U0276497

国家出版基金项目
NATIONAL PUBLICATION FOUNDATION

涡轮机械与推进系统出版项目
航空发动机技术出版工程

完全流程手册：

从流程建模到管理的知识体系
卷一（上）

The Complete Business Process Handbook
Body of Knowledge from Process Modeling to BPM
Volume I

〔法〕M. V. 罗辛（Mark von Rosing）

〔德〕A. W. 舍尔（August-Wilhelm Scheer） 著

〔瑞士〕H. V. 谢尔（Henrik von Scheel）

张玉金　王占学　等译

科学出版社

北京

图字:01-2021-1780号

内 容 简 介

本书从流程概念的发展与演变出发,汇集大量来自政府机构、标准组织、企业、大学、研究机构以及行业专家的杰出贡献,将涉及流程管理的相关知识概念与理论加以系统整合,构建了流程管理领域全面的知识体系。

本书采取理论与实践分析相结合的方法,通过对流程管理知识、模板、专家建议以及最佳实践的介绍,从什么是流程管理、流程管理的思维方式、工作方式、建模方式、实施与治理方式、培训与指导方式六部分进行编著,将每一个复杂主题拆解成便于理解的知识点,提供了有关实现流程管理的流程、框架、方法的所有内容,为商用航空发动机研制企业构建基于流程管理的自主研发体系提供了一份全面的实用指南。

本书可为企业中高级管理人员、流程管理业务人员、管理咨询行业从业人员、大学管理类专业相关人员提供全面的学习参考,也可作为高校管理类教师和研究生的参考书。

图书在版编目(CIP)数据

完全流程手册:从流程建模到管理的知识体系.卷一
= The Complete Business Process Handbook: Body
of Knowledge from Process Modeling to BPM Volume Ⅰ:
汉、英 / (法) 马克·冯·罗辛 (Mark von Rosing) 等
著;张玉金等译. — 北京:科学出版社,2021.12
(航空发动机技术出版工程)
国家出版基金项目 涡轮机械与推进系统出版项目
ISBN 978-7-03-068886-6

Ⅰ.①完… Ⅱ.①马…②张… Ⅲ.①航空发动机-
系统工程-流程-手册-汉、英 Ⅳ.①V23-65

中国版本图书馆 CIP 数据核字(2021)第101405号

责任编辑:徐杨峰 / 责任校对:谭宏宇
责任印制:黄晓鸣 / 封面设计:殷 靓

科学出版社 出版
北京东黄城根北街16号
邮政编码:100717
http:// www.sciencep.com
南京展望文化发展有限公司排版
广东虎彩云印刷有限公司印刷
科学出版社发行 各地新华书店经销

*

2021年12月第 一 版 开本:B5(720×1000)
2025年 2 月第五次印刷 总印张:91 3/4
总字数:1687 000
定价:600.00元
(如有印装质量问题,我社负责调换)

涡轮机械与推进系统出版项目
顾问委员会

主任委员
张彦仲

委 员
（以姓名笔画为序）

尹泽勇　乐嘉陵　朱　荻　刘大响　杜善义
李应红　张　泽　张立同　张彦仲　陈十一
陈懋章　闻雪友　宣益民　徐建中

航空发动机技术出版工程

专家委员会

航空发动机技术出版工程

编写委员会

涡轮机械与推进系统出版项目

序

　　涡轮机械与推进系统涉及航空发动机、航天推进系统、燃气轮机等高端装备。其中每一种装备技术的突破都令国人激动、振奋，但是由于技术上的鸿沟，使得国人一直为之魂牵梦绕。对于所有从事该领域的工作者，如何跨越技术鸿沟，这是历史赋予的使命和挑战。

　　动力系统作为航空、航天、舰船和能源工业的"心脏"，是一个国家科技、工业和国防实力的重要标志。我国也从最初的跟随仿制，向着独立设计制造发展。其中有些技术已与国外先进水平相当，但由于受到基础研究和条件等种种限制，在某些领域与世界先进水平仍有一定的差距。在此背景下，出版一套反映国际先进水平、体现国内最新研究成果的丛书，既切合国家发展战略，又有益于我国涡轮机械与推进系统基础研究和学术水平的提升。"涡轮机械与推进系统出版项目"主要涉及航空发动机、航天推进系统、燃气轮机以及相应的基础研究。图书种类分为专著、译著、教材和工具书等，内容包括领域内专家目前所应用的理论方法和取得的技术成果，也包括来自一线设计人员的实践成果。

　　"涡轮机械与推进系统出版项目"分为四个方向：航空发动机技术、航天推进技术、燃气轮机技术和基础研究。出版项目分别由科学出版社和浙江大学出版社出版。

　　出版项目凝结了国内外该领域科研与教学人员的智慧和成果，具有较强的系统性、实用性、前沿性，既可作为实际工作的指导用书，也可作为相关专业人员的参考用书。希望出版项目能够促进该领域的人才培养和技术发展，特别是为航空发动机及燃气轮机的研究提供借鉴。

张彦仲

2019 年 3 月

航空发动机技术出版工程

序

航空发动机被誉称为工业皇冠之明珠,实乃科技强国之重器。

几十年来,我国航空发动机技术、产品及产业经历了从无到有、从小到大的艰难发展历程,取得了显著成绩。在世界新一轮科技革命、产业变革同我国转变发展方式的历史交汇期,国家决策进一步大力加强航空发动机事业发展,产学研用各界无不为之振奋。

迄今,科学出版社于2019年、2024年两次申请国家出版基金,安排了"航空发动机技术出版工程",确为明智之举。

本出版工程旨在总结、推广近期及之前工作中工程、科研、教学的优秀成果,侧重于满足航空发动机工程技术人员的需求,尤其是从学生到工程师过渡阶段的需求,借此也为扩大我国航空发动机卓越工程师队伍略尽绵力。本出版工程包括设计、试验、基础与综合、前沿技术、制造、运营及服务保障六个系列,2019年启动的前三个系列近五十册任务已完成;后三个系列近三十册任务则于2024年启动。对于本出版工程,各级领导十分关注,专家委员会不时指导,编委会成员尽心尽力,出版社诸君敬业把关,各位作者更是日无暇晷、研教著述。同道中人共同努力,方使本出版工程得以顺利开展、如期完成。

希望本出版工程对我国航空发动机自主创新发展有所裨益。受能力及时间所限,当有疏误,恭请斧正。

2024 年 10 月修订

译 者 序

  航空发动机的研制是一项复杂的系统工程，长久以来，我国航空发动机受跟踪研发模式的影响，尚未建立起完整、统一的自主研发体系，存在数据不共享、标准不统一、管理"两张皮"、运行效率低等情况，明显阻碍型号项目研制顺利推进。要实现国产商用航空发动机的自主研制，必须遵循航空发动机发展的客观规律，建立面向航空发动机产品全生命周期完整统一的流程体系，为独立自主研制出先进可靠的航空发动机夯实基础。

  中国航发商用航空发动机有限责任公司自2009年成立以来，就明确提出"聚焦客户、流程主导、追求卓越、持续改进"的管理政策，以流程管理为主导，建设价值驱动型流程体系，以流程统领所有业务活动，以流程绩效评价业务结果，通过持续改进，不断推动业务高质量发展。商发公司矢志不渝地推进流程型企业的建设，围绕流程、组织和文化持续打造国际一流的企业运营系统：一是企业家面向系统求价值，管理者面向流程做改进，操作者面向作业做完善；二是企业作为价值创造系统，创新与变革是永远不变的主题，企业要始终动态地适应环境而改进，必须面对环境变化作战略取舍，面对客户需求优化产品价值链，面对资源需求优化供应链与资源保障；三是企业运营系统的核心要素：流程、人与文化，流程是主航道，必须持续地"清淤拓土筑基"，人与文化要素所赋予的知识资本、信息资本与组织资本必须嵌入流程鲜活灵动地动作。

  经过十余年单通道窄体干线客机发动机验证机研制走完全过程和全面推进产品研制的实践积累，商发公司在管理运营与产品研制等方面进行了全面探索与实践，初步形成了以产品研发体系为核心，涵盖公司20个业务领域的流程体系。同时，在不断地面向用户质量目标、聚焦价值创造、开展流程再造与数字化转型等创新变革活动中，逐步探索建立了面向商用航空发动机公司的体系运行管理知识体系，并尝试通过对供应商的管理体系延伸，进一步打造安全、可靠、稳定的供应链体系，建立产业链良好的演进生态，最终实现国产商用航空发动机的产品交付，达成客户满意。

本书总结吸纳了世界优秀企业实践，提供了一套流程优化剪裁模板、一套持续改进优化的方法论、一套文化重塑与习惯再造的利器。商发公司在流程体系建设实践中，借鉴了其中的原理与方法，特别是流程全生命周期以及实施与治理等部分，得到了很多启发和帮助。该书针对流程管理进行了全面的系统论述，并构建了较为完整的知识体系；该书作为流程管理的实用指南，详细揭示了流程管理中我们思考和使用流程的方式，深入论述了业务流程的本质，以及从流程建模到治理的完整知识。为了使广大的学习者和实践者能够准确掌握本书所包罗的理论与方法精髓，本译著将以中、英文逐页对照的方式出版发行。

企业管理就是以流程为核心、价值为目标、自我驱动的生命之旅，企业管理运作是一门技术，又是一门艺术，同时也是一门实践的系统工程。中国企业尤其是从事复杂系统工程与高端制造业的企业更应学习如何站在巨人的肩膀上不断成就自我，打造既有东方智慧又有西方商业理念优秀基因的一流企业。翻译和实践本书就是在汲取西方优秀的企业管理最佳实践，为我国航空制造企业提供参考借鉴。在此，要特别感谢支持本书翻译与校订工作，以及在过程中提供资料案例和提出宝贵建议的相关专家、同事，他们是黄博、陈楠、黄干明、陈天彧、项飞、黄飞、汤先萍、张澁滋、杨博文、吴帆、何宛文、陈婧怡、郑冰雷。希望这本译著能够进一步促进流程管理方法在我国航空制造企业乃至中国企业的管理变革中的实践，让东方智慧与西方商业理念的深度结合转化为企业治理效能，打造更多世界一流的中国企业，为国家打造央企"市营"新范式提供有益的借鉴和参考！

张玉金

2021 年 5 月

Foreword

This book has been put together to help you explore Business Process concepts and to understand what BPM really is all about.

We wrote this book for YOU—the individual. You may be a business executive, manager, practitioner, subject matter expert, student, or researcher. Or may be an ambitious career individual who wants to know more about business process concepts and/or BPM, what it is all about and how to apply it.

This, *The Complete Business Process Handbook*, provides a comprehensive body of knowledge written as a practical guide for you—by the authorities that have shaped the way we think and work with processes today. You hold the first of three books in the series in your hand.

- The first volume endows the reader with a deep insight into the nature of business process concepts and how to work with them. From BPM Ontology, semantics, and BPM Portfolio management, to the BPM Life Cycle, it provides a unique foundation within this body of knowledge.
- The second volume bridges theory and application of BPM in an advanced modeling context by addressing the subject of extended BPM.
- The third volume explores a comprehensive collection of real-world BPM lessons learned, best practices, and leading practices examples from award-winning industry leaders and innovators.

We wish you well on your Business Process journey and that is why we also have invested years putting this Handbook series together. To share the knowledge, templates, concepts, best and leading practices and to ensure high quality and standards, we have worked and coordinated with standard development organizations like International Organization for Standardization (ISO), Object Management Group (OMG), Institute of Electrical and Electronics Engineers (IEEE), North Atlantic Treaty Organization (NATO), Council for Scientific and Industrial Research (CSIR), MITRE—a Federally Funded Research and Development Center, European Committee for Standardization (CEN), The Security Forum, World Wide Web Consortium (W3C), and LEADing Practice.

We have also identified and worked with leading organizations and with their process experts/architects, and have described their practices. Among them are Lego, Maersk Shipping, Carlsberg, FLSmidth, the US Government, AirFrance, KLM, German Government, SaxoBank, Novozymes, the Canadian Government, US Department of Defense, Danish Defense, Johnson & Johnson, Dutch Railway, Australian Government, and many more. At last but not least the Global University Alliance consisting of over 400 universities, lecturers, and researchers have analyzed and examined what works, again and again (best practice), and what are the unique practices applied by these leading organizations (leading practices). They then identified common and repeatable patterns, which provide the basis for the BPM Ontology, BPM Semantics, the BPM standards, and the process templates found in this book.

原 书 序

本书已经整合在一起，可以帮助您探索业务流程概念，并了解业务流程管理（BPM）的真正含义。

我们为您编写的这本书。您可能是：业务主管、经理、工作者、某一领域专家、学生或研究员，也或许是一个雄心勃勃的职场人士，想要更多地了解业务流程概念和BPM是什么以及如何应用它。

这本《完全流程手册》，它为您提供一个全面的知识体系，作为一本实用指南，由那些塑造了我们今天思考和使用流程的方式的权威人士编写。您手里拿着的这本是这个系列的三卷书中的第一卷。

- 第一卷让读者深入了解业务流程概念的本质以及如何使用它们。从BPM本体论、语义、BPM项目组合管理，到BPM生命周期，它在这个知识体系中提供一个独特的基础。
- 第二卷通过处理扩展BPM主题，在高级建模背景中架起BPM理论和应用之间的桥梁。
- 第三卷探讨屡获殊荣的行业领导者和创新者的全面实际BPM经验教训、最佳实践和领导实践示例。

我们祝愿您在业务流程之旅中取得成功，这就是我们花费多年时间将本系列手册整合在一起的原因，分享知识、模板、概念、最佳和领导实践。为了确保高质量和高标准，我们与国际标准化组织（ISO）、对象管理组织（OMG）、电气和电子工程师协会（IEEE）、北大西洋公约组织（NATO）、科学与工业研究理事会（CSIR）、联邦资助的研究与发展中心（Federally Funded Research and Development Center, FFRDC）MITER、欧洲标准化委员会（CEN）、安全论坛（The Security Forum）、万维网联盟（W3C）和领导实践（LEADing Practice）等组织进行了协调。

我们还选择与领先的组织及其流程专家/架构师合作，并描述他们的实践。其中包括：乐高、马士基航运、嘉士伯、艾法史密斯（FLSmidth）、美国政府、法国航空、荷兰皇家航空、德国政府、盛宝银行、诺维信、加拿大政府、美国国防部（United States Department of Defense, DOD）、丹麦国防部、强生、荷兰铁路、澳大利亚政府等。最后但同样重要的是，由400多所大学的讲师和研究人员组成的全球大学联

We have worked years on this book, and as you just read, with contributions of standard bodies, governments, defense organizations, enterprises, universities, research institutes and individual thought leaders. We put these chapters and their subjects carefully together and hope you enjoy reading it—as much as we did writing, reviewing and putting it together.

Name	Organization
Mark von Rosing	Global University Alliance
August-Wilhelm Scheer	Scheer Group GmBH
Henrik von Scheel	LEADing Practices, Google Board
Adam D.M. Svendsen	Institute of Future Studies
Alex Kokkonen	Johnson & Johnson
Andrew M. Ross	Westpac
Anette Falk Bøgebjerg	LEGO Group
Anni Olsen	Carlsberg Group
Antony Dicks	NedBank
Asif Gill	Global University Alliance
Bas Bach	NS Rail
Bob J. Storms	LEADing Practices
Callie Smit	Reserve Bank
Cay Clemmensen	LEADing Practices
Christopher K. Swierczynski	Electrolux
Clemens Utschig-Utschig	Boehringer Ingelheim Pharma
Dan Moorcroft	QMR
Daniel T. Jones	Lean UK
David Coloma	Universitat Politècnica de Catalunya, Spain
Deb Boykin	Pfizer Pharmaceuticals
Dickson Hunja Muhita	LEADing Practices
Duarte Gonçalves	CSIR—Council for Scientific and Industrial Research
Ekambareswaran Balasubramanian	General Motors
Fabrizio Maria Maggi	University of Estonia
Fan Zhao	Florida Gulf Coast University
Fatima Senghore	NASA
Fatma Dandashi	MITRE
Freek Stoffel	LEADing Practices
Fred Cummins	OMG

盟（Global University Alliance）分析和检查哪些有效，以及这些领先组织独特的实践应用（领导实践）。然后，他们确定了常见和可重复的模式，这些模式为BPM本体、BPM语义、BPM标准和本书中的流程模板提供了基础。

正如您刚才所读，我们已经为编写这本书工作了多年，您将在本书中找到标准机构、政府、国防组织、企业、大学、研究机构和个人思想领袖的贡献。我们将这些章节及其主题精心放在一起，希望您就像我们写作、复习和整理一样喜欢阅读它。

姓名	组织
Mark von Rosing	Global University Alliance
August-Wilhelm Scheer	Scheer Group GmBH
Henrik von Scheel	LEADing Practices, Google Board
Adam D.M. Svendsen	Institute of Future Studies
Alex Kokkonen	Johnson & Johnson
Andrew M. Ross	Westpac
Anette Falk Bøgebjerg	LEGO Group
Anni Olsen	Carlsberg Group
Antony Dicks	NedBank
Asif Gill	Global University Alliance
Bas Bach	NS Rail
Bob J. Storms	LEADing Practices
Callie Smit	Reserve Bank
Cay Clemmensen	LEADing Practices
Christopher K. Swierczynski	Electrolux
Clemens Utschig-Utschig	Boehringer Ingelheim Pharma
Dan Moorcroft	QMR
Daniel T. Jones	Lean UK
David Coloma	Universitat Politècnica de Catalunya, Spain
Deb Boykin	Pfizer Pharmaceuticals
Dickson Hunja Muhita	LEADing Practices
Duarte Gonçalves	CSIR—Council for Scientific and Industrial Research
Ekambareswaran Balasubramanian	General Motors
Fabrizio Maria Maggi	University of Estonia
Fan Zhao	Florida Gulf Coast University
Fatima Senghore	NASA
Fatma Dandashi	MITRE
Freek Stoffel	LEADing Practices
Fred Cummins	OMG

Name	Organization
Gabriel von Scheel	LEADing Practices
Gabriella von Rosing	LEADing Practices
Gary Doucet	Government of Canada
Gert Meiling	Tommy Hilfiger
Gert O. Jansson	LEADing Practices
Hans Scheruhn	University of Harz, Gemany
Hendrik Bohn	Nedbank
Henk de Man	OMG, VeeBee
Henk Kuil	KLM, Air France
Henrik Naundrup Vester	iGrafx
Jacob Gammelgaard	FLSchmidt
James P. Womack	Cambridge University-Massachusetts Institute of Technology (MIT)
Jeanne W. Ross	Cambridge University-Massachusetts Institute of Technology (MIT)
Jeff Greer	Cardinal Health
Jens Theodor Nielsen	Danish Defense
John A. Zachman	Zachman International
John Bertram	Government of Canada
John Golden	iGrafx
John M. Rogers	Government of Australia
Jonnro Erasmus	CSIR—Council for Scientific and Industrial Research
Joshua Michael von Scheel	LEADing Practices
Joshua Waters	LEADing Practices
Justin Tomlinson	LEADing Practices
Karin Gräslund	RheinMain University-Wiesbaden Business School
Katia Bartels	Office Depot
Keith Swenson	Fujitsu
Kenneth Dean Teske	US Government
Kevin Govender	Transnet Rail
Klaus Vitt	German Federal Employment Agency
Krzysztof Skurzak	NATO ACT
LeAnne Spurrell	QMR
Lloyd Dugan	BPM.com
Lotte Tange	Carlsberg Group

姓名	组织
Gabriel von Scheel	LEADing Practices
Gabriella von Rosing	LEADing Practices
Gary Doucet	Government of Canada
Gert Meiling	Tommy Hilfiger
Gert O. Jansson	LEADing Practices
Hans Scheruhn	University of Harz, Gemany
Hendrik Bohn	Nedbank
Henk de Man	OMG, VeeBee
Henk Kuil	KLM, Air France
Henrik Naundrup Vester	iGrafx
Jacob Gammelgaard	FLSchmidt
James P. Womack	Cambridge University-Massachusetts Institute of Technology (MIT)
Jeanne W. Ross	Cambridge University-Massachusetts Institute of Technology (MIT)
Jeff Greer	Cardinal Health
Jens Theodor Nielsen	Danish Defense
John A. Zachman	Zachman International
John Bertram	Government of Canada
John Golden	iGrafx
John M. Rogers	Government of Australia
Jonnro Erasmus	CSIR—Council for Scientific and Industrial Research
Joshua Michael von Scheel	LEADing Practices
Joshua Waters	LEADing Practices
Justin Tomlinson	LEADing Practices
Karin Gräslund	RheinMain University-Wiesbaden Business School
Katia Bartels	Office Depot
Keith Swenson	Fujitsu
Kenneth Dean Teske	US Government
Kevin Govender	Transnet Rail
Klaus Vitt	German Federal Employment Agency
Krzysztof Skurzak	NATO ACT
LeAnne Spurrell	QMR
Lloyd Dugan	BPM.com
Lotte Tange	Carlsberg Group

Name	Organization
Mads Clausager	Maersk Group
Mai Phuong	Northrop Grumman Electronic Systems
Maria Hove	LEADing Practices
Maria Rybrink	TeliaSonera
Marianne Fonseca	LEADing Practices
Mark Stanford	iGrafx
Marlon Dumas	University of Tartu
Mathias Kirchmer	BPM-d
Maxim Arzumanyan	St. Petersburg University
Michael Tisdel	US Government, DoD
Michel van den Hoven	Philips
Mikael Munck	SaxoBank
Mike A. Marin	IBM Corporation
Mona von Rosing	LEADing Practices
Nathaniel Palmer	BPM.com, Workflow Management Coalition (WfMC)
Neil Kemp	LEADing Practices
Nils Faltin	Scheer Group GmBH
Partha Chakravartti	AstraZeneca
Patricia Kemp	LEADing Practices
Peter Franz	BPM-d
Philippe Lebacq	Toyota
Régis Dumond	French Ministry of Defense, NATO, ISO
Rich Hilliard	IEEE, ISO
Richard L. Fallon	Sheffield Hallam University
Richard N. Conzo	Verizon
Rod Peacock	European Patent Office
Rogan Morrison	LEAD Enterprise Architect Professional
Ronald N. Batdorf	US Government, DoD, Joint Staff
Sarel J. Snyman	SAP Solution Design
Scott Davis	Government of Canada
Simon Polovina	Sheffield Hallam University
Stephen White	IBM Corporation
Steve Durbin	Information Security Forum
Steve Willoughby	iGrafx

姓名	组织
Mads Clausager	Maersk Group
Mai Phuong	Northrop Grumman Electronic Systems
Maria Hove	LEADing Practices
Maria Rybrink	TeliaSonera
Marianne Fonseca	LEADing Practices
Mark Stanford	iGrafx
Marlon Dumas	University of Tartu
Mathias Kirchmer	BPM-d
Maxim Arzumanyan	St. Petersburg University
Michael Tisdel	US Government, DoD
Michel van den Hoven	Philips
Mikael Munck	SaxoBank
Mike A. Marin	IBM Corporation
Mona von Rosing	LEADing Practices
Nathaniel Palmer	BPM.com, Workflow Management Coalition (WfMC)
Neil Kemp	LEADing Practices
Nils Faltin	Scheer Group GmBH
Partha Chakravartti	AstraZeneca
Patricia Kemp	LEADing Practices
Peter Franz	BPM-d
Philippe Lebacq	Toyota
Régis Dumond	French Ministry of Defense, NATO, ISO
Rich Hilliard	IEEE, ISO
Richard L. Fallon	Sheffield Hallam University
Richard N. Conzo	Verizon
Rod Peacock	European Patent Office
Rogan Morrison	LEAD Enterprise Architect Professional
Ronald N. Batdorf	US Government, DoD, Joint Staff
Sarel J. Snyman	SAP Solution Design
Scott Davis	Government of Canada
Simon Polovina	Sheffield Hallam University
Stephen White	IBM Corporation
Steve Durbin	Information Security Forum
Steve Willoughby	iGrafx

Name	Organization
Sven Vollbehr	SKF
Thomas Boosz	German Government
Thomas Christian Olsen	NovoZymes
Tim Hoebeek	SAP
Tom Preston	Booz Allen Hamilton
Ulrik Foldager	LEADing Practices
Victor Abele	Government of Canada
Vincent Snels	Nationale Nederlanden
Volker Rebhan	German Federal Employment Agency
Wim Laurier	Université Saint-Louism Bruxelles
Ýr Gunnarsdottir	Shell
Yury Orlov	Smart Architects
Zakaria Maamar	Zayed University, United Arab Emirates

姓名	组织
Sven Vollbehr	SKF
Thomas Boosz	German Government
Thomas Christian Olsen	NovoZymes
Tim Hoebeek	SAP
Tom Preston	Booz Allen Hamilton
Ulrik Foldager	LEADing Practices
Victor Abele	Government of Canada
Vincent Snels	Nationale Nederlanden
Volker Rebhan	German Federal Employment Agency
Wim Laurier	Université Saint-Louism Bruxelles
Ýr Gunnarsdottir	Shell
Yury Orlov	Smart Architects
Zakaria Maamar	Zayed University, United Arab Emirates

Abbreviation Meaning

A2A	Application to application
AAIM	Agility adoption and improvement model
ACM	Adaptive case management
ADDI	Architect design deploy improve
API	Application programming interface
APQC	American productivity and quality center
B2B	Business to business
BAM	Business activity monitoring
BCM	Business continuity management
BEP	Break even point
BI	Business intelligence
BITE	Business innovation and transformation enablement
BOM	Business object management
BPA	Business process analysis
BPaaS	Business process as a service
BPCC	Business process competency center
BPD	Business process diagram
BPE	Business process engineering
BPEL	Business process execution language
BPEL4WS	Business process execution language for web services
BPG	Business process guidance
BPI	Business process improvement
BPM	Business process management
BPM CM	Business process management change management
BPM CoE	Business process management center of excellence
BPM LC	Business process management life cycle
BPM PM	Business process management portfolio management
BPMaaS	BPM as a service
BPMI	Business process management institute
BPMN	Business process model and notation
BPMS	Business process management system
BPO	Business process outsourcing
BPPM	Business process portfolio management
BPR	Business process reengineering
BRE	Business rule engine
BRM	Business rules management
CDM	Common data model
CE-BPM	Cloud-enabled BPM
CEAP	Cloud-enabled application platform
CEN	European committee for standardization
CEP	Complex event processing
CM	Configuration management
CMS	Content management system
COBIT	Control objectives for information and related technology

缩 略 词

A2A	应用到应用
AAIM	敏捷应用和改进模型
ACM	适应性案例管理
ADDI	架构师设计部署改进
API	应用程序编程接口
APQC	美国生产力和质量中心
B2B	业务到业务
BAM	业务活动监控
BCM	业务连续性管理
BEP	盈亏平衡点
BI	商务智能
BITE	业务创新和转型支持
BOM	业务对象管理
BPA	业务流程分析
BPaaS	业务流程即服务
BPCC	业务流程能力中心
BPD	业务流程图
BPE	业务流程工程
BPEL	业务流程执行语言
BPEL4WS	Web服务的业务流程执行语言
BPG	业务流程指导
BPI	业务流程改进
BPM	业务流程管理
BPM CM	业务流程管理变更管理
BPM CoE	业务流程管理卓越中心
BPM LC	业务流程管理生命周期
BPM PM	业务流程管理组合管理
BPMaaS	BPM即服务
BPMI	业务流程管理机构
BPMN	业务流程建模标记法
BPMS	业务流程管理系统

CPO	Chief process officer
CRM	Customer relationship management
CSF	Critical success factor
CSIR	Council for Scientific and Industrial Research
CxO	Chief x officer
DB	Database
DBMS	Database management system
DMS	Document management system
DNEAF	Domain neutral enterprise architecture framework
DSDM	Dynamic systems development method
EAI	Enterprise application integration
EITE	Enterprise innovation & transformation enablement
EMR	Enterprise-wide metadata repositories
EPC	Event-driven process chain
EPSS	Electronic performance support system
ERM	Entity relationship modeling
ERP	Enterprise resource planning
ESB	Enterprise service bus
FEAF	Federal enterprise architecture framework
FI	Financial
iBPM	Intelligent business process management
IDE	Integrated development environment
IE	Information engineering
IEEE	Institute of electrical and electronics engineers
ISO	International Organization for Standardization
ITIL	Information technology infrastructure library
KPI	Key performance indicator
L&D	Learning and development
LEADP	Layered enterprise architecture development and/or LEADing big in Practice
MDM	Master data management
NATO	North Atlantic Treaty Organisation
NIST	National Institute of Standards and Technology
OCM	Organizational change management
OLAP	Online analytic processing
OLTP	Online transaction processing
OMG	Object management group
PDC	Process data collection
PIM	Process instance management
PM	Portfolio management
PM	Project management
PMBOK	Project management body of knowledge
PMO	Project management offices
POA	Process oriented architecture
PPI	Process performance indicator
PPM	Project portfolios management
PPPM	Portfolio, program and project management
PRINCE	PRojects IN Controlled Environments
QM	Quality management

BPO	业务流程外包
BPPM	业务流程组合管理
BPR	业务流程再造
BRE	业务规则引擎
BRM	业务规则管理
CDM	通用数据模型
CE-BPM	支持云端的 BPM
CEAP	支持云的应用平台
CEN	欧洲标准化委员会
CEP	复杂事件处理
CM	配置管理
CMS	内容管理系统
COBIT	信息及相关技术控制目标
CPO	首席流程官
CRM	客户关系管理
CSF	关键成功因素
CSIR	科学和工业研究委员会
CxO	首席 x 官员
DB	数据库
DBMS	数据库管理系统
DMS	文件管理系统
DNEAF	领域中立的企业架构框架
DSDM	动态系统开发方法
EAI	企业应用程序集成
EITE	企业创新和转型支持
EMR	企业范围的元数据存储库
EPC	事件驱动的流程链
EPSS	电子绩效支持系统
ERM	实体关系建模
ERP	企业资源规划
ESB	企业服务总线
FEAF	联邦企业架构框架
FI	金融
iBPM	智能业务流程管理
IDE	集成开发环境
IE	信息工程
IEEE	电气和电子工程师协会
ISO	国际标准化组织
ITIL	信息技术基础架构库
KPI	关键绩效指标
L&D	学习和发展
LEADP	分层企业架构开发和/或领导实践
MDM	主数据管理

ROI	Return on investment
SBO	Strategic business objective
SCM	Supply chain management
SCOR	Supply chain operations reference model
SD	Sales and distribution
SNA	Social network analysis
SOA	Service oriented architecture
SPI	Service performance indicator
SRM	Supply relationship management
SW	Software
TCO	Total cost of ownership
TOGAF	The open group architecture framework
TQM	Total quality management
UI	User interface
ULM	Unified modeling language
USGAP	United States general accounting principles
VDML	Value delivery modeling language
VNA	Value network analysis
W3C	World Wide Web consortium
xBPMN	eXtended business process model and notation
XLM	Extensible markup language
XMI	Metadata interchange
XSD	XML schema definition

NATO	北大西洋公约组织
NIST	美国国家标准与技术研究院
OCM	组织变革管理
OLAP	联机分析处理
OLTP	联机事物处理
OMG	对象管理组织
PDC	过程数据收集
PIM	流程实例管理
PM	项目组合管理
PM*	项目管理
PMBOK	项目管理知识体系
PMO	项目管理办公室
POA	面向流程的体系结构
PPI	流程绩效指标
PPM	项目组合、项目集、项目管理
PPPM	投资组合、计划和项目管理
PRINCE	受控环境中的项目
QM	质量管理
ROI	投资回报率
SBO	战略业务目标
SCM	供应链管理
SCOR	供应链运作参考模型
SD	销售和分销
SNA	社会网络分析
SOA	面向服务的架构
SPI	服务绩效指标
SRM	供应关系管理
SW	软件
TCO	总拥有成本
TOGAF	开放组体系结构构框架
TQM	全面质量管理
UI	用户界面
UML	统一建模语言
USGAP	美国通用会计准则
VDML	价值交付建模语言
VNA	价值网络分析
W3C	万维网联盟
xBPMN	扩展业务流程模型和符号
XML	可扩展标记语言
XMI	元数据交换
XSD	XML模式定义

---

* 正文中出现PM缩写时请对照原文。

Introduction to the Book

Prof. Mark von Rosing, Henrik von Scheel, Prof. August-Wilhelm Scheer

It is not a new phenomenon that the markets are changing; however, the business environment in which firms operate lies outside of themselves and their control. So, while it is their external environment, which is always changing, most changes on the outside affect the need for innovation and transformation on the inside of the organization. The ability to change the business and to manage their processes is symbiotic, which is, among others, one of the reasons for such a high Business Process Management (BPM) adoption rate in the market. It is, however, important to note that unlike some analysts might claim, the size of the market and its adoption is in no way an indicator of maturity. As a matter of fact, the maturity of many of the BPM concepts can have a low maturity, even though the adoption is widespread. So while the high demand for BPM as a management method and a software solution, and the maturing BPM capabilities develop and unfold, the challenge quickly develops to provide concise and widely accepted BPM definitions, taxonomies, standardized, and integrated process templates, as well as overall frameworks, methods, and approaches.

Written as the practical guide for you—by the authorities that have shaped the way we think and work with process today. This handbook series stands out as a masterpiece, representing the most comprehensive body of knowledge published on business process. The first volume endows the reader with a deep insight into the nature of business process, and a complete body of knowledge from process modeling to BPM, thereby covering what executives, managers, practitioners, students, and researchers need to know about:

- Future BPM trends that will affect business
- A clear and precise definition of what BPM is
- Historical evolution of process concepts
- Exploring a BPM Ontology
- In-depth look at the Process Semantics
- Comprehensive Frameworks, Methods, and Approaches
- Process analysis, process design, process deployment, process monitoring, and Continuous Improvement
- Practical usable process templates
- How to link Strategy to Operation with value-driven BPM
- How to build BPM competencies and establish a Center of Excellence
- Discover how to apply Social Media and BPM
- Sustainable-Oriented process Modeling
- Evidence-based BPM
- Learn how Value and Performance Measurement and Management is executed
- Explore how to enable Process Owners

本 书 介 绍

Mark von Rosing, Henrik von Scheel, August-Wilhelm Scheer

市场的变化并不是一个新的现象，但是企业的经营环境是不受自身控制的。因此，虽然外部环境一直在变化，但大多数外部变化都会影响组织内部的创新和转型需求。改变业务和管理流程的能力是共生的，这也是BPM在市场上被广泛应用的原因之一。不过，值得注意的是，与一些分析人员发表的意见不同，市场的规模和应用程度绝不是BPM成熟的主要标志。事实上，尽管许多BPM概念被广泛应用，但是成熟度依然很低，因此，对BPM作为管理方法和软件解决方案的高需求以及日益成熟的BPM能力，两者不断发展和展现的同时，在提供简洁并广泛接受的BPM定义、分类法、标准化集成流程模板以及总体框架、方法和途径等方面的挑战也在快速发展。

作为实用指南，本书作者塑造了BPM中我们思考和使用流程的方式。本手册系列作为杰作脱颖而出，代表了当前业务流程文档中最全面的知识体系。第一卷使读者深入了解业务流程的本质，以及从流程建模到BPM的完整知识体系，涵盖高管、经理、工作者、学生和研究人员需要了解的内容：

- 影响未来业务的BPM趋势；
- 一个对BPM清晰而精确的定义；
- 流程概念的历史演变；
- 探索BPM本体；
- 深入研究流程语义；
- 综合框架、方法和途径；
- 流程分析、流程设计、流程部署、流程监控以及持续改进（continuous improvement，CI）；
- 实用的流程模板；
- 如何将战略与价值驱动的BPM运营联系起来；
- 如何建立BPM能力和卓越中心（center of excellence，CoE）；
- 了解如何应用社交媒体和BPM；

- BPM Roles and Knowledge Workers
- Discover how to develop information models within the process models
- Uncovering Process Life cycle
- BPM Maturity
- BPM Portfolio Management and BPM Alignment
- BPM Change Management and BPM Governance
- Learning a structured way of Thinking, Working, Modeling, and Implementing processes.

This book is organized into various chapters that have been thoughtfully put together to communicate many times a complex topic into a replicable and manageable structure—that you as a reader can apply. Furthermore, the book is structured into six parts with the intention to guide you in turning business processes into real assets.

In Part I, we introduce a comprehensive "history of process concepts" from Sun Tzu's, to Taylorism, to Business Process Reengineering to Lean and BPMN, providing the reader with an in-depth understanding of the evolution of process thinking, approaches, and methods: a fundamental insight to what has shaped and what is shaping process thinking.

In Part II, we introduce the "Way of Thinking" around Business Process with focus on the value of Ontology, and a comprehensive BPM Ontology—the essential starting point that creates the guiding principles.

In Part III, we establish a "Way of Working" with Business Processes—the critical discipline of translating both strategic planning and effective execution. Exploring the current and future process trends that you need to be aware of with a detailed practical guide on how to apply them in areas such as BPM Life cycle, BPM Roles, process templates, evidence-based BPM, and many more.

In Part IV, we provide the essential guidance to help you in a "Way of Modeling" in traditional Process Modeling concepts to BPMN and Value-Oriented Process Modeling, how to work with and model Business Processes variations, as well as how to interlink information models and process models.

In Part V, we focus on the "Way of Implementation" and "Way of Governance"— the approach the practitioner follows in order to apply and steer what exists, spanning issues ranging from BPM change management, agile BPM, business process outsourcing, and holistic governance to project, program, and portfolio alignment.

In Part VI, we focus on the "Way of Training and Coaching"—to provide insight into ideal process expert, process engineer and process architecture training, from online to class-based learning and coaching.

While this book certainly can be read cover to cover, depending on where you are in your Business Process journey, you may wish to choose a different path. If you are new to Business Process concepts, you might start at the beginning, with Part I. If you are beginning a BPM project, or it has already begun its journey, or you are looking for inspiration, we recommend using the book as a reference tool to access it by the topic of interest.

But no matter how you plan on building your knowledge, the book has been designed and architected to be a guide and a handbook able to create the right way of thinking, working, modeling implementation, and governance.

- 面向可持续发展的流程建模；
- 基于证据的BPM；
- 了解如何执行战略、绩效测量和管理；
- 探索如何启用流程责任人；
- BPM角色和知识工作者；
- 了解如何在流程模型中开发信息模型；
- 揭示流程生命周期；
- BPM成熟度；
- BPM组合管理和BPM协调机制；
- BPM变革管理和BPM治理；
- 学习思维、工作、建模和实施流程的结构化方法。

本书由不同的章节构成，这些章节经过深思熟虑被组合在一起，将一个复杂的主题分解成一个可复制和可管理的结构，便于读者应用。此外，本书分为六个部分，旨在指导您将业务流程转化为公司的宝贵资产。

第一部分，我们介绍从孙子到泰勒主义（Taylorism）、业务流程再造（BPR）、精益和业务流程建模标记法（BPMN）的流程概念的演变历史，有助于读者形成对进化流程思维、途径和方法的深入理解，建立对流程思维及形成流程的基本认识。

第二部分，我们介绍围绕业务流程的思维方法，重点介绍本体论的价值，以及一个全面的BPM本体论，这是创建指导原则的基本出发点。

第三部分，我们建立业务流程的工作方法，这是战略规划和有效执行的关键。我们协助您探索您需要了解的当前和未来流程趋势，并提供详细的实践指导，包括：如何将其应用于诸如BPM生命周期、BPM角色、流程模板、基于证据的BPM等领域。

第四部分，我们为您提供基本的指导，帮助您以传统流程建模概念的建模方法为基础，实现BPMN和价值导向的流程建模，包括：如何应对与建模业务流程有关的变化，以及如何将信息模型和流程模型相互链接等。

第五部分，我们将重点放在实施方法和治理方法——从业者遵循的方法方面，尤其是如何面对存在的问题，包括从BPM变革管理、敏捷BPM、业务流程外包（BPO）和整体治理到项目、计划和组合调整等方面的各种问题。

第六部分，我们将重点放在"培训和指导的方法"上，从在线、课堂学习到经验指导，我们将深入学习流程专家、流程工程专家和流程架构专家的培训内容与知识体系。

尽管这本书可以从头到尾详细地进行阅读，但您也可根据您在业务流程中所处的环节选择不同的章节学习。如果您对业务流程概念不熟悉，我们建议您可以从第一部分开始。如果您正在计划或者已经开始了一个BPM项目，我们建议您使用这本书作为参考工具，研究其中您感兴趣的主题。

但是，不管您打算如何构建您的知识体系，这本书在设计和架构方面都是一本能够创建正确的思考、工作、建模和治理方式的指南和手册。

目　录

第一部分

完全流程手册：

从流程建模到管理的知识体系
卷一（上）

The Complete Business Process Handbook

Body of Knowledge from Process Modeling to BPM
Volume I

Phase 1: Process Concept Evolution

Henrik von Scheel, Mark von Rosing, Marianne Fonseca, Maria Hove, Ulrik Foldager

INTRODUCTION

The term *process* comes from the Latin word *processus* or *processioat*, which translates as a performed action of something that is done, and the way it is done. A process is, therefore, a collection of interrelated tasks and activities that are initiated in response to an event which aims to achieve a specific result for the consumer of the process. Processes constantly occur and happen all around us, in all that we do throughout the course of the day. They are the basis of all actions that involve concepts such as time, space, and motion, and they shape and bend to the very reality in which we exist.

Imagine yourself reading this chapter. You glance briefly to the side of your table only to realize that your coffee cup is now empty. A process is then sparked and initialized, and you (1) get up from your chair, (2) lift your cup from the table, (3) walk into the kitchen, (4) pour yourself another cup of coffee, (5) you then return to your chair, and (6) sit down and continue reading. That, in itself, is a process by nature. This is just one example of a very simple and descriptive process so as to better illustrate its elusive concept.

A *business process*, however, is the same as a process, but with one major difference, namely with the emphasis on the word *business*. A business process is a collection of tasks and activities (business operations and actions) consisting of employees, materials, machines, systems, and methods that are being structured in such a way as to design, create, and deliver a product or a service to the consumer.

As such, a *business process* can be understood in the following way:

- It is a placeholder for the action (process area).
- An action is taking place (process group).
- A business task is taking place (business process).
- The location of the business task in the sequence (process step).
- The way the business task is carried out (process activity).

Business processes consist of nucleus tasks and activities that are connected with each other, and are categorized and grouped. High-level business processes occur in a far more abstract context, as they are, usually, utilized to illustrate how a business carries out many different sets of operations. The entire marketing department of a large corporation, for example, can be described as a process group, although it depends entirely on the process structure of each individual organization. A business process can also consist of minor activities within the business process itself, and in such a case, these minor activities are called subprocesses. One ought to view the

The Complete Business Process Handbook. http://dx.doi.org/10.1016/B978-0-12-799959-3.00001-X

第一部分

1.1　第1阶段：流程概念演变1.0

Henrik von Scheel, Mark von Rosing, Marianne Fonseca, Maria Hove,Ulrik Foldager

1.1.1　简介

"process" 这个词语来自拉丁语 "processus" 或 "processioat"，它被翻译成 "为完成某件事情而采取的操作过程，以及它的完成方式"。因此，流程是一个相互关联的任务和活动的集合，这些任务和活动旨在为流程的使用者实现特定结果。流程不断地发生在我们每天所做的一切之中。它们是所有涉及时间、空间和运动等概念的行动的基础，塑造并改变我们存在的真实世界。

读这一节时顺便想象一下：您瞥了一眼桌子上面的咖啡杯，发现咖啡杯已经空了，然后一个流程应机而生并开始初始化，您：① 从椅子上站起来；② 从桌子上拿起杯子；③ 走进厨房；④ 再给自己倒一杯咖啡；⑤ 然后回到椅子上；⑥ 坐下继续阅读。这本身就是一个很自然的行为过程，是一个非常简单和描述性流程的例子，这个例子能够帮我们更好地理解流程这个概念。

事实上，业务流程与流程基本相同，但有一个主要区别，即前者强调业务一词。业务流程是由员工、材料、机器、系统和方法组成的任务和活动（业务指令和行为）的集合，这些任务和活动的目的是设计、生产和向消费者提供产品或服务。

因此，可以通过以下几点更好地理解业务流程：
- 它是行为的体现（流程域）；
- 正在发生的行为（流程组）；
- 正在发生的业务（业务流程）；
- 业务任务序列中的环节（流程步骤）；
- 业务任务执行的方式（流程活动）。

processes in the big picture first (captured in the process map) since a business process can trigger many tasks and subprocesses but also initiate other processes. In that way, you often see a connection between the different processes (both the value-adding processes and the non-value-adding processes) that are involved in the servicing of a client. Business processes are usually illustrated by different readable business process diagrams—for example through the use of Business Process Modeling Notation[1] diagrams. Business Process Modeling Notation (BPMN) is a standardized, visual (graphical) modeling representation used to illustrate business process flows. It provides an easy to use, flow-charting notation that is independent of the implementation environment. Business processes are used to illustrate, document, and shape the way an organization carries out its business operations across all organizational levels, i.e., both the strategic, tactical, and operational business levels.

There are four major phases in the historical development of business processes. The first phase is launched with the introduction of Sun Tzu's *Art of War* in the era of Ancient China. In the *Art of War*, Sun Tzu describes military strategies and tactics where he would assign specific tasks to certain people and calculate the resources needed for the execution of these tasks. Thousands of years later, we share with you Adam Smith's observations of work processes, which eventually inspired Taylor's "Scientific Management." The main problem in the implementation of Scientific Management is that it does not integrate the person behind the machine, and this leads us to the second phase, in which Allan H. Mogensen, Frank Gilbreth, and Ben Graham involved the worker in optimizing the processes. Finally, the visualization and digitalization of processes leads to the third and fourth (present) phase, in which the processes are being implemented and, to some extent, executed through the use of information systems and technology (Figure 1).

PROCESS CONCEPT EVOLUTION

Sun Tzu

Sun Tzu (also rendered as Sun Zi) was a Chinese military general, strategist, and philosopher and is assumed to have lived his life from around 544–496 BC in ancient China. He is traditionally credited as the author of *The Art of War*[2]—an extremely influential ancient Chinese book on military strategy. Traditional accounts state that the general's descendant Sun Bin also wrote a treatise on military tactics, also titled *The Art of War*.

Sun Tzu's work has been praised and employed throughout East Asia since its composition. During the twentieth century, *The Art of War* grew in popularity and saw practical use in Western society as well. *The Art of War* is composed of 13 chapters, each of which is devoted to one aspect of warfare. It is commonly known to be the definitive work on military strategy and tactics of its time and has had an influence on Eastern and Western military thinking, business tactics, legal strategy, and beyond. In his work, Sun Tzu describes a subtle yet abstract use of process activities to fulfill specific goals, precisely as business processes today are used to fulfill the goals of a company. He describes carrying out specific sets of tasks and activities and then assigning resources to the execution of these tasks and their related activities in order to complete certain objectives, and thereby, fulfilling the strategies of warfare.

可见,业务流程由相互关联的核心任务和活动组成,并被分类和分组。高级业务流程发生在更复杂的环境中,因为它们通常用于说明如何将整体业务分解为许多不同子任务的操作集。例如,一个大公司的整个市场部门可以被描述为一个流程组,尽管它完全取决于每个组织的流程结构。业务流程也可以由业务流程本身中的分支活动组成,在这种情况下,这些分支活动称为子流程。因为业务流程可以分解为许多子任务和子流程,所以,应该首先查看流程图的整体流程。通过这种方式,您经常会看到参与客户端服务的不同流程(增值流程和非增值流程)之间的连接过程。借助业务流程图,业务流程通常由不同的业务流程图来说明。BPMN[1]是一种用于说明业务流程的标准化、可视化(图形化)的建模标记法,它提供了一个易于使用的、独立于现实环境的流程图符号体系。可见,业务流程主要用于说明、记录和塑造一个组织在所有组织级别(即战略、战术和运营业务级别)执行业务操作的方式。

业务流程的历史发展历经了四个主要阶段。第一阶段是随着中国古代《孙子兵法》的出现而展开的。孙子在《孙子兵法》中描述了一种军事战略和战术,即他将特定的任务分配给特定的人,并评估执行这些任务所需的资源。数千年后,我们将最终激发了 F. W. Taylor 的科学管理诞生的 A. Smith(亚当·斯密,1723—1790 年)对工作过程的领悟分享给您。科学管理实施中的主要问题是它并不整合机器背后的人,这导致我们进入第二阶段。在第二阶段中,A. H. Mogensen、F. B. Gilbreth 和 B. S. Graham 让工人参与优化流程。最后,流程的可视化和数字化导致了第三和第四(现在)阶段,在这一阶段中,流程正在被实施,并在某种程度上通过使用信息系统和技术来执行(图1)。

1.1.2　流程的演变历史

1. 孙子

孙子是中国一位军事将领、战略家和哲学家,被认为生活在公元前544—公元前496年左右。传统上他被认为是《孙子兵法》[2]——一本影响深远的中国古代军事战略书的作者。据传,孙子的后代孙膑还写了一本关于军事战术的著作,被称为《孙膑兵法》。

《孙子兵法》自问世以来,受到了东亚各国的赞扬和运用。20世纪,这本巨作开始流行,在西方社会也得到了实际应用。《孙子兵法》共有13章,每章都专门论述战争的一个方面。众所周知,它是当时军事战略和战术的权威性著作,对东西方军事思想、商业战略、法律战略等都产生了影响。正如今天的业务流程被用来实现公司的目标一样,孙子在他的著作中为实现特定的目标,描述了一种微妙而抽象的

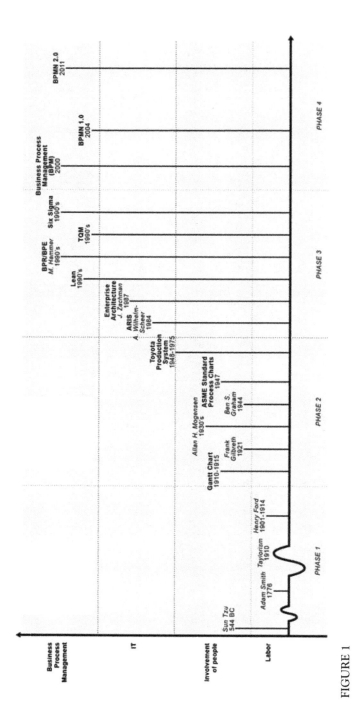

FIGURE 1

The historical evolution of processes over the course of time.[3]

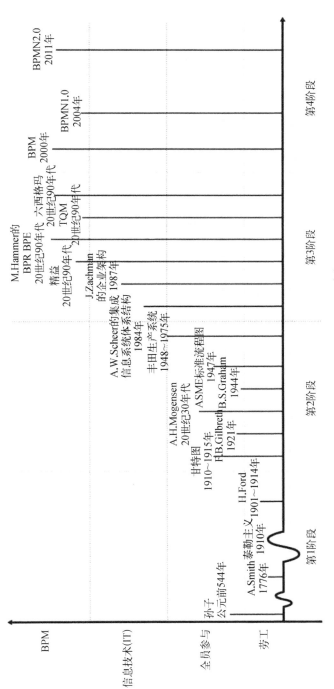

图 1　流程概念演变历史[3]

It should also be noted, however, that in 604 AD, Shotoku Taishi (573–621), the Prince of Holy Virtue, was a Japanese regent, statesman, and scholar who established a set of guidelines that served as the constitution of Japan at the time. In these documents, after studying the Chinese administration, he described the relations between Buddhism and what we would call public administration. These observations were later worked over and expanded and used as the foundation for the Japanese administration. Shotoku Taishi specified how tasks and responsibility best could be placed in the different branches of the administration. This is one of the earliest abstract views of the relation between organizational format and goals that the authors of this chapter have been able to identify.

Adam Smith

Adam Smith (1723–1790) was a Scottish moral philosopher and a pioneer of political economy. One of the key figures of the Scottish Enlightenment,[4] Smith is best known for two classic works: *The Theory of Moral Sentiments*[5] (1759), and *An Inquiry into the Nature and Causes of the Wealth of Nations*[6] (1776). While his exact date of birth is not known, Adam Smith's baptism was recorded on June 5, 1723, in Kirkcaldy, Scotland. He attended the Burgh School, where he studied Latin, mathematics, history, and writing. Smith entered the University of Glasgow when he was 14 years old and in 1740 he went to Oxford.[7]

In 1748, Adam Smith began giving a series of public lectures at the University of Edinburgh. Through these lectures in 1750, he met and became lifelong friends with Scottish philosopher and economist David Hume. This relationship led to Smith's appointment to the Glasgow University faculty in 1751.

In 1759, Smith published *The Theory of Moral Sentiments*, a book whose main contention is that human morality depends on sympathy between the individual and other members of society. On the heels of the book, he became the tutor of the future Duke of Buccleuch (1763–1766) and traveled with him to France, where Smith met with other prominent thinkers of his day, such as Benjamin Franklin[8] and French economist Anne-Robert-Jacques Turgot.[9]

After toiling for nine years, and with his work from 1776, *An Inquiry into the Nature and Causes of the Wealth of Nations* (usually shortened to *The Wealth of Nations*), Adam Smith changed the traditional way of viewing the production process. Smith's work is best known for his invisible-hand analogy of market pricing and self-regulation by the individual actors acting out of their own best interests. In this work, Smith uses the well-known example of the production of pins in which a specialized line of production can increase the production capability.[10] Furthermore, he is one of the first to do movement and time studies, and he suggests that the work processes can be split up in many parts. Smith's work is not only considered the first modern work of economics and the "father of modern economics"[11] but is also one of the main sources of inspiration for Taylorism.[12]

Smith's ideas are a reflection on economics in light of the beginning of the Industrial Revolution, and he states that free-market economies (i.e., markets based

流程项目的使用方法,包括确定特定的任务和活动,然后以完成某些目标为导向,为执行这些任务及其相关活动分配资源的方法,进而实现战争战略。

需要指出的是,公元604年,作为日本摄政王、政治家和学者的Shotoku Taishi(573—621年)制定了一套当时作为日本宪法的指导文献。在这些文献中,他在研究了中国的行政管理之后,论述了佛教与我们所谓的政府之间的关系。这些结论后来被研究和扩展,并被用作构建日本政府的理论基础。Shotoku Taishi详细说明了如何将任务和责任最好地安排在不同的行政部门之中。这也是本节作者能够确定的关于组织形式和目标之间关系的最早的著作之一。

2. A. Smith

A. Smith是苏格兰的伦理学家、政治经济学的先驱。同时也是苏格兰启蒙运动[4]的关键人物之一,A. Smith最著名的作品有两部:《道德情操论》[5](1759年),以及《国民财富的性质和原因研究》(通常简称为《国富论》)[6](1776年)。虽然A. Smith的确切出生日期不详,但他的洗礼(基督教的一种仪式)被记录是在1723年6月5日的苏格兰的Kirkcaldy。他在Burgh学校学习拉丁语、数学、历史和写作。A. Smith 14岁时进入Glasgow(格拉斯哥)大学,1740年去了牛津大学[7]。

1748年,A. Smith开始在Edinburgh(爱丁堡)大学进行一系列公开演讲。1750年,通过这些讲座,他认识了苏格兰哲学家和经济学家D. Hume,并与之成为终生的朋友。这种关系让A. Smith于1751年被任命为Glasgow大学的教师。

1759年,A. Smith出版了《道德情操论》,这本书的主要论点是人类的道德依赖于个人和社会其他成员之间的同情心。在这本书出版之后,他成为了未来的Buccleuch公爵(1763—1766年)*的导师,并与他一起前往法国,在那里A. Smith会见了当时其他著名的思想家,如B. Franklin(本杰明·富兰克林)、法国经济学家A. R. J. Turgot[9]。

A. Smith经过9年的创作,终于在1776年出版了他的著作《国富论》,从而改变了传统的生产流程观。A. Smith的作品最著名的是他用"看不见的手"来比喻市场定价和个体行为,人能够出于自己的最佳利益而进行自我调节。在这项研究中,A. Smith使用了著名的生产大头针的例子,在这个例子中,专业的生产线可以提高生产能力[10]。此外,他是最早对行为和时间的关系进行研究的人之一,他建议将工作流程分解为多个部分。因此,A. Smith被公认为"现代经济学之父",他的著作不仅被认为是第一部现代经济学著作[11],也是泰勒主义的主要灵感来源

---

*　此处年份为A. Smith担任Buccleuch公爵的导师时间。

on capitalism) are the most productive and beneficial to their societies. He goes on to argue for an economic system based on individual self-interest led by an "invisible hand," which would achieve the greatest good for all. In time, *The Wealth of Nations* won Smith a far-reaching reputation, and the work, considered a foundational work of classical economics, is one of the most influential books ever written.

In 1787, Smith was named rector of the University of Glasgow, and he died just three years later at the age of 67.

Taylorism[13]

Frederick Winslow Taylor (1856–1915), efficiency engineer and inventor, was born in Germantown, Philadelphia, Pa., the youngest child of Franklin and Emily Annette (Winslow) Taylor. He was a descendant of Samuel Taylor, who settled in Burlington, N. J., in 1677.[14]

His father was a lawyer, more interested, however, in literature than law; his mother was an ardent abolitionist and a coworker with Lucretia Mott [q.v.] in this cause. Taylor received his early education from his mother. In 1872, after two years of schooling in France and Germany, followed by 18 months of travel in Europe, he entered Phillips Exeter Academy at Exeter, N. H., to prepare for the Harvard Law School. Though he graduated with his class two years later, his eyesight had in the meantime become so impaired that he had to abandon further study. Between 1874 and 1878, he worked in the shops of the Enterprise Hydraulic Works, a pump-manufacturing company in Philadelphia, where he would learn the trades as a pattern-maker and machinist.

In the latter year, he joined the Midvale Steel Company, Philadelphia, as a common laborer. In the succeeding 12 years, he not only rose to be chief engineer (1884), but in 1883, by studying at night, obtained the degree of M.E. from Stevens Institute of Technology, Hoboken, N. J. In 1884, he married Louise M. Spooner of Philadelphia. His inventions during these years that affected improvements in machinery and manufacturing methods were many, the outstanding one being the design and construction of the largest successful steam hammer ever built in the United States (patent No. 424,939, Apr. 1, 1890). After three years (1890–93) as a general manager of the Manufacturing Investment Company, Philadelphia, operating large paper mills in Maine and Wisconsin, he began a consulting practice in Philadelphia—his business card read "Systematizing Shop Management and Manufacturing Costs a Specialty"—which led to the development of a new profession.

Behind this lay Taylor's years of observation and study of manufacturing conditions and methods. From these he had evolved a theory that, by scientific study of every minute step and operation in a manufacturing plant, data could be obtained as to the fair and reasonable production capacities of both man and machine, and that the application of such data would, in turn, abolish the antagonism between employer and employee and bring about increased efficiencies in all directions. He had, in addition, worked out a comprehensive system of analysis, classification, and symbolization to be used in the study of every type of manufacturing organization.

之一[12]。

A. Smith的思想是从工业革命角度对经济学的一种诠释,他指出自由市场(即基于资本主义的)市场)经济是最有生产力和对他们的社会最有益的。他主张建立一个以"看不见的手"为首的个人利益为基础的经济体系,这将为所有人带来最大的好处。随着时间的推移,《国富论》给A. Smith赢得了深远的声誉,这本被认为是古典经济学基础的著作,是有史以来最具影响力的著作之一。

1787年,A. Smith被任命为Glasgow大学校长,三年后去世,享年67岁。

3. 泰勒主义[13]

F. W. Taylor(1856—1915年),效率工程师和发明家,出生于宾夕法尼亚州费城的德国城,是Franklin和E. A. (Winslow)Taylor最小的孩子。他也是S. Taylor的后裔,1677年他定居在新泽西州伯灵顿[14]。

他的父亲是一名律师,然而,他对文学比法律更感兴趣;他的母亲是一名热心的废奴主义者,也是L. Mott在这方面的合作者。F. W. Taylor从母亲那里接受了早期教育。1872年,他在法国和德国接受了两年的教育,随后在欧洲旅行了18个月,进入位于新罕布什尔州埃克塞特的菲利普斯埃克塞特学院,为进入哈佛法学院做准备。尽管两年后他随班毕业,但他的视力在这段时间变得很差,不得不放弃继续深造。1874 ~ 1878年,他在费城的一家泵制造公司"企业液压厂"的商店工作,在那里他学习做一名制模工和机械师。

之后的一年,他加入了Midvale(费城米德维尔)钢铁公司,成为一名普通工人。在随后的12年里,他不仅晋升为总工程师(1884年),而且在1883年,通过夜间学习,获得了新泽西州霍博肯的史蒂文斯技术学院的工程学硕士学位。1884年,他与费城的L. M. Spooner结婚。这些年来,他的发明影响了机械和制造方法的改进,其中最突出的是美国有史以来最成功的蒸汽锤的设计和建造(专利号424939,1890年4月1日)。在费城制造投资公司(the Manufacturing Investment Company)担任总经理三年(1890 ~ 1893年),在缅因州和威斯康星州经营大型造纸厂之后,他在费城开始了一项咨询业务,他的名片上写着"系统化车间管理和制造成本是一项专业",这促进了新专业的发展。

这背后是F. W. Taylor多年来对制造条件和方法的观察和研究。由此,他发展了一种理论,即通过对制造厂每一个步骤和每一步操作的科学研究,可以获得有关人和机器公平合理生产能力的数据,而这些数据的应用,反过来又会消除雇主和雇员之间的对立,并带来各个方面效率的提升。此外,他还制定了一套全面的分析、分类和标准化系统,用于研究各种类型的制造组织。五年来,他成功地将他的

For five years, he successfully applied his theory in a variety of establishments, administrative and sales departments, as well as shops. In 1898, he was retained exclusively for that purpose by the Bethlehem Steel Company in Bethlehem, Pa. In the course of his work there, he undertook with J. Maunsel White a study of the treatment of tool steel, which led to the discovery of the Taylor–White process of heat treatment of tool steel, yielding increased cutting capacities of 200–300 percent. This process and the tools treated by it are now used in practically every machine shop of the world. While he was at Bethlehem, too, Taylor's ideas regarding scientific management took more concrete form. Being convinced of the results that would be attained if these principles should be generally adopted throughout the industrial world, he resigned from the Bethlehem Steel Company in 1901, returned to Philadelphia, and devoted the remainder of his life to expounding these principles, giving his services free to anybody who was sincerely desirous of carrying out his methods. While he met with many unbelievers among both employers and employees, he lived to see his system widely applied. In 1911, the Society to Promote the Science of Management (after his death renamed the Taylor Society) was established by enthusiastic engineers and industrialists throughout the world to carry on his work and legacy.

Among Taylor's contributions to the technical journals were "A Piece-Rate System"[15] (Transactions of the American Society of Mechanical Engineers, vol. XVI, 1895), an exposition of the principles on which his system of management was subsequently based, and "Shop Management," which was translated and published in almost every country of Europe. As an active member of the American Society of Mechanical Engineers, he served as vice-president in 1904–05 and as president in 1906, when he delivered as his presidential address an exhaustive monograph *On the Art of Cutting Metals*.

In 1911, Frederick Winslow Taylor published his work *Principles of Scientific Management*.[16] The purpose of this method, which is also known as Taylorism, is to identify the optimum method to achieve a certain goal. By categorizing people as a machine that can be manipulated, the factory processes could be streamlined and optimized. Taylor was among those who did the most to systematize the production technique and make it more scientific. Before the turn of the century, work processes and organization were based on experience, tradition, and, as seen from a production point of view, sheer chance. Taylorism meant a development of time and method studies and job analysis. The jumping-off point was the way the best and fastest workers performed a certain job. Taylor put a lot of emphasis on the standardization and specialization of the work—for both workers, foremen, and planners—and on the importance of increased quality control. Along with the practician Henry Ford, the car manufacturer, Taylor was among those who contributed the most to the development of the capitalist large-scale industry.[17]

Scientific Management is based on objective data, like measuring the time required for a certain job, which helped you set expected quotas for the workers. At the same time, the system was supposed to show the value of the work, and thus ensures the worker a reasonable salary for a job well done. Taylor was convinced that all interests were considered, which has not always been the case in practice. Taylorism was also

理论应用于各种企业、行政机构、销售部门以及商店。1898年,他被宾夕法尼亚州伯利恒市的伯利恒钢铁公司(Bethlehem Steel Company)专门为这一目的而雇用。在伯利恒钢铁公司工作的过程中,他与 J. M. White 一起对工具钢进行了处理,并进行了研究,发明了工具钢热处理的"泰勒-怀特(Taylor-White)流程",使切割能力增长了200% ~ 300%。这种流程和方法现在几乎世界上的每一个机器厂都在使用。F. W. Taylor 在伯利恒钢铁公司的时候,他关于科学管理的思想也有了更具体的形式,他深信如果这些原则在整个工业界得到普遍采用,将会取得丰硕成果。1901年,他从伯利恒钢铁公司辞职回到费城,并倾其余生致力于阐述这些原则,将他的服务免费提供给任何真诚希望执行他的方法的人。虽然他在雇主和雇员中遇到了许多不信赖他的人,但他在有生之年看见了他的理论得到广泛的应用。1911年,促进管理科学学会(他去世后改名为泰勒学会)由全世界热情的工程师和实业家建立,以继承他的工作和遗产。

F. W. Taylor 对学术的贡献还包括"计件工资制"[15][美国机械工程师协会(American Society of Mechanical Engineers, ASME)汇刊,第16卷,1895年],其中阐述了他的管理体制所依据的原则以及车间管理,这几乎在欧洲的每个国家都被翻译和出版。作为 ASME 的积极成员,他在1904 ~ 1905年担任副会长,1906年担任会长,在发表会长致辞时,他论述了关于金属切削技术的详尽专论。

1911年,F. W. Taylor 出版了他的《科学管理原理》[16],这种方法也称为泰勒主义,其目的是找出实现某一目标的最佳方法。通过将人的动作分解为一台可以操作的机器,工厂流程被简化和优化。F. W. Taylor 是那些把生产技术系统化并使其更科学的人中做得最多的。在世纪之交(19世纪与20世纪)之前,工作流程和组织是基于经验、传统的方法,从生产的角度来看,则是纯粹的随机。泰勒主义意味着时间、方法研究和工作分析的发展,起点是最好和最快的工人完成某项工作的方式。F. W. Taylor 非常重视工人、工头和规划者工作的标准化和专业化,并注重加强质量的把控性。F. W. Taylor 与汽车制造商 H. Ford(亨利·福特),都在对资本主义大型工业发展贡献最大的人之列[17]。

科学管理是建立在客观数据的基础上的,如测量某项工作所需的时间,这有助于为工人设定任务配额。同时,该系统还可体现工作的价值,从而确保工人在完成一项工作时获得合理的工资。F. W. Taylor 深信所有的利益都得到了考虑,但这一点在实际情况中并非总是如此。泰勒主义也是 H. Ford 的起点,而他创造了非常

the jumping-off point for Henry Ford, who created his own very successful production machinery. Taylor based his theory on the assumption that industrial growth required a large-scale influx of workers from the primary trades to the industry. These were workers with no industrial experience. This made it expedient to ensure a strong specialization and strict control, which would leave nothing to chance. But Taylor also claimed that most workers are lazy and stupid and solely interested in their salary; therefore, they will be better off with no responsibility and only simple jobs.

Taylorism has a bad reputation these days since it contributed to creating a self-fulfilling prophecy. Based on the assumption that workers are lazy and stupid, they created boring jobs that left no room for growth. Naturally, the workers reacted with apathy in a situation like that. This cemented the employer's belief that workers simply lacked skills, and that increased the strict control. The kind of work that these guidelines created has contributed to the alienation within the industry.

Frederick Winslow Taylor died in Philadelphia of pneumonia, survived by his widow and three adopted children.

Henry Ford[18]

Famed automobile manufacturer Henry Ford was born on July 30, 1863, on his family's farm in Wayne County, near Dearborn, Michigan. When Ford was 13 years old, his father gifted him a pocket watch, which the young boy promptly took apart and reassembled. Friends and neighbors were impressed and requested that he fixed their timepieces too. Unsatisfied with farm work, Ford left home the following year, at the age of 16, to take an apprenticeship as a machinist in Detroit. In the years that followed, he would learn to skillfully operate and service steam engines and would also study bookkeeping.

In 1888, Ford married Clara Ala Bryant and briefly returned to farming to support his wife and son, Edsel. But three years later, he was hired as an engineer for the Edison Illuminating Company. In 1893, his natural talents earned him a promotion to chief engineer. All the while, Ford developed his plans for a horseless carriage, and in 1896, he constructed his first model, the Ford Quadricycle. Within the same year, he attended a meeting with Edison executives and found himself presenting his automobile plans to Thomas Edison. The lighting genius encouraged Ford to build a second, better model.

After a few trials building cars and companies, in 1903, Henry Ford established the Ford Motor Company. Ford introduced the Model T in October of 1908, and for several years, the company posted 100 percent gains. However, more than for his profits, Ford became renowned for his revolutionary vision: the manufacture of an inexpensive automobile made by skilled workers who earn steady wages. Inspired by Taylorism, Ford started his mass production[17] and revolutionized the production process. In 1914, his plant in Highland Park, Michigan, could produce a complete frame every 93 min thanks to innovative production techniques. This was a remarkable improvement in comparison with the former production time of 728 min.

成功的生产机械化。F. W. Taylor的理论基于这样一个假设：工业增长需要大量的工人从初级行业涌入工业，他们是没有工业经验的工人，这对确保强大的专业化和严格的控制是有利的，从而把事情做得四平八稳，不留漏洞。但F. W. Taylor也声称，大多数工人都懒惰、愚蠢，只对自己的工资感兴趣，因此，他们会在无责任和简单工作的情况下企图过得更好。

泰勒主义现在的名声并不好，因为它促成了一个自我实现的预言。基于工人懒惰和愚蠢的假设，他们创造了没有增长空间和无聊的工作。自然而然，工人们在这种情况下变得反应漠然。这巩固了雇主的信念，即：工人只是缺乏技能，需要进行严格的控制。这些指导方针所创造的工作类型促成了行业内部的异化。

F. W. Taylor在费城病逝于肺炎，留下他的遗孀和三个收养的孩子。

4. H. Ford[18]

著名的汽车制造商H. Ford于1863年7月30日出生在密歇根州迪尔伯恩（Dearborn）附近韦恩（Wayne County）的一个家庭农场。当H. Ford 13岁时，他的父亲送给他一块怀表，这个小男孩很快就把它拆开重新组装了。朋友和邻居对他印象深刻，并请他也修理他们的钟表。H. Ford不满足于只做农活，16岁离开家，在底特律当了一名机械师学徒。在接下来的几年里，他学会了熟练操作和维修蒸汽机，还学会了记账。

1888年，H. Ford与C. A. Bryant结婚，并暂返农场，与他的妻子和儿子埃德塞尔一起生活。但三年后，他被爱迪生照明公司聘为工程师。1893年，他的天赋使他晋升为总工程师。一直以来，H. Ford都在发展他的"无马马车计划"（horseless carriage），1896年，他建造了第一个福特四轮车模型。同年，他参加了一次爱迪生照明公司高管的会议，向T. Edison（托马斯·爱迪生）展示了自己的汽车计划。照明天才T. Edison鼓励H. Ford制造第二个更好的车型。

1903年，H. Ford在几次汽车建造和公司建立的试验之后，成立了福特汽车公司。福特汽车公司在1908年10月推出了T型车，几年后，福特汽车公司实现了100%的增长。然而，福特汽车公司不仅因为利润，还因为其革命性的愿景而出名，即：生产一种价廉的汽车，这种汽车由熟练的工人制造，由此他们挣得稳定的工资。在泰勒主义的启发下，H. Ford开始了他的大规模生产，并彻底改变了生产流程。1914年，由于创新的生产技术，他在密歇根州高地公园的工厂每93分钟就可以生产一个完整的车架。与以前728分钟的生产时间相比，这是一个显著的改进。通过使用传送带，分解工作流程，仔细协调和操作，福特汽车公司通过提升

Through the use of a conveyer belt, the breaking down of work processes and careful coordination and operations, Ford gained serious profits in productivity.

In 1914, Ford started paying his employees five dollars a day, which was almost double the salary paid by other manufacturers. He cut the working hours from nine to eight hours for a three-shift working day at the factory. Ford's mass production techniques would finally make it possible to build a Model T every 24 s. Ford's affordable Model T and the way it was produced has changed American society for good. As more Americans became car owners, the patterns of city life changed. As the United States saw a traffic increase in the suburbs, a highway system was established, and the population could transport itself anywhere it wanted to. From a social perspective, Henry Ford was marked by seemingly contradictory viewpoints. In business, Ford offered profit sharing to select employees who stayed with the company for six months and, most importantly, who conducted their lives in a respectable manner.

The company's "Social Department" looked into an employee's drinking, gambling, and otherwise uncouth habits to determine eligibility for participation. Ford was also an ardent pacifist and opposed World War I, and even funded a peace ship to Europe. Later, in 1936, Ford and his family established the Ford Foundation to provide ongoing grants for research, education, and development. But despite these philanthropic leanings, Ford was also a committed anti-semite, going as far as to support a weekly newspaper, The Dearborn Independent, which furthered such views.

Henry Ford died of a cerebral hemorrhage on April 7, 1947, at the age of 83, near his Dearborn estate, Fair Lane. Ford, considered one of America's leading businessmen, is credited today for helping to build America's economy during the nation's vulnerable early years. His legacy will live on for decades to come.[19]

End Notes

1. BPMN: The business process modeling notation, Patrice Briol, s. l. s. n. 2008.
2. The Art of War, Sun Tzu (author), B. H. Liddell Hart (Foreword), Samuel B. Griffith (Translator), Oxford University Press, 1971.
3. Process Evolution Timeline Model, LEADing Practice Business Process Reference Content #LEAD-ES20005BP.
4. Great Thinkers of the Scottish Enlightenment, BBC, from http://www.bbc.co.uk/history /scottishhistory/enlightenment/features_enlightenment_enlightenment.shtml.
5. The Theory of Moral Sentiments, Adam Smith, 1759 from http://www.earlymoderntexts. com/pdfs/smith1759.pdf.
6. An Inquiry into the Nature and Causes of the Wealth of Nations, Adam Smith, 1776 from http://www.gutenberg.org/files/3300/3300-h/3300-h.htm#link2HCH0028.
7. Smith A., The Biography.com website, (2014), Retrieved 11:52, August 11, 2014, from http://www.biography.com/people/adam-smith-9486480.
8. Benjamin Franklin from http://www.biography.com/people/benjamin-franklin-9301234.
9. Anne-Robert-Jacques Turgot from http://www.econlib.org/library/Enc/bios/Turgot.html.
10. The History of Management Thought, Daniel A. Wren, Wiley, 2005.
11. Davis William L., Figgins B., Hedengren D., and Klein D.B., "Economic Professors' Favorite Economic Thinkers, Journals, and Blogs," Econ Journal Watch 8, no. 2 (May 2011): 126–146.

生产效率获得了可观的利润。

1914年，H. Ford开始向员工每天支付5美元的工资，这几乎是其他制造商工资的两倍。他把工厂的工作时间从9个小时缩短到8个小时，三班制工作日。福特汽车公司的大规模生产技术最终使每24分钟生产一辆T型车成为可能。福特汽车公司这种大众消费得起的T型车和它的生产方式已经彻底改变了美国社会。随着越来越多的美国人拥有汽车，城市生活的模式发生了变化。美国政府看到郊区的交通量增加后，建立了高速公路系统，人们可以自行去任何他想去的地方。从社会的角度来看，H. Ford的观点似乎矛盾。在商业上，福特汽车公司提供利润分成给选择在公司工作六个月以上的员工，最重要的是：他们能以体面的方式生活。

福特汽车公司的公共部门会经常调查员工的饮酒、赌博和其他粗俗习惯及行为，以确定他们是否有资格留在公司工作。H. Ford也是一个热心的和平主义者，反对第一次世界大战，甚至资助了一艘前往欧洲的和平船。后来，在1936年，H. Ford和他的家人建立了福特基金会，为研究、教育和发展提供持续的补助金。但是，尽管有这些慈善的倾向，H. Ford还是一个坚定的反犹分子，甚至为发表他的观点而支持一家名为《迪尔伯恩独立报》的周报。

1947年4月7日，H. Ford在他位于费尔莱恩的迪尔伯恩庄园附近因脑出血病逝，享年83岁。H. Ford被认为是美国最成功的商人之一，至今仍被赞颂在美国脆弱的早期帮助重塑了美国经济。他的遗产也将在未来几十年中继续运作[19]。

参考文献

[1] BPMN: The business process modeling notation, Patrice Briol, s. l. s. n. 2008.

[2] The Art of War, Sun Tzu (author), B. H. Liddell Hart (Foreword), Samuel B. Griffith (Translator), Oxford University Press, 1971.

[3] Process Evolution Timeline Model, LEADing Practice Business Process Reference Content #LEAD-ES20005BP.

[4] Great Thinkers of the Scottish Enlightenment, BBC, from http: //www.bbc.co.uk/history/scottishhistory/enlightenment/features_enlightenment_enlightenment.shtml.

[5] The Theory of Moral Sentiments, Adam Smith, 1759 from http: //www.earlymoderntexts.com/pdfs/smith1759.pdf.

[6] An Inquiry into the Nature and Causes of the Wealth of Nations, Adam Smith, 1776 from http: //www.gutenberg.org/files/3300/3300-h/3300-h.htm#link2HCH0028.

[7] Smith A., The Biography.com website, (2014), Retrieved 11: 52, August 11, 2014, from http: //www.biography.com/people/adam-smith-9486480.

[8] Benjamin Franklin from http: //www.biography.com/people/benjamin-franklin-9301234.

[9] Anne-Robert-Jacques Turgot from http: //www.econlib.org/library/Enc/bios/Turgot.html.

[10] The History of Management Thought, Daniel A. Wren, Wiley, 2005.

[11] Davis William L., Figgins B., Hedengren D., and Klein D.B., "Economic Professors' Favorite Economic Thinkers, Journals, and Blogs," Econ Journal Watch 8, no. 2 (May 2011): 126-146.

12. The History of Management Thought, Daniel A. Wren, Wiley, 2005.

13. American genesis: A century of invention and technological enthusiasm, 1870–1970, Thomas Parke Hughes, University of Chicago Press, 2004.

14. *Dictionary of American Biography Base Set*, Frederick Winslow Taylor, American Council of Learned Societies, 1928–1936.

15. The visible hand: The managerial revolution in American business, Alfred Dupont Chandler, Belknap Press, 2002.

16. The Principles of Scientific Management, Frederick Winslow Taylor, Courier Dover Publications, 1998.

17. Henry Ford, mass production, modernism and design, Ray Batchelor, Manchester University Press, 1994.

18. The People's Tycoon: Henry Ford and the American Century, Steven Watts, Vintage Books, 2006.

19. Ford H., The Biography.com website, (2014), Retrieved 11:34, August 11, 2014, from http://www.biography.com/people/henry-ford-9298747.

[12]　The History of Management Thought, Daniel A. Wren, Wiley, 2005.

[13]　American genesis: A century of invention and technological enthusiasm, 1870−1970, Thomas Parke Hughes, University of Chicago Press, 2004.

[14]　Dictionary of American Biography Base Set, Frederick Winslow Taylor, American Council of Learned Societies, 1928−1936.

[15]　The visible hand: The managerial revolution in American business, Alfred Dupont Chandler, Belknap Press, 2002.

[16]　The Principles of Scientific Management, Frederick Winslow Taylor, Courier Dover Publications, 1998.

[17]　Henry Ford, mass production, modernism and design, Ray Batchelor, Manchester University Press, 1994.

[18]　The People's Tycoon: Henry Ford and the American Century, Steven Watts, Vintage Books, 2006.

[19]　Ford H., The Biography.com website, (2014), Retrieved 11: 34, August 11, 2014, from http: //www. biography.com/people/henry-ford-9298747.

Phase 2: Process Concept Evolution

Henrik von Scheel, Mark von Rosing, Maria Hove, Marianne Fonseca, Ulrik Foldager

INTRODUCTION

Throughout the first phase of the historical development of processes, we elaborated on some of the key figures who are widely regarded as pioneers—both in academia as well as in business—and laid the foundation for the upcoming—as well as today's—process studies and the use of them in organizations across the world. The second phase of the historical evolution of process flows incorporates a far more extensive use of human involvement, however, and tells the story of human interaction around processes, process flows, tasks, and activities, as well as the necessary application and consumption of resources that are associated with carrying out each of these tasks and activities.

Upon entering the second phase, we are introduced to Henry Laurence Gantt, an American engineer who is best known for his planning methodology.[1] This methodology helped him realize major infrastructure projects including the construction of the Hoover Dam in the United States. His manager, Frederick W. Taylor, involved Gantt in a number of large infrastructure projects. Together with Taylor, Henry Gantt applied different scientific management principles in order to implement these projects successfully.

While closing in on the twentieth century, Frank B. Gilbreth introduces us to much more efficient bricklaying techniques, as he studies the sequence and number of physical movements associated with laying down each brick upon the construction of a building. By studying and documenting the movements associated with physical labor, he found that one would be able to reduce the number of movements required in order to carry out a specific task. This enabled him to effectively reduce the number of physical movements associated with that particular task, and thus, improve the flow of the process by shortening the time consumption while increasing productivity. He would also be the first person ever to document the flow of a process through the use of a so-called Process Chart, as he coined it.

Later on in the early 1930s, Allan H. Mogensen evolved upon this methodology through the study of motion pictures, which enabled him to further improve upon the sequence of activities that was involved with each task. He created process diagrams—a more detailed and elaborate version of Frank Gilbreth's process charts—to document each task that enabled him to understand and visualize the activities that were involved with that task. This would allow him to reduce the number of motions required to carry out a particular task, and reducing the time needed to complete it. These techniques were used in hospitals at the time and would thereby

The Complete Business Process Handbook. http://dx.doi.org/10.1016/B978-0-12-799959-3.00002-1

1.2　第2阶段：流程概念演变2.0

Henrik von Scheel, Mark von Rosing, Maria Hove, Marianne Fonseca, Ulrik Foldager

1.2.1　介绍

我们第1阶段详细阐述的一些被广泛认为是学术界和商业领域先驱者的关键人物对整个流程的历史发展所做的研究，为今后世界范围内各组织的使用奠定了基础。流程演变的第2阶段包含了人员与流程之间、流程之间、任务和活动之间的互动，以及执行这些任务和活动所必需的应用和资源投入。

在第2阶段，我们将介绍以擅长项目规划方法而闻名[1]的美国工程师H. L. Gantt（1861—1919年）。这种方法帮助他完成了重大的基础设施项目，包括建造美国胡佛水坝。他的经理F. W. Taylor参与了他的许多大型基础设施项目。为了成功实施这些项目，F. W. Taylor和H. L. Gantt一起运用了科学管理的多种原理。

在20世纪初的时候，F. B. Gilbreth向我们展现了更有效的砌砖技术，他研究了每一块砖在建筑施工时的铺设顺序和搬运次数。通过研究和记录与体力劳动相关的活动，他发现人们可以减少执行特定任务所需的动作次数。这使人们能够有效地减少与该特定任务相关的物理移动的次数，从而通过缩短时间消耗和提高生产力来改善流程的步骤。他也是第一个创造流程图并通过使用所谓的流程图来记录流程的人。

20世纪30年代早期，A. H. Mogensen通过对电影的研究，对这种方法进行了改进（如改进与每项任务有关的活动顺序），他创建了流程图，这是F. B. Gilbreth的流程图的一个更详细和复杂的版本，用来记录每个任务，使他能够理解和可视化与该任务相关的活动。这将使他能够减少执行特定任务所需的动作次数，并减少完成该任务所需的时间。这些技术当时已在医院使用，证明通过减少不必要的动作和活动能够提高患者在手术期间存活的概率。

be used to increase the likelihood of survival of patients during surgery by reducing the need for unnecessary movements and activities.

At the end of the 1930s, William E. Boeing would then take these practices, and the knowledge thereof, and learn how to effectively break down a process into smaller bits, thereby enabling him to simplify and improve processes that were involved with the building of both civilian as well as military aircrafts. Through the use of these enhanced process techniques, Boeing would become able to supply the American Air Force with the B-17 Flying Fortresses and B-29 Superfortresses during the Second World War that were able to carry heavy military equipment, such as nuclear bombs and other material. It would become one of the first very good examples of process documentation for process flows and breakdown.

In the 1940s, Ben S. Graham would then take these process practices that he had acquired from Allan H. Mogensen and Lillian Gilbreth (the wife of Frank B. Gilbreth) into office work in order to simplify paperwork and work more efficiently with information at an office desk. He used his own techniques to map out the processes of the employees that worked with paperwork—nonphysical and non-manufacturing labor—which would later be used across the world today. It is also known as Business Process Improvement, although the term was not coined as such at that time.

In 1947, the American Society of Mechanical Engineers—or ASME—would become the first organization to develop and establish an international standard of process symbols, and document how to use them in process charts and process diagrams. Shortly thereafter, namely in the 1950s, Functional Flow Block Diagrams began to surface that would commonly be used to describe processes in development and production system environments.

In 1957, the PERT methodology was introduced and is utilized to illustrate and analyze program-related tasks. This was particularly helpful, as it would give employees and managers involved with programs the ability to set up timelines, estimated durations, and the following efficiency evaluation—with a focus on time consumption—in each program.

With the introduction of the Data Flow Diagram, engineers would have a tool to produce a graphical illustration (process diagrams) that would place information blocks into the flow of a process, giving them the ability to see how and where data would be stored in the process, as well as which kind of input would be delivered to what and the consequential output of this behavior. Around nearly the same time, we see the arrival of IDEF (Integrated Definition), which was developed by Knowledge Based Systems, Inc. (KBSI). IDEF would introduce methods of function modelling, information modelling, data modelling, and object-oriented design, as well as a method for capturing process and ontology descriptions.

In the 1960s, the theory of Zero Defects was established and expressed the idea that *a product that meets the requirements of the consumer is the right product*. In 1979, Philip Crosby penned *Quality Is Free: The Art of Making Quality Certain*,[2] which would introduce a 14-step quality improvement program designed to not only improve quality but also reduce an unnecessary waste of resources in production facilities.

20世纪30年代末，W. E. Boeing采用这些实践和相关知识，尝试如何有效地将一个过程分解为更小的部分，从而使他能够简化和改进民用和军用飞机制造流程。通过使用这些细分的工艺技术，波音公司在第二次世界大战期间向美国空军提供了能够携带如核弹和其他武器等重型军事装备的B-17飞行堡垒和B-29超级堡垒，这成为了以流程记录和分解为目标的流程文档应用的第一个非常好的案例。

20世纪40年代，B. S. Graham将从A. H. Mogensen和L. Gilbreth（F. B. Gilbreth的妻子）那里获得的这些流程实践应用到办公室工作中，以简化文书工作并更有效地处理办公桌上的文件。他利用自己掌握的技术来规划员工涉及文书工作、无形劳动和非制造劳动的工作流程，而这种办法如今已在世界各地使用，它也被称为业务流程改进（BPI，尽管当时这个术语并没有被创造出来）。

1947年，ASME成为第一个制定和建立国际流程符号标准的组织，并记录如何在工艺图和流程图中使用这些符号。此后不久，即在20世纪50年代，用于描述开发和生产系统环境的功能流程框图（functional flow block diagram，FFBD）开始出现。

1957年，计划评估评审技术（program evaluation review technique，PERT）方法被引入，并用它来说明和分析程序相关的任务。这一点特别有用，因为这将使参与计划的员工和管理者能够制定时间表、评估持续时间，并对每个计划的时间消耗进行有效评估。

随着数据流程图（data flow diagram，DFD）的引入，工程师有了一个绘成图解（流程图）的工具，它将信息块放入流程图中，使他们能够了解数据在流程中的存储方式和位置，以及流程行为输入哪种信息和输出什么结果。几乎在同一时间，我们看到了由知识系统公司（Knowledge Based Systems Inc.，KBSI）开发的集成定义（integrated definition，IDEF）的到来。IDEF方法不仅包括流程构建和本体描述，还包括功能建模、信息建模、数据建模和面向对象设计的方法。

在20世纪60年代，零缺陷（zero defects）理论被确立并表达了这样一种观点：满足消费者需求的产品才是正确的产品。1979年，P. B. Crosby（菲利浦·克劳士比）笔下的《质量免费：确定质量的艺术》[2]引入了一个14步质量改进计划，旨在不仅提高质量，而且要减少生产设施中不必要的资源浪费。

At the end of the second phase, Taichi Ohno and Shigeo Shingo developed the Toyota Production System (TPS) between 1948 and 1975. In 1988, Taichi Ohno explained the philosophies and concepts behind TPS through the publication of *Just-in-Time for Today and Tomorrow*.[3] TPS focuses mainly around two different concepts: (1) *Jidoka*, which means that when a problem occurs, the equipment stops immediately, preventing defective products from being produced, and (2) *Just-in-Time*, in which each process produces only what is needed by the next process within the continuous flow of production processes.

GANTT CHART

Henry Laurence Gantt, AB, ME, (1861–1919) was an American engineer and famous management consultant who is best known for his planning methodology. This methodology helped him realize major infrastructure projects, including the construction of the Hoover Dam in the United States. Henry Gantt graduated with a bachelor's degree (AB) from McDonogh School (United States) in 1878. Then he went on to the Stevens Institute of Technology (New Jersey, United States) to obtain his master's degree in engineering (ME). After obtaining this degree, Gantt worked as a teacher until 1887. From 1887, Henry Gantt chose a new challenge and joined Midvale Steel Company in Philadelphia, United States. His manager, Frederick W. Taylor, involved Gantt in a number of large infrastructure projects. Together with Taylor, Henry Gantt applied different scientific management principles in order to implement these projects successfully.

Henry Gantt worked for Midvale Steel Company until 1893. He continued his career as a management consultant and developed his famous planning methodology. Henry Gantt also developed a task and bonus system of wage payment and measurement instruments to provide an insight into worker efficiency and productivity. In 1916, inspired by Thorsten Veblen, Gantt set up a trade association that was aimed at the development of industrial efficiency within political processes. In addition, he called into question the industrial system under control of managers and Polakov's analysis[4] of inefficiency within the industrial sector.

While Ford implemented Taylorism in practice, Henry Gantt chose to illustrate the process work behind it all. This way of keeping track of all processes is utilized all over the world today. The Gantt chart that we know today was created from 1910 to 1915, and it was later published in the entire western world.[5] A Gantt chart is a type of bar chart that illustrates a project schedule. Gantt charts illustrate the start and finish dates of the terminal elements and summary elements of a project. Terminal elements and summary elements comprise the work breakdown structure of the project. Modern Gantt charts also show the dependency (i.e., precedence network) relationships between activities. Gantt charts can be used to show current schedule status using percent-complete shadings and a vertical "TODAY" (the "TODAY" indicator describes what happens on that particular day or at this particular time) line as shown in (Figure 1).

Gantt charts (see Figure 1) have become the common method to present the phases and activities in a project's work breakdown structure thus making it comprehensible to a wide audience. Although now regarded as a common charting technique, Gantt

在第二阶段结束时，Taichi Ohno 和 Shigeo Shingo 在 1948 ～ 1975年开发了丰田生产方式（Toyota production system，TPS）。1988年，Taichi Ohno 通过《准时化生产的今天与明天》的出版，解释了TPS背后的哲学和概念[3]，TPS主要聚焦于两个不同的概念：① 自働化*，这意味着当出现问题时，设备立即停止，防止生产出有缺陷的产品；② 准时制，在生产过程的连续流程中，每个流程只生产下一个流程所需的产品。

1.2.2　甘特（Gantt）图

美国著名的工程师兼管理顾问 H. L. Gantt（学士，工程学硕士）以其规划方法论而闻名。这种方法帮助他完成了主要的基础设施项目，包括美国胡佛大坝的建设。H. L. Gantt 于1878年毕业于美国麦克多诺学院（McDonogh School），获得学士学位。随后他继续到史蒂文斯技术学院（美国新泽西州）攻读工程学硕士学位。在获得这个学位后，H. L. Gantt 一直担任教师直到1887年。从1887年开始，H. L. Gantt 选择了一个新的挑战，加入了美国费城的 Midvale 钢铁公司。他的经理 F. W. Taylor 参与了他的一些大型基础设施项目。为了成功实施这些项目，F. W. Taylor 和 H. L. Gantt 一起创造了科学管理原理体系。

H. L. Gantt 在 Midvale 钢铁公司工作到1893年，然后以管理顾问的身份继续他的职业生涯，并发展他的著名的规划方法。H. L. Gantt 还开发了一项度量"任务、工资支付和奖金制度"的方法，能够对工人效率和生产力进行深入了解。1916年，受 Thorsten Veblen 的启发，H. L. Gantt 成立了一个贸易协会，旨在提升制度流程中的工业效率。此外，他还对管理者控制下的工业体系提出了质疑，并对 Polakov（波拉科夫）对工业部门效率低下的分析提出了质疑[4]。

当 H. Ford 在实践中实施泰勒主义时，H. L. Gantt 开始研究流程背后的原理。这种跟踪所有流程的方法在当今世界各地都得到了应用。我们今天所知道的甘特图是在1910 ～ 1915年创建的，后来在整个西方世界得以应用[5]。甘特图是一种用于说明工程项目的条形图，说明了项目的终端元素和概要元素的开始和结束日期。终端元素和概要元素构成了项目的工作分解结构。现代甘特图还显示了活动之间的依赖关系（即前导网络）。甘特图可以用来显示当前进度状态使用的完成百分比阴影和一个垂直的"TODAY"（"TODAY"指标描述了发生在某一天里或在这个特殊时间的事件），线型如图1所示。

甘特图（图1）的结构使它能为广大观众所理解，并已成为在项目的工作分解结构中显示阶段和进程的常用方法。虽然甘特图现在被认为是一种常用的制图技

---

* 此处自働化不同于自动化，是日语 Jidoka 译文。

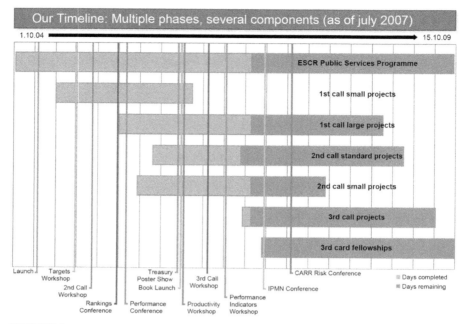

FIGURE 1

Seen as a whole, a Gantt chart is a useful tool to get a general idea of the process work and to be able to control it, but it does have its shortcomings that you need to take into consideration.[6]

charts were considered extremely revolutionary when first introduced. This chart is also used in information technology to represent data that have been collected.

Even though a Gantt diagram is useful and valuable to smaller projects, which fits nicely on a single sheet or on the screen, they tend to become a bit too overwhelming for projects that includes a lot of activities. Larger Gantt charts are unsuitable for most computer screens, and they do not give the intended easy overview. Gantt charts are also criticized for not having more than a little information in each area of the screen. Projects are often more complicated that can be illustrated effectively in a Gantt chart. Since the horizontal bars in a Gantt chart have a fixed height, they cannot be representative of the work-load in the time phase (resource requirements) of a project. In the model below, all activities have the same size, but in reality the work-load itself can be much different. Another criticism states that activities in a Gantt chart show a planned work-load as a constant.

FRANK B. GILBRETH

Born in Fairfield, ME, on July 7, 1868, Gilbreth broke into the construction industry as a bricklayer shortly after his high school graduation. In the course of his work, Gilbreth observed that each bricklayer approached his job differently, some seemingly

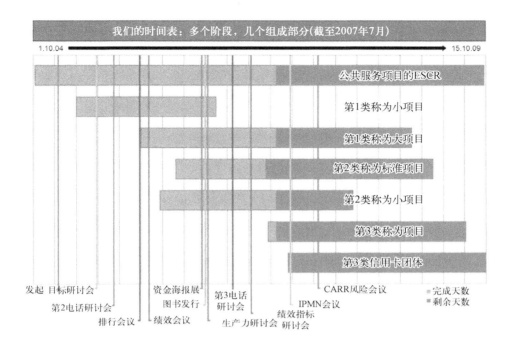

图1　从整体上看,甘特图是一个很有用的工具,它可以让您对流程工作有一个大致的了解,并能够对其进行控制,但它确实有一些不足之处需要您加以考虑[6]

术,但在最初引入甘特图时,它被认为是一种革命性的技术。甘特图还用于信息技术(information technology,IT)领域表示已收集的数据。

尽管甘特图对于较小的项目(非常适合放在单个工作表或控制板上)非常有用和有价值,但对于包含大量活动的项目来说,它们往往会变得有些难以承受。较大的甘特图不适用于大多数计算机屏幕,并且不能给出预期的简单概述。甘特图也因为在屏幕的每个区域没有太多的信息而受到批评。项目通常比在甘特图中的表示要复杂得多。此外,由于甘特图中的水平条具有固定的高度,它们不能代表项目的时间阶段(资源需求)中的工作负荷。在下面的模型中,所有活动都有相同的大小,但实际上,工作负荷本身可能会有很大的不同。另一种批评认为,甘特图中的活动将计划的工作量显示为固定量也不太合适。

1.2.3　F. B. Gilbreth

1868年7月7日,F. B. Gilbreth出生于缅因州费尔菲尔德,高中毕业后不久,他进入建筑业,做了一名砖匠。在他的工作过程中,F. B. Gilbreth观察到,每一个砖匠对待自己的工作都不一样,有些似乎比其他人更有效率。然后他开始分析他们

more efficient than others. He then began analyzing their motions to determine which approach to bricklaying was the best. Hence, his pioneering work in motion analysis and how it was applied to the workforce was under way. Over the years, he developed many improvements in bricklaying. He invented a scaffold that was easily adjusted to allow the worker to be at the most advantageous level at all times. He created a system whereby bricks were stacked on the scaffold in such a manner that the worker could easily pick up a brick in one hand and mortar in the other. By using some of the previously established best pracitces that he had observed, and some that he made up himself, he improved his performance by 190%. This led Gilbreth to improve his theories on how to find the best practices.[7,8] He used different methods to gain a better understanding of the movements that people performed while doing a certain job—first he filmed them, and later, he divided the hand movements into 17 minor standard movements. To monitor the processes and these movements in a work situation, he developed the "process chart" and the "flow diagram." The flow diagram is a diagram that shows the separate parts of a process.[9]

Gilbreth learned every trade in the construction business and advanced to superintendent without the typical three years of apprentice work. At the age of 27, Gilbreth started his own contracting firm, where he patented many inventions, including a concrete mixer and concrete conveyor system. He adopted the slogan "Speed Work" for his company and expressed his goals as the elimination of waste, the conservation of ability, and the reduction of cost. He was lauded for the application of these principles in the rapid construction of the Augustus Lowell Laboratory of Electrical Engineering for the Massachusetts Institute of Technology. His company was involved in a variety of construction projects, including dams, canals, houses, factory buildings, and an industrial facility. He eventually expanded his business to England.[10]

Gilbreth had the good fortune to meet Lillian Evelyn Moller, and they married in 1904. In addition to raising 12 children and being the subject of the Hollywood movie *Cheaper by the Dozen*, they became one of the great husband–wife teams of science and engineering. Together, they collaborated on the development of a micromotion study as an engineering and management technique and introduced the application of psychology to industrial management. They saw the need to improve worker satisfaction, which would in turn improve overall job performance and worker efficiency. Gilbreth designed systems to ease worker fatigue and increase productivity by studying each movement a worker made, and in doing so, document the best way to perform the task. They also considered the physical comfort of the worker and their innovations in office furniture led to the study of ergonomics.

In 1907, Gilbreth met engineer and inventor Frederick Winslow Taylor and became a proponent of the Taylor System of Time Study.[11] Frank and Lillian were instrumental in creating the Taylor Society. In 1912, the Gilbreths left the construction industry to focus their efforts on scientific management consulting. They broke with Taylor in 1914 and formed their own scientific management company with the intent to focus on the human element of management, as well as the technical.

的动作,以确定哪种砌砖方法是最好的。因此,他在动作分析方面的开创性工作以及如何将其应用于劳动力方面的工作由此展开。多年中,他在砌砖方面的研究取得了许多进步。他发明了一种易于调整的脚手架,使工人能一直发挥最好的水平。他建立了一个系统,在脚手架上堆放砖块,这样工人一只手就可以轻松地拿起一块砖,另一只手就可以很容易地拿起砂浆。通过使用观察到的一些最佳实践,以及他自己填补的一些方法,他将自己的绩效提高了190%。这使F. B. Gilbreth得以改进他的如何找到最佳实践的理论[7,8]。他用不同的方法来更好地了解人们在做某项工作时所做的动作。他先拍摄这些动作,然后将手部动作分成17个小的标准动作。为了监控工作环境中的流程和这些动作,他开创了"流程路径图"和"流程分解图"。"流程分解图"是显示流程各个部分的图解[9]。

F. B. Gilbreth学习了建筑业的每一个业务,在没经历一贯的三年学徒工作的情况下晋升为主管。27岁时,F. B. Gilbreth创办了自己的承包公司,并获得了许多发明专利,包括混凝土搅拌机和混凝土输送系统。他为公司采用了"加快工作"("Speed Work")的口号,并表示他的目标是消除浪费、提升能力和降低成本。他因将这些原则应用于麻省理工学院(Massachusetts Institute of Technology, MIT)奥古斯都·洛厄尔电气工程实验室的快速建设而受到称赞。他的公司参与了各种建设项目,包括水坝、运河、住宅、厂房和工业设施。他最终把生意扩展到了英国[10]。

F. B. Gilbreth有幸遇见了L. E. Moller,他们于1904年结婚,抚养了12个孩子,这一事情还成为好莱坞电影《儿女一箩筐》的题材,他们还成为科学和工程领域最伟大的夫妻团队之一。他们共同合作开发了一项微动作研究,作为一项工程和管理技术,并将心理学应用于工业管理。他们看到了提高员工满意度对提高整体工作绩效和员工效率的重要性。F. B. Gilbreth通过研究工人所做的每一个动作设计了一个缓解工人疲劳、提高生产力的系统,并以此来作为完成任务的最佳标准。他们还考虑了员工的身体舒适度,他们在办公家具上的创新引发了人体工程学的研究。

1907年,F. B. Gilbreth会见了工程师和发明家F. W. Taylor,并成为泰勒时间管理理论研究的支持者[11]。F. B. Gilbreth和L. Gilbreth在创建泰勒协会方面发挥了重要作用。1912年,F. B. Gilbreth一家离开建筑业,把精力集中在科学管理咨询上。他们于1914年与F. W. Taylor分道扬镳,成立了自己的科学管理公司,致力于关注管理的人为因素和技术因素。他们认为F. W. Taylor的"秒表"方法主要是为

They felt Taylor's "stop-watch" approach was primarily concerned with reducing process times, whereas the Gilbreths focused on making processes more efficient by reducing the motions involved. The Gilbreths continued their micromotion studies in other fields, pioneering the use of motion pictures for studying various aspects of work and workers.

In the early months of World War I, Gilbreth studied industrial processes and machinery in Germany. As wounded soldiers began returning home, Gilbreth applied his principles to improving surgical procedures and was the first to use the motion-picture camera in the operating room for educational purposes. He was also the first to propose that a surgical nurse serve as a "caddy" to the surgeon by handing surgical instruments to the surgeon during a procedure. He also helped rehabilitate injured soldiers by developing ways to help them manage their daily activities.

In 1920, ASME instituted its Management Division, which Gilbreth had helped to establish. He became one of the most widely known engineers in the US and Europe and reaped financial rewards and many professional honors. At an ASME conference in 1921, Gilbreth became the first to present a structured method to document process flow. He called his presentation "Process Charts—First Steps in Finding the One Best Way."[12] Seen as a whole, the purpose of Gilbreth's work was to develop the worker's full potential through effective training, new work methods, improved working environment and tools, and a healthy psychological outlook on life.[13]

Frank B. Gilbreth suggested the inaugural international management congress that was held in Prague in 1924. He was stricken with a heart attack shortly after the conference and died on June 14, 1924, while traveling from his home in Montclair, NJ, to New York City.

ALLAN H. MOGENSEN

Allan H. Mogensen, also known as Mogy, (Pennsylvania, May 1901–March 1989), was an American industrial engineer and authority in the field of work simplification and office management. He is noted for popularizing flow charts in the 1930s and is remembered as the "father of work simplification."

In the 1920s, Mogensen received his BA in Industrial Engineering at Cornell University, where he had studied the methods of Frank Gilbreth. Afterward, he started as an industrial engineering consultant, among other places at Eastman Kodak. During his consultancy practice, he experienced that improvements made by employees on the work floor were the most successful. Mogensen's career started at a time when there was a dramatic increase in productivity. This was due to Scientific Management, which Frank Gilbreth had further improved on by increasing the efficiency of the analysis and the adding of techniques, which made it possible for people to make an increased output with a slightly improved effort. Even so, some of the workers saw this as a new way of taking advantage of

了缩短过程时间,而Gilbreth则专注于通过减少所涉及的动作来提高过程的效率。F. B. Gilbreth继续在其他领域进行微动作研究,开创性地将动作分解图片用于研究工作和工人的各个方面。

在第一次世界大战的头几个月里,F. B. Gilbreth在德国研究工业流程和机械。当受伤的士兵开始回家时,F. B. Gilbreth运用他的理论改进了外科手术,并且成为第一个在手术室使用电影摄影机进行教学的人,他也是第一个提出将外科护士作为外科医生的助手、在手术过程中把手术器械交给外科医生的人,他还为帮助受伤士兵康复开发了管理日常活动的方法。

1920年,F. B. Gilbreth帮助ASME成立了管理部门,他因此成为美国和欧洲最知名的工程师之一,并获得了经济奖励和许多专业荣誉。在1921年的一次ASME会议上,F. B. Gilbreth成为第一个提出结构化方法来记录工艺流程的人。他称自己的陈述为"流程图是找出最佳方法的第一步。"[12]从整体上看,F. B. Gilbreth的工作目的是通过有效的培训、新的工作方法、改善的工作环境和改进工具以及健康的心理生活观来开发工人的全部潜能[13]。

1924年,F. B. Gilbreth建议在布拉格召开首届国际管理大会。会议结束后不久,他突发心脏病,1924年6月14日在从新泽西州蒙特克莱尔的家中前往纽约市的途中去世。

1.2.4　A. H. Mogensen

A. H. Mogensen,也被称为Mogy(宾夕法尼亚州,1901年5月—1989年3月),是美国工业工程师,在工作简化和办公室管理领域具有权威性。20世纪30年代,他以推广流程图而闻名,并被誉为"工作简化之父"。

20世纪20年代,A. H. Mogensen在康奈尔大学获得工业工程学士学位,在那里他学习了F. B. Gilbreth的方法。后来,他开始在伊士曼柯达(Eastman Kodak)担任工业工程顾问。在咨询过程中,他体会到只有员工在工作场所所做的改进才是最成功的。A. H. Mogensen的职业生涯始于生产力大幅提高的时期,由于科学管理思想的应用,F. B. Gilbreth通过提高分析效率和添加技术进一步改进了这一点,这使得人们只要稍加努力就可使提高产量成为可能。尽管如此,一些工人还是把这看作一种利用他们的新方法[14],因此有时这些做法会导致强烈的反对,从而破

them,[14] and at times it lead to such an intense opposition as to destroy the entire process.

In the 1930s, Mogensen further experimented with time and motion studies using motion pictures. In 1932, Allan H. Mogensen invented the "Work Simplification", which is defined as the organized utilization of common sense.[15] Mogensen made use of process diagrams to organize and study the work, and he tapped into the common sense of the one performing the work in order to develop ideas for improvement.[16] One of Allan H. Mogensen's main points was that the one performing the task was the one best suited to improve it. What sets his method apart from earlier methods is the fact the he sees the people performing the work as the most vital resource when it comes to an ongoing improvement of the work processes.[17] By involving each individual worker in the process, he achieves two things: (1) he reduces the opposition to the implementation of the scientific management methods and (2) he gets input to improvement of the processes. In 1937, he had organized the process enough to be able to start arranging his conferences.

During the early 1940s, Mogensen was noted for making movies of operations in hospitals, where he discovered that surgeons could work faster by avoiding lost motions, and in doing so reduce the mortality rate. At Lake Placid, Mogensen kept organizing the Work Simplification Conferences for almost 50 years.

BOEING B17

Boeing is an American plane manufacturer producing planes for both civilian and military use. The company was founded in Seattle in the United States by William E. Boeing and George Conrad Westervelt on July 15, 1916, and was named B&W after their initials. Later, the name was changed to Pacific Aero Products, and in 1917, it became Boeing Airplane Company. William E. Boeing had been educated at Yale University, and originally he had earned a lot of money in the forest industry. He acquired a certain knowledge of a breakdown of work processes, simplification, and improvement in dealing with the work that was being done, and thus, the workflow and the context of processes. He put this knowledge to use in the design and production of airplanes.

Toward the end of the 1930s, Boeing constructed the world's largest passenger plane, the Boeing 314 Clipper, which could hold 90 passengers during daytime flights and 40 passengers during night flights. The plane was designed in cooperation with Pan American World Airways and was used for long-distance international flights. It was a seaplane that could take off and land on the water.

During the Second World War, the company manufactured aircraft for the military and is known for its bombers. The production rate went up dramatically, and they produced about 350 planes each month. These planes were primarily B-17 Flying Fortresses and B-29 Superfortresses. It was planes of the latter type that carried the nuclear bombs to Hiroshima and Nagasaki in Japan, thus contributing to ending the Second World War through the pacification of Japan.

坏整个过程。

20世纪30年代，A. H. Mogensen进一步利用电影进行了时间和动作的研究。1932年，A. H. Mogensen发明了工作简化方法，即有组织地运用常识[15]。A. H. Mogensen利用流程图解来组织和研究这项工作，他利用执行这项工作的人的常识来提出和改进意见[16]。A. H. Mogensen的主要观点之一是执行任务的人最适合改进它，这使他的方法与早期方法不同的是，当工作流程持续改进时，他认为执行工作的人员是最重要的资源[17]。通过让每个工人参与到这一过程中，他实现了两件事：① 减少了工人对实施科学管理方法的反对；② 对流程的改进进行投入。1937年，他已经组织了足够的流程来开始他的会议。

在20世纪40年代初，A. H. Mogensen以拍摄医院手术的电影而闻名，他发现外科医生可以通过避免动作缺失来更快地工作，这样做可以降低死亡率。在普莱西德湖，A. H. Mogensen组织了近50年的工作简化会议。

1.2.5　波音B17

波音是一家生产民用和军用飞机的美国飞机制造商。1916年7月15日，这家公司由W. E. Boeing和G. C. Westervelt在美国西雅图成立，并以其首字母命名为B&W。后来，该公司更名为太平洋航空产品公司，1917年成为波音飞机公司。W. E. Boeing曾在耶鲁大学接受教育，最初他在森林产业赚了很多钱。他对工作流程的分解、简化和改进有了一定的了解，从而对工作流程和流程节点有了一定的认识。最终，他把这些知识应用到飞机的设计和生产中。

20世纪30年代末，波音公司建造了当时世界上最大的客机"波音314飞剪船"，白天可搭载90名乘客，夜间可搭载40名乘客。这架飞机是其与泛美航空公司合作设计的，用于长途国际航班。这是一架可以在水面上起飞降落的水上飞机。

在第二次世界大战期间，波音公司因为军方制造轰炸机而闻名，生产速度急剧上升，每月生产约350架飞机，这些飞机主要是B-17飞行堡垒和B-29超级堡垒。正是后一种类型的飞机将核弹带到了日本的广岛和长崎，从而使得日本投降并结束了第二次世界大战。

It was a great technological triumph, but the greatest innovation was the handling of the complexity of the job. They started working with checklists. Checklists are not just useful for making sure that every detail is carried out, but also in controlling the context in which it is done.[18] Nowadays, checklists are used in many places, and not just in aviation, but also for precision tasks such as surgery and laser carving. It is one of the first good examples of process documentation for complicated workflows and the breakdown of processes.

BENJAMIN S. GRAHAM

Benjamin S. Graham, Sr. (1900–1960) was a pioneer in the development and use of scientific management and industrial improvement techniques for office work. He is to this day recognized as the founder of "paperwork simplification." He saw a growing need for improvement in dealing with information back in the 1940s when leaders and office staff still constituted a minority of the workforce.[19] Benjamin S. Graham's work sprang from Mogensen's, and focused on the simplification of the work processes.

He was trained in "Work Simplification" by Allan Mogensen and Lillian Gilbreth, later, and adapted Gilbreth's flow process diagrams which had been used to improve the rather complex business processes in factory work. He worked at the Standard Register Company of Dayton, Ohio, and his job was to work out processes so that the employees could study them and streamline them. Graham was invited to participate in Mogensen's conferences to teach simplification of paperwork. His technique for the mapping of processes is used all over the world for business process improvement. While Mogensen's work focused on factory work, Graham adapted the process to integrate the indirect workforce, which he called "Paperwork." He defines this in the light of two terms: "The desired purpose or end result" and "paperwork includes absorption, transmission, analysis, communication, and storing of information/facts".[20]

Graham finds the above information very important, since it will be used as management tools for the employees. He said, "Earlier the purpose was to help the leadership in their decision-making, but it seemed that the message was not being conveyed to the employees. Therefore, the employees constantly tried to "increase production, improve quality, and reduce costs, but in vain, since they had not been given any guidelines." Graham felt that the employees should be involved in the accumulated knowledge.

ASME: AMERICAN SOCIETY OF MECHANICAL ENGINEERS

The American Society of Mechanical Engineers—also known as ASME[21]—is a not-for-profit membership organization that enables collaboration, knowledge sharing, career enrichment, and skills development across all engineering disciplines towards the goal of helping the global engineering community develop

这是一个巨大的技术胜利，但最大的创新是处理工作的复杂性。他们开始使用检查表这个工具。检查表不仅有助于确保每一个细节都得到执行，而且也有助于控制完成它的整体环境[18]。如今，检查表在许多地方都被使用，不仅在航空业，而且还被用于诸如外科手术和激光雕刻等精确的任务。对于复杂的工作流和流程分解，它是流程文档的第一个好例子。

1.2.6　B. S. Graham

B. S. Graham（1900—1960年）是开发和使用科学管理和工业改进技术到办公室工作中的先驱。直到今天，他仍然被公认为是纸上作业简化的创始人。他发现，在20世纪40年代，当领导和办公室工作人员仍占劳动力的少数时，对处理信息的需求却越来越大[19]。B. S. Graham 的工作源于 A. H. Mogensen 的工作，重点是简化工作流程。

后来，他接受了 A. H. Mogensen 和 L. Gilbreth 的工作简化培训，并改造了 F. B. Gilbreth 的流程图，该流程图用于改进工厂工作中复杂的业务流程。他在俄亥俄州代顿的标准注册公司（Standard Register Company）工作，他的工作是制定流程，使员工能够学习并简化流程。B. S. Graham 应邀参加 A. H. Mogensen 的会议，教授简化纸上作业。他的流程地图技术被世界各地用于BPI。虽然 A. H. Mogensen 的工作重点是工厂业务，但 B. S. Graham 调整了这一流程，以整合间接劳动力，他称之为纸上作业。他根据两个术语来定义这一点："预期目的或最终结果" 和 "纸上作业包括信息/事实的吸收、传递、分析、沟通和存储"[20]。

B. S. Graham 认为上述信息非常重要，因为它将被用作员工的管理工具。他说："早些时候的目的是帮助领导层做出决策，但似乎这一信息没有传达给员工。因此，员工不断地试图增加生产、提高质量、降低成本，但因为没有得到任何指导方针，所以徒劳。"B. S. Graham 认为，员工应该参与知识的积累。

1.2.7　ASME：美国机械工程师协会

美国机械工程师协会也称为ASME[21]，是一个非营利性的会员组织，能够跨所有工程学科开展协作、共享知识、丰富职业和发展技能，以帮助全球工程界建立解决方案。ASME成立于1880年，由一小群企业家组成，经过几十年的发展，它在

solutions to benefit lives and livelihoods. Founded in 1880 by a small group of leading industrialists, ASME has grown through the decades to include more than 130,000 members in 158 countries. Circa thirty-thousand of these members are students.

From college students and early-career engineers to project managers, corporate executives, researchers, and academic leaders, ASME's members are as diverse as the engineering community itself. ASME serves this wide-ranging technical community through quality programs in continuing education, training and professional development, codes and standards, research, conferences and publications, government relations, and other forms of outreach.

A process chart is one of the simplest forms of mapping procedures. Process charts are a common "language" between different groups of people and across several branches of an organizational environment. A set of different process diagrams are designed to meet the needs at a certain level or step in the analysis, and they can be used at a very detailed level (for the monitoring of the activities at a certain place of work) but also in a much larger system, process, or procedure level. The different forms of process diagrams share a common set of symbols, but some have other symbols for specific and specialized steps within the process. The most common and well-known process symbols (of which there are only five) were first announced by the American Society of Mechanical Engineers and are known as the ASME symbols[22] (see Figure 2).

Operation: The Operation symbol represents the handling of goods, preparing and putting away, loading and unloading, and all sorts of activities that do not include the exchange of information. This includes physical "paper carrying" and "electronic paper carrying" activities (i.e., the typing of electronic documents and the use of software applications), and usually there are more of these symbols than any other on the chart.

Checking: The Checking symbol represents the checking of an object to see if it is "all right." The Checking symbol is not used when dealing with normal checking routines within a given work. The purpose of this symbol is to check if a certain job has been done correctly. The Checking symbol is often followed by correction routines.

Transport: The Transport symbol represents a movement from one area of work to another. It is not used for minor movements within a given area of work. The purpose of this symbol is to illustrate the movements that separate the employees from one another and, thus, requires time and resources when the employees have to cooperate across the organizations. These movements are often time-consuming and expensive.

Storage: The Storage symbol represents a period where nothing happens at the step being monitored. We should show the occurrence of storage, when it involves a period of neglect of an existing project, and thus is of importance for the subsequent processes.

Delay: The Delay symbol represents a period which is extended. How short or how long this period is depends on the situation. We should show the occurrence of delays when they involve the consumption of time and when they are of importance to the subsequent processes.

158个国家拥有约13万名成员,其中大约有3万名学生。

从大学生和起步期的工程师,再到项目经理、企业高管、研究人员和学术领袖,ASME的成员和工程界本身一样多种多样。ASME通过持续教育、培训和专业发展、规范和标准、研究、会议和出版物、政府关系和其他形式的外联服务于广泛的技术界。

流程图是程序地图中最简单的形式之一。流程图是不同人群之间以及组织环境的多个分支之间的通用语言。一组不同的流程图是为了满足分析中某一级别或步骤的需求而设计的,它们可以在非常详细的级别(用于监控某一工作地点的活动)使用,也可以在更大的系统、流程或流程级别使用。不同形式的流程图共享一组通用的符号,但有些流程图中的特定步骤和专用步骤具有其他符号。ASME首次发布了最常见和最著名的工艺符号(其中只有五个),称为ASME符号[22](图2)。

操作:操作符号表示货物的处理、准备和存放、装卸以及各种不包括信息交换的活动。这包括物理的"携带纸张"和"携带电子纸张"活动(即电子文档的打印和软件应用的使用),通常这些符号比图表上的任何其他符号都多。

检查:检查符号表示检查对象是否正常。在处理给定工作中的正常检查例程时,通常不使用检查符号。此符号的目的是检查某项工作是否已正确完成。检查符号后面通常跟着修正程序。

转移:转移符号表示从一个工作区域到另一个工作区域的移动。它不用于特定工作区域内的微小移动。这个符号的目的是说明员工之间相互分离的动作,因此,当员工必须在组织间进行合作时,需要时间和资源。这些动作往往费时且昂贵。

存储:存储符号表示在被监视的步骤中没有发生任何事情的时段。当涉及对现有项目的一段时间的忽视时,我们应该说明存储的发生,这对于后续的过程是很重要的。

延迟:延迟符号表示延长的周期。这段时间有多长取决于具体情况。当延迟涉及时间的消耗以及它们对后续过程的重要性时,我们应该显示延迟的发生。

Operation
A complex action or process (possibly described elsewhere), often changing something.

Transport
Movement of people or things. May be accompanied by a distance measurement.

Delay
Idle time of people or machines, or temporary storage of materials.

Storage
Longer-term storage of materials or other items.

Inspection
Checking of items to ensure correct quality or quantity.

FIGURE 2

ASME symbols (Ref. 23).

FUNCTIONAL FLOW BLOCK DIAGRAM OF PERT
Functional Flow Block Diagrams

Functional Flow Block Diagrams (FFBD) surfaced in the 1950s. They are a way of illustrating and describing the processes in a development and production system environment and are a multitiered, time-sequenced, step-by-step flow diagram of a system's functional flow. The FFBD notation (see Figure 3) is widely used in classical systems engineering and is one of the classic business process modeling methodologies, along with flow charts, data flow diagrams, control flow diagrams, Gantt charts, PERT diagrams, and IDEF.[24] FFBDs are also referred to as Functional Flow Diagrams, functional block diagrams, and functional flows.[25]

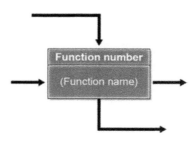

FIGURE 3

FFBD Functional block (Ref. 26).

操作
一种复杂的动作或过程(可能在别处描述)，经常改变某件事。

转移
人或事物的运动。可能伴随着距离度量。

延迟
人员或机器的闲置时间，或材料的临时储存。

存储
材料或其他物品的长期储存。

检查
检查项目以确保正确的质量或数量。

图2　ASME符号[23]

1.2.8　PERT功能流程框图

1. 功能流程框图

功能流程框图出现在20世纪50年代。它是一种说明和描述开发和生产系统环境中的流程的方法，是一个多层次、有时间顺序、逐步的系统功能流程图。FFBD表示法(图3)广泛应用于经典系统工程中，是经典的业务流程建模方法之一，与流程图、DFD、控制流程图、甘特图、PERT图和IDEF一起使用[24]。FFBD也被称为功能流程图、功能框图和功能流程[25]。

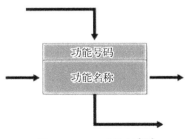

图3　FFBD功能块[26]

The purpose of this method is to show the sequential relations between all the functions in a given system. FFBD illustrates the chronological sequence of functional occurrences. The amount of time spent on each function or the time period between two functions are not indicated. FFBD has its focus on functions and shows only which function is to be carried out, but not how this is to be done.

The first structured method for documenting process flow was introduced by Frank B. Gilbreth to members of the American Society of Mechanical Engineers (ASME) in 1921 in his presentation "Process Charts—First Steps in Finding the One Best Way." Gilbreth's tools quickly found their way into industrial engineering curricula. In the early 1930s, Allan H. Mogensen began training business people in the use of some of the tools of industrial engineering at his Work Simplification Conferences at Lake Placid, New York.

A 1944 graduate of Mogensen's class, Art Spinanger, took the tools back to Procter and Gamble where he developed their Deliberate Methods Change Program.[27] Another 1944 graduate, Benjamin S. Graham, Director of Formcraft Engineering at Standard Register Industrial, adapted the flow process chart to information processing with his development of the multiflow process chart to display multiple documents and their relationships. In 1947, ASME adopted a symbol set as the ASME Standard for Operation and Flow Process Charts, derived from Gilbreth's original work. The modern FFBD was developed by TRW Incorporated, a defense-related business, in the 1950s. In the 1960s, it was exploited by NASA to visualize the time sequence of events in space systems and flight missions. FFBDs became widely used in classical systems engineering to show the order of execution of system functions.

PERT: Program Evaluation Review Technique[28]

PERT is a method to analyze the tasks involved in completing a given program, especially the time needed to complete each task, and to identify the minimum time needed to complete the entire program. PERT was developed primarily to simplify the planning and scheduling of large and complex programs and was developed for the United States Navy Special Projects Office in 1957 to support the United States Navy's Polaris nuclear submarine program.[29] It was able to incorporate uncertainty by making it possible to schedule a program while not knowing precisely the details and durations of all the activities. It is more of an event-oriented technique rather than being start- and completion-oriented, and it is mostly used in projects where time is the major factor rather than the cost. It is often applied to large-scale, one-time, complex, nonroutine infrastructure and research and development programs.

An example of this was for the 1968 Winter Olympics in Grenoble, which applied PERT from 1965 until the opening of the 1968 Games. This program model was the first of its kind, a revival for scientific management, founded by Frederick Winslow Taylor (Taylorism) and later refined by Henry Ford (Fordism).

这种方法的目的是展示系统中所有功能节点之间的顺序关系。FFBD说明了功能出现的时间顺序。每个功能所花费的时间量或两个功能之间的时间段不会显示出来。FFBD将重点放在功能上,只展示要执行的功能,而不展示如何执行。

1921年,F. B. Gilbreth向ASME的成员介绍了第一种记录工艺流程的结构化方法,即"流程图:找到最佳方法的第一步"。F. B. Gilbreth的工具很快就进入了工业工程课程。20世纪30年代初,A. H. Mogensen在纽约普莱西德湖举行的工作简化会议上开始培训商界人士使用工业工程的这些工具。

1944年,A. H. Mogensen班的毕业生Art Spinanger把这些工具带回了宝洁公司,在那里他开发了他们的深思熟虑的方法改变计划[27]。另一名1944年毕业的毕业生B. S. Graham是标准注册公司(Standard Register Industrial)的模板工艺工程总监,他开发了多流程图来显示多个文档及其关系,并将流程图应用于信息处理。1947年,ASME采用了一套符号作为ASME操作和流程图的标准,其源自F. B. Gilbreth的原著。现代FFBD是由防务相关企业天合公司在20世纪50年代开发的。20世纪60年代,美国国家航空航天局(National Aeronautics and Space Administration, NASA)利用它来可视化航天系统和飞行任务中事件的时间序列。FFBD在经典系统工程中得到了广泛的应用,用来表示系统功能的执行顺序。

2. PERT:计划评估评审技术[28]

PERT是计划评估评审技术,简单地说,PERT是利用网络分析制定计划以及对计划予以评价的技术。它能协调整个计划的各道工序,合理安排人力、物力、时间、资金,加速计划的完成。在现代计划的编制和分析手段上,PERT被广泛地使用,是现代项目管理的重要手段和方法。该方法专门评估每个任务所需的时间,并确定完成整个程序所需的最短时间。PERT的开发主要是为了简化大型和复杂项目的规划和调度,并于1957年为美国海军特别项目办公室开发,以支持美国海军的北极星核潜艇项目[29]。它能够包容不确定性,使其能够在不准确了解所有活动的细节和持续时间的情况下安排计划。它更像是一种面向事件的技术,而不是面向开始和完成的技术,并且它主要用于主要因素是时间而不是成本的项目中。它通常应用于大规模的、一次性的、复杂的、非常规的基础设施和研发项目。

其中一个例子就是1968年格勒诺布尔冬奥会,从1965年到1968年冬奥会开幕,冬奥会一直采用PERT。这种计划模式是同类中的首创,也是科学管理的复兴,其由F. W. Taylor创立(泰勒主义),后来由H. Ford完善(福特主义)。

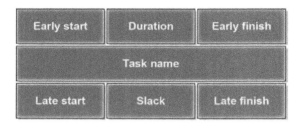

FIGURE 4

The PERT diagram (Ref. 30).

Figure 4 shows one way to describe a PERT diagram. Early start (ES) is the earliest possible starting point. Early finish (EF) is the earliest possible finishing point. EF−ES = duration of the task. The latest starting point (LS) and the latest finishing point (LF) is stated as well.

PERT uses terms like Slack or Float, Lead Time, Lag Time, Critical Way, Fast Tracking, and Crashing. Slack (or Float) states the amount of time that a given task can be delayed without resulting in a delay of the total program. Slack = LF−EF. Lead Time describes when a subsequent task can start before the previous task has been completed. In contrast to this is Lag Time, which describes the waiting time between two tasks (drying time). The Critical Way is the way through the project where you do not experience any Slack. If a task on the Critical Way is delayed, this will result in a total delay if you are using either Fast Tracking or Crashing. Fast Tracking means that you re-plan (if at all possible) and perform several tasks simultaneously on the Critical Way. The advantage of Fast Tracking, as opposed to Crashing, is that there is no increased consumption of resources as a result of the re-planning. Crashing is an attempt to catch up or accelerate activities on the Critical Way by speeding them up. This almost always results in an increased use of resources.

DATA FLOW DIAGRAMS AND IDEF

Data Flow Diagram

A Data Flow Diagram (DFD) is a graphical representation of the "flow" of data through an information system (as shown on the DFD flow chart Figure 5), modeling its process aspects. Often it is a preliminary step used to create an overview of the system that can later be elaborated. DFDs can also be used for the visualization of data processing (structured design) and show what kind of information will be input to and output from the system, where the data will come from and go to, and where the data will be stored. It does not show information about the timing of processes or information about whether processes will operate in sequence or in parallel.

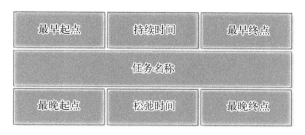

图4　PERT图[30]

图4显示了描述PERT图的一种方法。最早起点(ES)是最早可能的开始点，最早终点(EF)是最早可能的完成点，EF − ES=持续时间。同时图4还说明了最晚起点(LS)和最晚终点(LF)。

PERT使用松弛或浮动、提前期、延迟时间、关键方式、快速跟踪和赶工等术语。松弛(或浮动)表示给定任务可以延迟而不会导致整个程序延迟的时间量，松弛=LF − EF。提前期描述的是在上一个任务完成之前，后续任务何时可以开始。与此相反的是延迟时间，它描述了两个任务之间的等待时间(容许时间)。关键方式是在项目中不经历任何松弛的方式。如果关键路径上的任务被延迟，那么如果您正使用快速跟踪或赶工中任意一个，这将导致完全延迟。快速跟踪意味着您重新计划(如果可能的话)并以关键的方式同时执行几个任务。与赶工相比，快速跟踪的优势在于重新规划不会增加资源消耗。赶工是一种试图通过加速来赶上或加速关键的活动，这几乎总是导致使用的资源增加。

1.2.9　DFD和IDEF

1. DFD

DFD是通过信息系统建模展示其流程各方面的数据流的图示(如图5 DFD流程图所示)，通常，它是创建系统概览的一个初步步骤，稍后可以详细说明。DFD还可用于数据处理的可视化(结构化设计)，并显示系统将输入和输出何种类型的信息、数据的来源和去向以及数据的存储位置。它不显示有关流程时间安排的信息，也不显示有关流程是按顺序运行还是并行运行的信息。

FIGURE 5

An example of a Data Flow Diagram (Ref. 31).

It is common practice to draw the context-level data flow diagram first, which shows the interaction between the system and external agents that act as data sources and data sinks. This helps to create an accurate drawing in the context diagram. The system's interactions with the outside world are modelled purely in terms of data flows across the system boundary. The context diagram shows the entire system as a single process and gives no clues as to its internal organization.

This context-level DFD is next "exploded" to produce a Level 1 DFD that shows some of the detail of the system being modeled. The Level 1 DFD shows how the system is divided into subsystems (processes), each of which deals with one or more of the data flows to or from an external agent, and which together provide all of the functionality of the system as a whole. It also identifies internal data stores that must be present in order for the system to do its job and shows the flow of data between the various parts of the system.

Integrated Definition

IDEF is an abbreviation of Integrated Definition. It is a modulation method that is used in system and software design. It is a broad-spectrum model, and it covers a lot of applications within data modulation, simulation, object-oriented analysis/design, and gathering of information (Figure 6).

IDEF contains a lot of different variables, of which the most commonly used is IDEF0, which is the only model described here. IDEF0 is focused on describing decisions, actions, and activities in an organization or a system. Just like DFD, IDEF0 shows the data flow between various functions.

IDEF0[32] Overview: Function Modeling Method

IDEF0 is a method designed to model the decisions, actions, and activities of an organization or system. IDEF0 was derived from a well-established graphical language, the Structured Analysis and Design Technique (SADT). The United States Air Force commissioned the developers of SADT to develop a function modeling method for analyzing and communicating the functional perspective of a system. Effective IDEF0 models help to organize the analysis of a system and to promote good communication between the analyst and the customer. IDEF0 is useful in establishing the scope of an analysis, especially for a functional analysis.

图5   DFD示例[31]

通常的做法是首先绘制具有前后关系的DFD,它展示系统内外数据输出与数据输入之间的交互,这有助于创建准确的流程图。系统内外的交互完全按照跨系统边界的数据流进行建模。流程图将整个系统展示为一个整体的流程,并没有提供关于其流程内部的信息。

这个前后关系级别的DFD接下来将被分解以生成一个第1级DFD,该DFD显示了正在建模的系统的一些细节。第1级DFD显示了如何将系统划分为子系统(流程),每个子系统处理一个或多个进出的数据流,这些子系统一起提供了整个系统的所有功能。它还标识系统执行其工作所必须存在的内部数据存储,并显示系统各个部分之间的数据流。

2. 集成定义

IDEF是集成定义的缩写。它是一种用于系统和软件设计的调制方法。它是一种广谱模型,在数据调制、仿真、面向对象的分析/设计、信息采集等方面有着广泛的应用(图6)。

IDEF包含许多不同的变量,其中最常用的是IDEF0,它是这里描述的唯一模型。IDEF0的重点是描述组织或系统中的决策、行动和活动。与DFD一样,IDEF0展示了各种功能之间的数据流。

3. IDEF0[32]概述:功能建模方法

IDEF0是一种用于对组织或系统的决策、行动和活动进行建模的方法,其基本格式如图6所示。IDEF0源于成熟的图形化语言、结构化分析与设计技术(structured analysis and design technique,SADT),是美国空军委托SADT的开发人员开发的一种功能建模方法,用于分析和展示系统的功能透视图。有效的IDEF0模型有助于组织系统的分析,并促进分析师和客户之间的良好沟通。IDEF0在确定分析的范围时很有用,特别是对于功能分析。

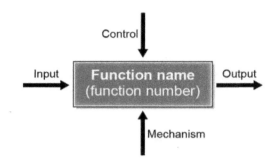

FIGURE 6

IDEF0 basis format (Ref. 33).

IDEF1[34] Overview: Information Modeling Method

IDEF1 was designed as a method for both analysis and communication in the establishment of requirements. IDEF1 is generally used to (1) identify what information is currently managed in the organization, (2) determine which of the problems identified during the needs analysis are caused by lack of management of appropriate information, and (3) specify what information will be managed in the TO-BE implementation.

IDEF1 captures the information that exists about objects within the scope of an enterprise. The IDEF1 perspective of an information system includes not only the automated system components, but also non-automated objects such as people, filing cabinets, telephones, etc. IDEF1 was designed as a method for organizations to analyze and clearly state their information resource management needs and requirements. Rather than a database design method, IDEF1 is an analysis method used to identify the following:

- Information collected, stored, and managed by the enterprise.
- Rules governing the management of information.
- Logical relationships within the enterprise reflected in the information.
- Problems resulting from the lack of good information management.

IDEF1X[35] Overview: Data Modeling Method

IDEF1X is a method for designing relational databases with a syntax designed to support the semantic constructs necessary in developing a conceptual schema. A conceptual schema is a single integrated definition of the enterprise data that is unbiased toward any single application and independent of its access and physical storage. Because it is a design method, IDEF1X is not particularly suited to serve as an AS-IS analysis tool, although it is often used in that capacity as an alternative to IDEF1. IDEF1X is most useful for logical database design after the information requirements are known and the decision to implement a relational database has been made. Hence, the IDEF1X system perspective is focused on

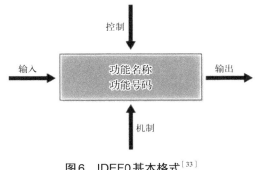

图6　IDEF0基本格式[33]

4. IDEF1[34]概述：信息建模方法

IDEF1被设计成在建立需求时分析和沟通的一种方法。IDEF1通常用于：
① 确定组织中当前管理的信息；② 确定需求分析过程中发现的哪些问题是由缺乏适当的信息管理造成的；③ 指定在未来的实施中将管理哪些信息。

IDEF1获得有关企业范围内对象的信息。信息系统的IDEF1透视图不仅包括自动化系统组件，还包括人、文件柜、电话等非自动化对象。IDEF1被设计为一种组织分析并清楚说明其信息资源管理需求和要求的方法。IDEF1不是一种数据库设计方法，而是一种用于识别以下内容的分析方法：

- 企业收集、存储和管理信息；
- 制订管理信息的规则；
- 反映在信息中企业内部的逻辑关系；
- 缺乏准确的信息管理导致的问题。

5. IDEF1X[35]概述：数据建模方法

IDEF1X是一种设计关系型数据库的方法，其语法设计常用于支持开发概念框图所需的语义构造。概念框图是企业数据的单一集成定义，它对任何单个应用程序都没有偏见，并且独立于其访问和物理存储。由于IDEF1X是一种设计方法，它并不特别适合用作现有的分析工具，尽管它经常被用作IDEF1的替代方案。IDEF1X在了解信息需求并决定实现关系数据库之后，对逻辑型数据库的设计非常有用。因此，IDEF1X系统透视图主要关注关系型数据库中的实际数据元素。如果

the actual data elements in a relational database. If the target system is not a relational system, for example, an object-oriented system, IDEF1X is not the best method.

IDEF3[36] Overview: Process Description Capture Method

The IDEF3 Process Description Capture Method provides a mechanism for collecting and documenting processes. IDEF3 captures precedence and causality relations between situations and events in a form natural to domain experts by providing a structured method for expressing knowledge about how a system, process, or organization works. IDEF3 descriptions can do the following:

- Record the raw data resulting from fact-finding interviews in systems analysis activities.
- Determine the impact of an organization's information resource on the major operation scenarios of an enterprise.
- Document the decision procedures affecting the states and life-cycle of critical shared data, particularly manufacturing, engineering, and maintenance product definition data.
- Manage data configuration and change control policy definition.
- Make system design and design trade-off analysis.
- Provide simulation model generation.

IDEF4[37] Overview: Object-Oriented Design Method

The intuitive nature of object-oriented programming makes it easier to produce code. Unfortunately, the ease with which software is produced also makes it easier to create software of poor design, resulting in systems lacking re-usability, modularity, and maintainability. The IDEF4 method is designed to assist in the correct application of this technology.

IDEF5[38] Overview: Ontology Description Capture Method

The IDEF5 method provides a theoretically and empirically well-grounded method specifically designed to assist in creating, modifying, and maintaining ontologies. Standardized procedures, the ability to represent ontology information in an intuitive and natural form, and higher quality results enabled through IDEF5 application also serve to reduce the cost of these activities.

ZERO DEFECTS

The theory of Zero Defects was first brought into use in the space and defense industry in the United States in 1960. The term Zero Defects expresses the idea that a product which meets the requirements of the consumer is the right product. If the product can do more or less than needed, it is a waste of resources.

目标系统不是关系型系统,如面向对象系统,IDEF1X则不是最佳方法。

6. IDEF3[36]概述:过程描述获取方法

IDEF3过程描述获取方法提供了一种收集和记录过程的机制。IDEF3通过提供一种结构化的方法来展示有关系统、流程或组织如何工作的知识,领域专家以自然的方式捕获情况和事件之间的优先级和因果关系。IDEF3的描述可以做到以下几点:

- 记录系统分析活动中开展业务调查访谈产生的原始数据;
- 确定组织信息资源对企业主要运营场景的影响;
- 记录影响关键共享数据状态和生命周期的决策程序,尤其是制造、工程和维护产品方面的数据;
- 制订管理数据配置和更改控制策略;
- 进行系统设计和设计得失分析;
- 提供仿真模型生成。

7. IDEF4[37]概述:面向对象的设计方法

面向对象的编程使代码的生成更加容易。然而,代码的容易生成也使得软件变得更糟糕,从而导致系统缺乏可重用性、模块化和可维护性。IDEF4方法的设计是为了帮助这项技术的正确应用。

8. IDEF5[38]概述:本体描述捕获方法

IDEF5方法提供了一种在理论和经验上专门设计用于帮助创建、修改和维护本体的很好的方法。标准化的过程、以直观自然的形式表示本体信息的能力以及通过IDEF5应用程序实现的更高质量的结果,这些都有助于降低这些活动的成本。

1.2.10　零缺陷

零缺陷理论于1960年首次在美国航天和国防工业中得到应用。零缺陷一词表达了这样一种观点:满足消费者要求的产品就是正确的产品。如果产品能做的比需要的多或少,那就是浪费资源。因此,如果满足客户的期望,一个便宜的产品可

Thus, a cheap product can be just as good as an expensive one if it meets the expectations of the client.

In 1979, quality expert Philip Crosby penned *Quality Is Free: The Art of Making Quality Certain*, which preserved the idea of Zero Defects in a 14-step quality improvement program and the concept of the "Absolutes of Quality Management." Zero Defects seeks to directly reverse the attitude that the amount of mistakes a worker makes does not matter since inspectors will catch them before they reach the customer. This stands in contrast to activities that affect the worker directly, such as receiving a paycheck in the correct amount. Zero Defects involves reconditioning the worker to "take a personal interest in everything he does by convincing him that his job is just as important as the task of the doctor or the dentist."

According to Crosby, there are four Absolutes:

1. **Quality is conformance to requirements**

   Every product or service has a requirement: a description of what the customer needs. When a particular product meets that requirement, it has achieved quality, provided that the requirement accurately describes what the enterprise and the customer actually need. This technical sense should not be confused with more common usages that indicate weight or goodness or precious materials or some absolute idealized standard. In common parlance, an inexpensive disposable pen is a lower-quality item than a gold-plated fountain pen. In the technical sense of Zero Defects, the inexpensive disposable pen is a quality product if it meets requirements: it writes, does not skip or clog under normal use, and lasts the time specified.

2. **Defect prevention is preferable to quality inspection and correction**

   The second principle is based on the observation that it is nearly always less troublesome, more certain, and less expensive to prevent defects than to discover and correct them. It saves a lot of human power and cost of inspection and correction. For example, if a person changes the poor condition brake shoes of his bike before next riding, then it will save a lot of the rider's energy and reduce the risk of accident on the road, not to mention the generation of new defects in the bike due to poorly conditioned brake shoes, which when observed later will have a higher cost of repair.

3. **Zero Defects is the quality standard**

   The third is based on the normative nature of requirements: if a requirement expresses what is genuinely needed, then any unit that does not meet requirements will not satisfy the need and is no good. If units that do not meet requirements actually do satisfy the need, then the requirement should be changed to reflect reality. Furthermore, the idea that mistakes are inevitable is rejected out of hand. Just as the CEO wouldn't accept "mistakenly" not getting paid occasionally, his/her chauffeur "mistakenly" driving them to the wrong business, or their spouse "mistakenly" sleeping with someone else, so the company should not take the attitude that they'll "inevitably" fail to deliver

以和一个昂贵的产品一样好。

1979年，质量专家P. B. Crosby撰文《质量免费：确定质量的艺术》，这一理论保留了14步质量改进计划中零缺陷的理念和"质量管理的绝对性"的概念。零缺陷旨在直接扭转一种态度，即员工所犯错误的数量并不重要，因为检查员会在错误到达客户之前发现它们。这与直接影响员工的活动形成了对比，比如获得正确数额的工资。零缺陷让员工"对自己做的每件事都感兴趣，让他相信自己的工作和医生或牙医的任务一样重要"。

根据Crosby的说法，有以下四种绝对原则。

1. 质量符合需求

每个产品或服务都有一个需求：对客户需求的描述。当一个特定的产品满足该需求时，只要该需求准确地描述了企业和客户实际需要什么，那么它就已经达到了质量。这种技术意义不应与表示重量、质量、贵重材料或某种绝对理想化的标准相混淆。一般来说，便宜的一次性钢笔比镀金钢笔质量差，但是从零缺陷的技术意义上讲，廉价的一次性钢笔如果能满足要求，就是一种高质量的产品：它能写字，在正常使用下不会中断或堵塞，而且能在有效的时间内使用。

2. 预防缺陷优于质量检查和纠正

第二个原则基于这样的观察：预防缺陷几乎总是比发现和纠正缺陷更简单、更确定、更便宜。它节省了大量的人力和检验及校正费用。例如，如果一个人在下一次骑车前更换了状况不佳的自行车刹车蹄片，那么它将节省骑车人大量的能量，并降低道路上发生事故的风险，更不用说由于刹车蹄片状况不佳而在自行车上产生新的缺陷，而且长远来看，这将带来更高的维修成本。

3. 零缺陷是质量标准

第三个原则是基于需求的规范性：如果一个需求表达的是真正需要的东西，那么任何不符合需求的单元就不能满足需要、就不是好的。如果不满足需求的单元确实满足了需求，那么应该更改需求以反映现实。此外，"错误是不可避免的"这一观点被全盘否定。正如首席执行官（chief executive officer, CEO）不会错误地接受偶尔得不到报酬，他/她的司机不会错误地开车送他们去错误的公司，或者他们的配偶不会错误地与别人发生性关系，所以公司不应该采取这样的态度，即他们不可避免

what was promised from time to time. Aiming at an "acceptable" defect level encourages and causes defects.

4. **Quality is measured in monetary terms—the Price of Nonconformance (PONC)**

The fourth principle is key to the methodology. Phil Crosby believes that every defect represents a cost, which is often hidden. These costs include inspection time, rework, wasted material and labor, lost revenue, and the cost of customer dissatisfaction. When properly identified and accounted for, the magnitude of these costs can be made apparent, which has three advantages. First, it provides a cost justification for steps to improve quality. The title of the book *Quality is Free* expresses the belief that improvements in quality will return savings more than equal to the costs. Second, it provides a way to measure progress, which is essential to maintaining management commitment and to rewarding employees. Third, by making the goal measurable, actions can be made concrete, and decisions can be made on the basis of relative return.

Another principle that Crosby stresses is that it is much cheaper and more efficient to prevent errors than to check and remedy errors later. Crosby formulated the idea that Zero Defects was what was expected, and one defect or error means that the product does not meet the expectations. In cases where the product meets the needs but not the expectations, the expectations will have to be adapted to the need.

Crosby thought that every error or defect represented an expense, which is often camouflaged. Whether it is control, lost materials, reparations, or extra labor, this expense has to be made visible. By mapping these factors, you would gain three advantages:

1. An expense that can justify the quality improvement effort.
2. A valuable measuring tool that allows you to monitor the progress and motivate or reward the employees.
3. A valuable foundation for decisions on where to put in an extra effort.

Crosby's principles had their renaissance in the 1990s when the battered American car industry needed to make cuts in the budget. By increasing the requirements to the suppliers, they could downsize several quality control functions and, thus, save a lot of money.

TOYOTA PRODUCTION SYSTEM

Toyota's production system (TPS) was developed between 1948 and 1975 by Taichi Ohno, Shigeo Shingo and the grandchild of the founder of Toyota, Eiji Toyoda. One of the most popular books written by the founders themselves is *Just-In-Time for Today and Tomorrow* (Ohno et al., 1988), which accounts for the philosophy behind TPS. TPS is based on a scientific production method.[39] When Eiji Toyoda

地不能按时兑现承诺。力求达到可接受的缺陷级别会鼓励并导致更大、更多的缺陷。

4. 质量是用金钱来衡量的：质量代价（price of nonconformance，PONC）

第四项原则是方法论的关键。P. B. Crosby 认为，每一个缺陷都代表着一种成本，这种成本往往是隐藏的。这些成本包括检查时间、返工、材料和人工的浪费、收入的损失以及客户不满的成本。如果正确地确定和说明这些费用的规模，就可以清楚地看到这有三个优点。首先，它为质量改进的活动提供了节约成本的理由。《质量免费：确定质量的艺术》一书的书名表达了这样一种信念，即质量的提高所带来的节省将大于成本。其次，它提供了一种衡量进展的方法，这对于维持管理承诺和奖励员工至关重要。最后，通过使目标可测量，行动可以具体化，决策可以建立在相对回报的基础上。

P. B. Crosby 强调的另一个原则是，预防错误要比以后检查和纠正错误便宜得多，效率也更高。P. B. Crosby 提出了零缺陷是预期的想法，一个缺陷或错误意味着产品不符合预期。在产品满足需求而不是期望的情况下，期望必须适应需求。

P. B. Crosby 认为每一个错误或缺陷都代表着一种代价，而这种代价往往被掩盖。无论是控制、丢失的材料、修理或额外的劳动力，这些代价都必须是可见的。通过这些因素，您将获得三个优势：

（1）能够确定质量改进工作的费用；

（2）一个有价值的测量工具，可以让您监控进度、激励或奖励员工；

（3）这是决定在何处付出额外努力的宝贵基础。

P. B. Crosby 的原则在20世纪90年代得到了复兴，当时遭受重创的美国汽车业需要削减预算。通过增加对供应商的要求，他们可以缩小几个质量控制功能，从而节省大量资金。

1.2.11　TPS

TPS是由 Taiichi Ohno、Shigeo Shingo 和丰田（Toyota）汽车公司（简称丰田）创始人 Eiji Toyoda 的孙子在1948 ～ 1975年开发的。创始人自己写的最受欢迎的书之一就是《准时化生产的今天与明天》（Taichi Ohno et al., 1988），它解释了TPS背后的哲学。TPS以科学的生产方法为基础[39]，当 Eiji Toyoda 从美国旅行回来，向

returned from a trip to the United States to learn about mass production from Ford, the factory manager at Toyota, Taiichi Ohno, was given the task of improving the production processes at Toyota to achieve the same level of productivity as Ford. Toyota had only limited funds and, therefore, had to find a way to increase flexibility and efficiency.

This lead to the Toyota Production System (TPS). The TPS concept focuses on a set of main ideas. The first is the elimination of all waste. Waste is to be understood as anything superfluous that does not add any value (i.e., defect components, unnecessary processes, downtime, etc.). The other two main ideas are a constant improvement of processes as well as people (we respect others) and teamwork (stimulate personal and professional growth).

The Toyota Production System (TPS) was established based on two concepts: The first is called "jidoka" (which can be loosely translated as "automation with a human touch"), which means that when a problem occurs, the equipment stops immediately, preventing defective products from being produced; The second is the concept of "Just-in-Time," in which each process produces only what is needed by the next process in a continuous flow.

Jidoka: Highlighting/Visualization of Problems

Quality must be built in during the manufacturing process!

If equipment malfunction or a defective part is discovered, the affected machine automatically stops, and operators cease production and correct the problem. For the Just-in-Time system to function, all of the parts that are made and supplied must meet predetermined quality standards. This is achieved through jidoka.

1. Jidoka means that a machine safely stops when the normal processing is completed. It also means that, should a quality/equipment problem arise, the machine detects the problem on its own and stops, preventing defective products from being produced. As a result, only products satisfying quality standards will be passed on to the following processes on the production line.

2. Since a machine automatically stops when processing is completed or when a problem arises and is communicated via the "andon" (problem display board), operators can confidently continue performing work at another machine, as well as easily identify the problem's cause to prevent its recurrence. This means that each operator can be in charge of many machines, resulting in higher productivity, while continuous improvements lead to greater processing capacity.

Just-In-Time: Productivity Improvement

Making only "what is needed, when it is needed, and in the amount needed!"

Just-in-Time means producing quality products efficiently through the complete elimination of waste, inconsistencies, and unreasonable requirements on the

H. Ford了解了大规模生产的情况时，丰田的工厂经理Taiichi Ohno被赋予了改善丰田生产流程以达到与福特汽车公司相同的生产力水平的任务。当时，丰田只有有限的资金，因此必须找到提高灵活性和效率的方法。

这就由此产生了TPS。TPS概念包括一系列主要思想。第一是消除一切浪费。浪费被理解为不增加任何价值的任何多余的东西（即缺陷组件、不必要的过程、停机时间等）。另外两个主要思想是不断改进流程，以及个人（尊重他人）和团队合作（促进个人和职业发展）。

TPS的建立基于两个概念：第一个概念是自働化（jidoka），这意味着当出现问题时，设备立即停止，防止生产出有缺陷的产品；第二个概念是准时化生产（just-in-time，JIT），其中每个流程仅在连续流程中生成下一个流程所需的内容。

1. 自働化：突出显示/可视化问题

质量必须在制造过程中建立起来！

如果发现设备故障或零件有缺陷，受影响的机器应该自动停止，操作员停止生产并纠正问题。为了使准时制系统发挥作用，所有制造和供应的零件必须达到预定的质量标准，这是通过自働化实现的。

（1）自働化是指机器在正常加工完成后安全停机。它还意味着，如果出现质量/设备问题，机器会自动检测到问题并停止工作，防止生产出有缺陷的产品。因此，只有符合质量标准的产品才会在生产线上传递到下一工序。

（2）由于一台机器在处理完成或出现问题时自动停止，并通过"andon"（"问题显示板"）进行沟通，操作人员可以放心地在另一台机器上继续工作，也可以轻松识别问题的原因，防止问题的再次发生。这意味着每个操作员可以管理许多机器，从而提高生产率，而不断的改进促成更大的处理能力。

2. 准时化生产：提高生产率

只做"需要什么，什么时候需要，需要多少"！

准时化生产是指通过完全消除生产线上的浪费、不一致和不合理的要求，高效地生产高质量的产品。为了能尽快交付客户订购的车辆，我们会在最短的时间内

production line. In order to deliver a vehicle ordered by a customer as quickly as possible, the vehicle is efficiently built within the shortest possible period of time by adhering to the following:

1. When a vehicle order is received, a production instruction must be issued to the beginning of the vehicle production line as soon as possible.
2. The assembly line must be stocked with the required number of all needed parts so that any type of ordered vehicle can be assembled.
3. The assembly line must replace the parts used by retrieving the same number of parts from the parts-producing process (the preceding process).
4. The preceding process must be stocked with small numbers of all types of parts and produce only the numbers of parts that were retrieved by an operator from the next process.

Based on the basic philosophies of jidoka and Just-in-Time, the TPS can efficiently and quickly produce vehicles of sound quality, one at a time, that fully satisfy customer requirements.

The entire Toyota "house" is shown in Figure 7. The roof can only be held up by Just-In-Time (JIT) and Jidoka, which is achieved through people's efforts and the reduction of waste. The idea of minimizing waste led to the Lean principle, which states that any activity that does not add value must be cut off.

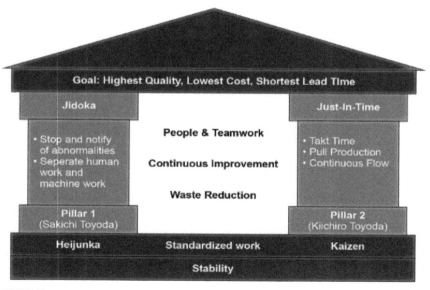

FIGURE 7

The TPS house (Ref. 40).

有效地制造车辆,并遵守以下规定:

(1)当收到车辆订单时,必须尽快向车辆生产线的起点发出生产指令;

(2)装配线必须存放所需数量的所有零件,以便装配任何类型的订购车辆;

(3)装配线必须通过从零件生产过程(前一个过程)中获取相同数量的零件来替换所使用的零件;

(4)前面的流程必须存储少量所有类型的部件,并且只生成操作人员从下一个流程检索到的部件的数量。

基于自働化和准时化生产的基本理念,TPS能够高效、快速地生产出一次一辆的同质车辆,完全满足客户的需求。

TPS架构屋如图7所示,屋顶通过准时化生产和自働化来支撑,这是通过人们的努力和减少浪费来实现的。减少浪费的想法形成了精益原则,即任何不增加价值的活动都必须被切断。

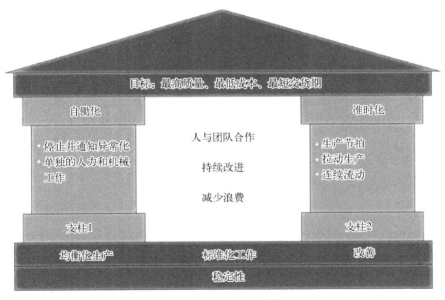

图7　TPS架构屋[40]

Toyota identified seven kinds of waste that had to be eliminated: overproduction, waiting time, unnecessary transport, over processing or misprocessing, superfluous storage, unnecessary movement, defects, and unutilized employee creativity. The entire Toyota concept states that only by combining JIT and Jidoka can you achieve the goal of the best quality, lowest costs, shortest lead time, the greatest safety, and high morale.

The 14 Principles of TPS

All in all, there are 14 principles in TPS, which are divided into four main areas[41] that we will describe in the following sections. These four sections are illustrated as a pyramid in Figure 8.

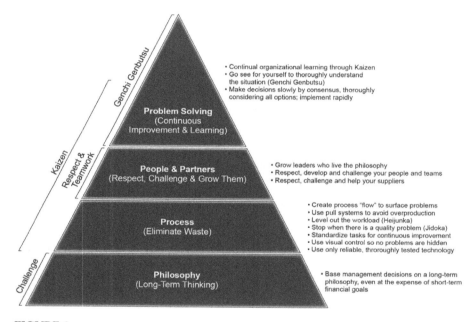

FIGURE 8

The four overriding sections of TPS are divided into Problem Solving, People & Partners, Process, and Philosophy.

Long-Term Philosophy

Principle 1: Base your management decisions on a long-term philosophy, even at the expense of short-term financial goals.

- Have a philosophical sense of purpose that supersedes any short-term decision making. Work, grow, and align the whole organization toward a common purpose that is bigger than making money. Understand your place in the history of the company and work to bring the company to the next level. Your philosophical mission is the foundation for the other principles.

丰田指出了必须消除的7种浪费：生产过剩、等待时间、不必要的运输、过度加工或误加工、多余的库存、不必要的移动、缺陷和未被利用的员工创造力。丰田的整个理念是：只有将准时化生产和自働化结合起来，才能达到质量最好、成本最低、交货期最短、安全性最高、士气高昂的目标。

3. TPS的14个原则

总而言之，TPS有14条原则，分为4个主要领域[41]，我们将在下面的章节中描述。这4个部分见图8。

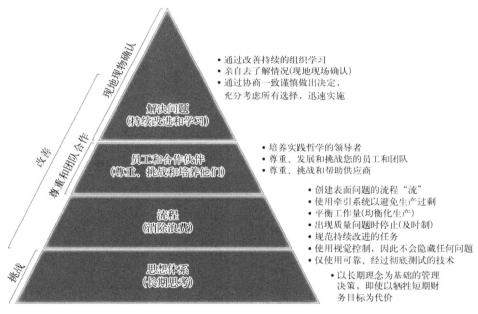

图8　TPS的4个主要部分分为问题解决、人员与合作伙伴、流程和哲学

（1）长期的哲学。

原则1：把您的管理决策建立在一个长期的哲学基础上，即使是以牺牲短期的财务目标为代价。

- 具有超越任何短期决策的哲学意义。我们应该为了一个比赚钱更重要的共同目标而工作、推动和调整整个组织成长。您的使命就是了解您在公司历史上的位置，努力把公司提升到一个新的水平，并把该使命作为其他原则的基础。

- Generate a value of the customer, society, and the economy—it is your starting point. Evaluate every function in the company in terms of the ability to achieve this.
- Be responsible. Strive to decide your own fate. Act with self-reliance and trust in your own abilities. Accept responsibility for your own conduct and maintain to improve the skills that enable you to produce added value.

The Right Process Will Produce the Right Results

Principle 2: Create continuous process flow to bring problems to the surface.

- Redesign work processes to achieve high-value-added, continuous flow. Strive to cut back to zero the amount of the time that any work project is sitting idle or waiting for someone to work on it.
- Create flow to move material and information fast, as well as to link processes and people together so that problems surface right way.
- Make flow evident throughout your organizational culture. It is the key to a true continuous improvement process and to developing people.

Principle 3: Use "Pull" system to avoid overproduction.

- Provide your downline customers in the production process with what they want, when they want it, and in the amount they want it. Material replenishment is initiated by consumption in the basic principle of just-in-time.
- Minimize your work in process and warehousing of inventory by stocking small amounts of each product and frequently restocking based on what the customer actually takes away.
- Be responsive to the day-by-day shifts in customer demand rather than relying on computer schedules and systems to track wasteful inventory.

Principle 4: Level out the workload (heijunka).

- Eliminating waste is just one-third of the equation for making Lean successful. Eliminating overburden to people and equipment and eliminating unevenness in the production schedule is just as important—yet, generally not understood at companies attempting to implement Lean principles.
- Work to level out the workload of all manufacturing and service processes as an alternative to the start/stop approach of working on projects in batches that is typical at most companies.

Principle 5: Build a culture of "stopping to fix problems to get quality right the first time".

- Quality of the customer drives your value proposition.
- Use all the modern quality assurance methods available.
- Build into your equipment the capability of detecting problems and stopping itself. Develop a visual system to alert team or project leaders that a machine or process needs assistance. Jidoka (machines with human intelligence) is the foundation for "building in" quality.

- 为客户、社会和经济创造价值,这是您的出发点。根据实现这一目标的能力评估公司的每一项职能。
- 负责。努力决定自己的命运。自力更生,相信自己的能力。对自己的行为负责,并持续提高技能,使您能够创造附加值。

(2)正确的流程将产生正确的结果。

原则2:通过创造连续的流程不断地揭示问题。

- 为了实现高附加值、连续流程而重新设计工作流程,努力将任何闲置的工作项目或等待有人处理的时间减少到零。
- 创建流程以快速转移材料和信息,并将流程和人员联系在一起,以便问题以正确的方式呈现。
- 使整个组织文化中的流程清晰可见。这是持续改进流程和发展人才的关键。

原则3:用拉动式生产制度避免生产过剩。

- 在生产过程中为您的下线客户提供他们想要的产品、他们想要的时间以及他们想要的数量。物料补货是在准时制的基本原则下由消耗发起的。
- 仅为每类产品库存较少数量,并根据客户实际购买的产品重新进货,达到最小工作量和库存仓储。
- 响应客户每日需求,而不是依靠计算机时间表和系统来跟踪库存。

原则4:均衡化生产(heijunka)。

- 消除浪费只是实现精益成功的三分之一。消除人员和设备负担过重与消除生产计划的不均匀性同样重要,但在试图实施精益原则的公司中,人们通常不理解这一点。
- 努力使所有制造和服务流程的工作量均衡化,以替代在大多数公司通常采用的项目批量开始/停止的方法。

原则5:建立一种“停下来解决问题,第一时间把质量做好”的文化。

- 客户的质量决定您的价值主张。
- 使用所有可用的现代质量保证方法。
- 在您的设备中内置检测问题并自动停止的功能。开发一个可视化系统,来提醒团队或项目负责人机器或流程需要帮助。自働化是构建质量的基础。

- Build into your organization support systems to quickly solve the problems and put in place countermeasures.
- Build into your culture the philosophy of stopping or slowing down to get quality right the first time to enhance productivity in the long run.

Principle 6: Standardized tasks are the foundation for continuous improvements and employee empowerment.

- Use stable, repeatable methods everywhere to maintain the predictability, regular timing, and regular output of your processes. It is the foundation for the flow and pull.
- Capture the accumulated learning about a process up to a point in time by standardizing today's best practices. Allow creative and individual expression to improve upon the standard; then incorporate it into the new standard so that when a person moves on, you can hand off the learning to the next person.

Principle 7: Use Visual Control so no problems are hidden.

- Use simple visual indicators to help people determine immediately whether they are in standard condition or deviating from it.
- Avoid using a computer screen when it moves the worker's focus away from the workplace.
- Design a simple visual system at the workplace where the work is done to support flow and pull.
- Reduce your reports to one piece of paper whenever possible, even for your most important financial decisions.

Principle 8: Use only reliable, thoroughly tested technology that serves your people and process.

- Use technology to support people, not to replace people. Often it is best to work out the process manually before adding technology to support the people.
- New technology is often unreliable and difficult to standardize and, therefore, endangers "flow." A proven process that works generally takes precedence over new and untested technology.
- Conduct actual tests before adopting new technology in business processes, manufacturing systems, or products.
- Reject or modify technologies that conflict with your culture or that might disrupt stability, reliability, and predictability.
- Nevertheless, encourage your people to consider new technologies when looking into new approaches to work. Quickly implement a thoroughly considered technology if it has been proven in trials and if it can improve the flow in your processes.

Add Value to Your Organization by Developing Your People and Partners
Principle 9: Grow leaders who thoroughly understand the work, live philosophy, and teach it to others.

- 建立组织支持系统,快速解决问题并制定对策。
- 在您的企业文化中树立这样一种理念:第一时间停止或放慢速度以保证质量,从长远来看能提高生产率。

原则6:标准化的任务是持续改进和员工授权的基础。

- 在任何地方使用稳定的、可重复的方法来维护流程的可预测性、定时和定期输出,它是流程和拉动式的基础。
- 将当前的最佳实践标准化,及时积累关于流程的知识。允许创造性和个性化的改进标准,然后把它融入新的标准中。这样,当一个人离开时,您就可以把关于流程的知识传给下一个人。

原则7:使用可视化控制,确保不会隐藏任何问题。

- 使用简单的视觉指示器帮助人们立即确定他们是处于标准状态还是偏离标准状态。
- 避免使用电脑屏幕,因为它会转移员工的注意力。
- 在工作场所设计一个简单的视觉系统来支持流程和动力。
- 尽可能将您的报告简化为一张纸,即使是最重要的财务决策。

原则8:为您的员工和流程服务仅使用可靠的、经过完整测试的技术。

- 使用技术来支持人,而不是取代人。通常,在添加技术以支持人员之前,最好先手动制定流程。
- 新技术通常不可靠,难以标准化,因此会危及流程。一个行之有效的流程通常优先于新技术和未经测试的技术。
- 在业务流程、制造系统或产品中采用新技术之前,进行实际测试。
- 拒绝或修改与您的文化冲突或可能破坏稳定性、可靠性和可预测性的技术。
- 鼓励您的员工在寻找新的工作方法时考虑新技术。如果已经在试验中得到验证,并且能够改善使用中的流程,那么可以快速实施经过充分考虑的技术。

(3)通过培养员工和合作伙伴为您的组织增加价值。

原则9:培养完全理解工作、生活哲学并将其传授给他人的领导者。

- Grow leaders within, rather than buying them from outside the organization.
- Do not view the leader's job as simply accomplishing tasks and having good people skills. Leaders must be role models for the company's philosophy and the way of doing business.
- A leader must understand the daily work in great detail so that he or she can be the best teacher of your company's philosophy.

Principle 10: Develop exceptional people and teams who follow your company's philosophy.

- Create a strange, stable culture, in which company values and beliefs are widely shared and lived out over a period of many years.
- Train exceptional individuals and teams to work within the corporate philosophy to achieve exceptional results. Work hard to reinforce the culture continually.
- Use cross-functional teams to improve quality and productivity and enhance flow by solving difficult technical problems. Empowerment occurs only when people use the company's tools to improve the company.
- Make an ongoing effort to teach individuals how to work together as teams toward common goals. Teamwork is something that has to be learned.

Principle 11: Respect your extended network of partners and suppliers by challenging them and helping them to improve.

- Have respect for your partners and suppliers and treat them as an extension of your business.
- Challenge your outside business partners to grow and develop. It shows that you value them. Set challenging targets and assist your partners in achieving them.

Continuously Solving Root Problems Drives Organizational Learnings

Principle 12: Go and see for yourself to thoroughly understand the situation (genchi genbutsu).

- Solve problems and improve processes by going to the source and personally observing and verifying data, rather than theorizing on the basis of what other people or the computer screen tell you.
- Think and speak based on personally verified data.
- Even high-level managers and executives should go and see things for themselves, so they will have more than a superficial understanding of the situation.

Principle 13: Make a decision slowly by consensus, thoroughly considering all options; implement decisions rapidly.

- Do not pick a single direction and go down that one path until you have thoroughly considered alternatives.
- Nemawashi is the process of discussing problems and potential solutions with all of those affected, to collect their ideas and get agreement on a path forward. This consensus process, though time-consuming, helps broaden the

- 在内部培养领导者,而不是从组织外部招聘。
- 不要将领导者的工作视为简单地完成任务和拥有良好的人际交往技能,领导者必须是公司理念和经营方式的榜样。
- 领导者必须非常详细地了解日常工作,这样他或她才能成为公司理念的最佳倡导者。

原则10:培养符合公司理念的优秀人才和团队。

- 创造一种强大、稳定的文化,在这种文化中,公司价值观和信仰被广泛分享,并在许多年的时间里得以延续。
- 培训优秀的个人和团队在公司理念范围内工作,以取得卓越的成果。努力不断地加强文化建设。
- 利用跨职能团队来提高质量和生产力,并通过解决困难的技术问题来增强流程。只有当人们使用公司的工具来改进公司时,授权才会发生。
- 持续努力,教个人如何以团队的形式合作,实现共同目标。团队合作是必须学习的内容。

原则11:尊重您的合作伙伴和供应商,通过目标帮助他们提升。

- 尊重合作伙伴和供应商,并将其视为业务的延伸。
- 挑战您的外部业务合作伙伴,使其成长和发展,这表明您重视他们。制定具有挑战性的目标,并协助合作伙伴实现这些目标。

(4)持续解决根本问题推动组织学习。

原则12:自己亲自去看现场,彻底了解情况(现地现物)。

- 通过访问数据源并亲自观察和验证数据来解决问题并改进流程,而不是根据其他人或计算机屏幕告诉您的内容进行理论化。
- 根据个人验证的数据进行思考和发言。
- 即使是高层管理人员和主管也应该亲自去看现场,确保他们对情况的了解不仅仅是表面的。

原则13:以协商一致的方式缓慢做出决定,充分考虑所有选择,迅速执行决定。

- 备选方案在考虑充分之前,不要选择一个方向并沿着这条路径前进。
- 根回(原指在移植树木时小心翼翼地将所有根须都包缠起来)是指与所有受影响的人讨论问题和潜在解决方案的过程,收集他们的想法,并就前进道路达成一致。这种共识过程虽然耗时,但有助于扩大对解决方案的搜索

search for solutions, and once a decision is made, the stage is set for rapid implementation.

Principle 14: Become a learning organization through relentless reflection (hansei) and continuous improvements (kaizen).

- Once you have established a stable process, use continuous improvement tools to determine the root cause of inefficiencies and apply effective countermeasures.
- Design processes that require almost no inventory. This will make wasted time and resources visible for all to see. Once waste is exposed, have employees use a continuous improvement process (kaizen) to eliminate it.
- Protect the organizational knowledge base by developing stable personnel, slow promotion, and very careful succession systems.

End Notes

1. ToolsHero.com: Henry L. Gantt from http://www.toolshero.com/henry-gantt.
2. Philip B. Crosby *Quality Is Free: The Art of Making Quality Certain: How to Manage Quality–So That It Becomes A Source of Profit for Your Business*, (1979).
3. Taiichi Ohno and Setsuo Mito., *Just-In-Time for Today and Tomorrow*, (1988).
4. *Mastering Power Production: The Industrial, Economic and Social Problems Involved and Their Solution*, (Walter Nicholas Polakov, 1923).
5. Gantt H.L. et al., published by The Engineering Magazine, New York, 1910; republished as Work, Wages and Profits, Easton, Pennsylvania, Hive Publishing Company, 1974.
6. Gantt Chart from http://www.gantt.com/.
7. *The One Best Way, Frederick Winslow Taylor and the Enigma of Efficiency.*
8. The Ben Graham Corporation, "Business Process Improvement Methodology for Graham Process Charting Software", 57–58.
9. http://pascal.computer.org/sev_display/search.action;jsessionid=C022AFA5E46051D302 D9E138E7A90DAF. D. 30. October 2009 –kl. 20.00 (Software and System Engineering Vocabulary).
10. Frank Bunker Gilbreth, Tom Ricci, http://www.ASME.org, 2012.
11. F. W. Taylor: Critical Evaluations in Business and Management, Volume 2, John Cunningham Wood, Michael C. Wood, Taylor & Francis, 2002.
12. Dr. Ben S. Graham. Jr, *Flowchart.*
13. George C.S., *The history of management thought*, 100–101.
14. Dr. Ben S. Graham Jr., Allan Mogensen and his Legacy, p. 5.
15. Mogensenm A.H., *Common sense applied to motion and time study*, (McGraw-Hill, 1932).
16. Ben G, Rediscover Work Simplification, 1–2.
17. Allan H. Mogensen, Carry Out a Methods Improvement Program, 1949 at http://www.nickols.us/Mogensen.pdf.
18. Schamel J., *How the Pilot's Checklist Came About.* (FAA Flight Service Training, Retrieved 12 January 2007). "The idea of a pilot's checklist spread to other crew members, other Air Corps aircraft types, and eventually throughout the aviation world".
19. http://en.wikipedia.org/wiki/Benjamin_S._Graham 30.09.2009 21:34.
20. Ben S., Graham Sr., *Paperwork Simplification.*
21. https://www.asme.org/.

范围,一旦做出决定,就为快速实施做好了准备。

原则14:通过不断反思(hansei)和持续改进(kaizen)成为学习型组织。

- 一旦您建立了一个稳定的流程,就使用持续改进工具来确定效率低下的根本原因,并采取有效的对策。

- 设计流程几乎不需要库存,这将使所有人都能看到浪费的时间和资源。一旦浪费暴露出来,就让员工使用持续改进流程来消除它。

- 通过发展稳定的人员、缓慢的晋升和非常谨慎的继任制度来保护组织知识库。

参考文献

[1] ToolsHero.com: Henry L. Gantt from http://www.toolshero.com/henry-gantt.

[2] Philip B. Crosby Quality Is Free: The Art of Making Quality Certain: How to Manage Quality—So That It Becomes A Source of Profit for Your Business, (1979).

[3] Taiichi Ohno and Setsuo Mito., Just-In-Time for Today and Tomorrow, (1988).

[4] Mastering Power Production: The Industrial, Economic and Social Problems Involved and Their Solution, (Walter Nicholas Polakov, 1923).

[5] Gantt H.L. et al., published by The Engineering Magazine, New York, 1910; republished as Work, Wages and Profits, Easton, Pennsylvania, Hive Publishing Company, 1974.

[6] Gantt Chart from http://www.gantt.com/.

[7] The One Best Way, Frederick Winslow Taylor and the Enigma of Efficiency.

[8] The Ben Graham Corporation, "Business Process Improvement Methodology for Graham Process Charting Software", 57−58.

[9] http://pascal.computer.org/sev_display/search.action;jsessionid=C022AFA5E46051D302D9E138E7A90D AF. D. 30. October 2009 −kl. 20.00 (Software and System Engineering Vocabulary).

[10] Frank Bunker Gilbreth, Tom Ricci, http://www.ASME.org, 2012.

[11] F. W. Taylor: Critical Evaluations in Business and Management, Volume 2, John Cunningham Wood, Michael C. Wood, Taylor & Francis, 2002.

[12] Dr. Ben S. Graham. Jr, Flowchart.

[13] George C.S., The history of management thought, 100−101.

[14] Dr. Ben S. Graham Jr., Allan Mogensen and his Legacy, p. 5.

[15] Mogensen A.H., Common sense applied to motion and time study, (McGraw-Hill, 1932).

[16] Ben G, Rediscover Work Simplification, 1−2.

[17] Allan H. Mogensen, Carry Out a Methods Improvement Program, 1949 at http://www.nickols.us/Mogensen. pdf.

[18] Schamel J., How the Pilot's Checklist Came About. (FAA Flight Service Training, Retrieved 12 January 2007). "The idea of a pilot's checklist spread to other crew members, other Air Corps aircraft types, and eventually throughout the aviation world".

[19] http://en.wikipedia.org/wiki/Benjamin_S._Graham 30.09.2009 21:34.

[20] Ben S., Graham Sr., Paperwork Simplification.

[21] https://www.asme.org/.

22. *A.S.M.E. standard operation and flow process charts*, developed by the A.S.M.E. Special committee on standardization of therbligs, process charts, and their symbols, (1947).

23. Henzold G., *Geometrical Dimensioning and Tolerancing for Design, Manufacturing and Inspection*. 2nd ed., (Oxford, UK: Elsevier, 2006).

24. Thomas Dufresne and James Martin, *Methods for Information Systems Engineering: Knowledge Management and E-Business* (2003).

25. *Task Analysis Tools Used Throughout Development* (FAA, 2008).

26. Graham B. B., *Detail Process Charting: Speaking the Language of Process* (2004).

27. Motivating People to Work: The Key to Improving Productivity, Warren C. Hauck, Industrial Engineering and Management Press, 1984.

28. Tim Weilkiens., *Systems Engineering with SysML/UML: Modeling, Analysis, Design*, 287.

29. Malcolm D.G., et al., *Application of a Technique for Research and Development Program Evaluation Operations Research* vol. 7, no. 5, (September–October 1959), 646–669.

30. Tim Weilkiens., *Systems Engineering with SysML/UML: Modeling, Analysis, Design* (2008), 287.

31. John Azzolini., *Introduction to Systems Engineering Practices*. (July 2000).

32. http://www.idef.com/IDEF0.htm.

33. *ICAM Architecture Part II, Volume V–Information Modeling Manual (IDEF1)*, AFWAL-TR-81–4023, Materials Laboratory, Air Force Wright Aeronautical Laboratories, Air Force Systems Command, (Wright-Patterson Air Force Base, Ohio 45433, June 1981).

34. http://www.idef.com/IDEF1.htm.

35. http://www.idef.com/IDEF1x.htm.

36. http://www.idef.com/IDEF3.htm.

37. http://www.idef.com/IDEF4.htm.

38. http://www.idef.com/IDEF5.htm.

39. Ohno T., *Toyota Production System: Beyond Large-scale Production* (Productivity Press Inc, 1995).

40. Liker J.K., Chapter 3 "The TPS House Diagram" *The Toyota Way: 14 Management Principles from the World's Greatest Manufacturer* (McGraw Hill, 2004).

41. Jeffrey Liker *The Toyota Way: 14 Management Principles from the World's Greatest Manufacturer*, First edition, McGraw-Hill (2003).

［22］ A.S.M.E. standard operation and flow process charts, developed by the A.S.M.E. Special committee on standardization of therbligs, process charts, and their symbols, (1947).

［23］ Henzold G., Geometrical Dimensioning and Tolerancing for Design, Manufacturing and Inspection. 2nd ed., (Oxford, UK: Elsevier, 2006).

［24］ Thomas Dufresne and James Martin, Methods for Information Systems Engineering: Knowledge Management and E-Business (2003).

［25］ Task Analysis Tools Used Throughout Development (FAA, 2008).

［26］ Graham B. B., Detail Process Charting: Speaking the Language of Process (2004).

［27］ Motivating People to Work: The Key to Improving Productivity, Warren C. Hauck, Industrial Engineering and Management Press, 1984.

［28］ Tim Weilkiens., Systems Engineering with SysML/UML: Modeling, Analysis, Design, 287.

［29］ Malcolm D.G., et al., Application of a Technique for Research and Development Program Evaluation Operations Research vol. 7, no. 5, (September–October 1959), 646–669.

［30］ Tim Weilkiens., Systems Engineering with SysML/UML: Modeling, Analysis, Design (2008), 287.

［31］ John Azzolini., Introduction to Systems Engineering Practices. (July 2000).

［32］ http://www.idef.com/IDEF0.htm.

［33］ ICAM Architecture Part Ⅱ, Volume V–Information Modeling Manual (IDEF1), AFWAL-TR-81-4023, Materials Laboratory, Air Force Wright Aeronautical Laboratories, Air Force Systems Command, (Wright-Patterson Air Force Base, Ohio 45433, June 1981).

［34］ http://www.idef.com/IDEF1.htm.

［35］ http://www.idef.com/IDEF1x.htm.

［36］ http://www.idef.com/IDEF3.htm.

［37］ http://www.idef.com/IDEF4.htm.

［38］ http://www.idef.com/IDEF5.htm.

［39］ Ohno T., Toyota Production System: Beyond Large-scale Production (Productivity Press Inc, 1995).

［40］ Liker J.K., Chapter 3 "The TPS House Diagram" The Toyota Way: 14 Management Principles from the World's Greatest Manufacturer (McGraw Hill, 2004).

［41］ Jeffrey Liker, The Toyota Way: 14 Management Principles from the World's Greatest Manufacturer, First edition, McGraw-Hill (2003).

Phase 3: Process Concept Evolution

Mark von Rosing, August-Wilhelm Scheer, John A. Zachman, Daniel T. Jones, James P. Womack, Henrik von Scheel

INTRODUCTION

In this book, we have until now covered key figures like Adam Smith, Frederick Winslow Taylor, Henry Ford, and many others who have dramatically impacted the way of thinking of, working with, and how we model processes. This chapter of the book will elaborate on the move from process-centric work around labor and people to the coupling of information technology (IT) aspects with process modeling aspects. We witness a much wider and more profound involvement of IT with the incorporation of processes in business and organizations across geographical borders and cultural differences.

With the entry of Phase 3 at the beginning of the 1980s, Professor August-Wilhelm Scheer founded his company, IDS Scheer, in Germany. IDS Scheer is a consulting and development firm focusing on business process management (BPM)[1] and is widely regarded as a frontrunner and pioneer of today's BPM industry. Professor August-Wilhelm Scheer would later launch the Architecture of Integrated Information Systems (ARIS) concept in 1991—a system concept that would allow company data to be associated with information flows, function, and control while simultaneously be supervised by different management viewpoints for the sake of business clarity and a better control of circumstances and process execution within the organization.

In 1987, John Zachman (born December 16, 1934), an American business and IT consultant and early pioneer of Enterprise Architecture[2], published his initial framework named "A Framework for Information Systems Architecture"[3] in an article in the IBM Systems Journal. This would lead to the ongoing development of what is now known as the Zachman Framework for Enterprise Architecture—a conceptual development of a framework that intends to establish principles for and to document how an enterprise is constructed and designed through the use of accountabilities, responsibilities, interrogatives, functions, control, information and process flows, infrastructure and technology, etc. This framework would be the very first concept to relate an entire enterprise with information systems and business processes and—although not discussed in this chapter—would then spark the development of many more frameworks in the future.

After World War II, Taiichi Ohno and Shingeo Shingo created the "Just In Time," "Waste Reduction," and "Pull System" concepts for Toyota, which, together with other flow management techniques, resulted in the Toyota Production System (TPS).[4] Ever since its initial conception during the early 1970s, the TPS has undergone industrial

1.3 第3阶段：流程概念演变3.0

Mark von Rosing, August-Wilhelm Scheer, John A. Zachman, Daniel T. Jones, James P. Womack, Henrik von Scheel

1.3.1 介绍

在这本书中，我们到目前为止已经介绍了一些关键人物，如A. Smith、F. W. Taylor、H. Ford，以及许多对我们的思维方式、工作方式和如何进行流程建模产生巨大影响的其他人。本书的这一节将详细阐述从"围绕工作和人员的流程中心"到"将IT方面与流程建模等各方面耦合"的过程。我们将讨论在存在地域边界和文化差异的前提下，IT技术如何更广泛、更深入地参与业务和组织的流程整合。

第3阶段始于20世纪80年代初，A. W. Scheer（奥古斯特·威廉·舍尔）教授在德国创立了IDS Scheer公司（当前公司网站为：www.scheer-group.com）。IDS Scheer是一家专注于BPM咨询和开发的公司[1]，该公司被认为是当今BPM行业的建立者和领导者。A. W. Scheer教授随后于1991年推出了集成信息系统体系结构（architecture of integrated information systems, ARIS）的概念，这个概念的定义是：一个允许公司数据与信息流、功能和控制相关联，同时由不同的管理主体进行监督，以确保业务清晰、更好地控制组织内的环境和流程执行的系统。

1987年，J. A. Zachman（出生于1934年12月16日），一位美国的商业和IT顾问，也是企业体系结构理论的早期创立者[2]，在IBM系统杂志的一篇文章中发表了他的"信息系统体系结构框架"的初始框架[3]，该框架被称为"Zachman企业结构框架的雏形"，旨在通过使用职责、责任、问询、功能、控制、信息和流程、基础设施和技术等来建立和记录企业构建和设计的过程。该框架是第一个将整个企业与信息系统和业务流程联系起来的概念，尽管本节没有讨论，但它在未来必将引发更多框架的开发。

第二次世界大战后，Taiichi Ohno和Shigeo Shingo提出了准时化生产、减少浪费和"拉动系统"的概念，与其他流程管理技术一起，形成了TPS[4]。自20世纪70年代初的最初构想提出以来，TPS经历了产业多次演进，并在不断得到行业和学术

evolution and matured dramatically while being constantly improved upon by industry professionals as well as throughout the academic world. In 1990, James Womack summarized the TPS concepts to create Lean Manufacturing[5] at a time when Japanese expertise was spreading to the West, and the success achieved by companies applying these principles and techniques became undeniable. Then, in 2005, James P. Womack and Daniel T. Jones published an article in the *Harvard Business Review* describing a new theory called "Lean Consumption."[6] While "Lean Manufacturing" sets out ways to streamline manufacturing processes, Lean Consumption *minimizes customers' time and effort by delivering exactly what they want, when, and where they want it.*

Perhaps one of the biggest ideas in the 1990s was conceived by Dr Michael Hammer, an American engineer, management author, and a former professor of computer science at the Massachusetts Institute of Technology (MIT), as he introduced the business world to a concept he called business process reengineering (BPR).[7] BPR involves the redesign of core business processes to achieve improvements in productivity, cycle times, and quality. In BPR, companies start with a blank sheet of paper and rethink existing processes to deliver more value to the customer. The terms *process improvement, process excellence,* and *process innovation* were all coined by Dr Michael Hammer.

In the 1980s to the 1990s, a new phase of quality control and management began. This became known as total quality management (TQM).[8] Having observed Japan's success of employing quality management, western companies started to introduce their own quality initiatives.[9] TQM developed as a catchall phrase for the broad spectrum of quality-focused strategies, programs, and techniques during this period, and became the center of focus for the western quality movement.[10]

A typical definition[11] of TQM includes phrases such as the following:

- Customer focus
- The involvement of all employees
- Continuous improvement
- The integration of quality management throughout the entire organization

Although the definitions were all similar, there was confusion. It was not clear what sort of practices, policies, and activities needed to be implemented to fit the TQM definition. Some of the tools involved with the implementation of the TQM concept in organizations were such as:

- House of Quality or Quality Function Deployment (QFD)[12]
- Taguchi techniques[13]
- Pareto diagrams[14]
- Process diagrams
- Cause and effect analysis diagrams—also known as root cause analysis
- Statistical process control (SPC)[15]
- The Plan, Do, Check, Act (PDCA) Cycle[16]
- The EFQM Excellence Model[17] or simply "business excellence"

In the early and mid-1980s, with Chairman Bob Galvin at the front, Motorola engineers decided that the traditional quality levels—measuring defects in

界专业人士的改进的同时,得到了巨大的发展。1990年,J. P. Womack总结了TPS概念,在将日本的专业知识体系向西方传播的过程中创建了精益制造的概念[5],从效果来看,大部分应用这些原则和技术的公司取得了巨大的成功。然而,在2005年,J. P. Womack和D. T. Jones在《哈佛商业评论》上发表了一篇文章,描述了一种称为精益消费的新理论[6]。虽然精益制造提出了简化制造流程的方法,但精益消费可以根据客户想要的产品和时间、地点来最大限度地减少客户的时间和精力。

20世纪90年代,美国工程师、管理学作家、麻省理工学院前计算机科学教授M. Hammer博士提出的BPR[7]也许是当前最伟大的想法之一。BPR涉及提高生产率、生命周期和质量的核心业务流程。在BPR中,公司从一张空白的纸开始,重新思考现有的流程,以便为客户提供更多的价值。术语流程改进、流程优化和流程创新都是由M. Hammer博士创造的。

20世纪80年代至90年代,以质量管理和控制为目标的全面质量管理(TQM)[8]思想迅速发展,这源于西方公司在观察到日本采用质量管理的成功后,开始引入符合自身特点的质量举措[9]。在这一时期,作为以质量为中心的战略、计划和技术的TQM逐渐发展起来,并成为西方质量运动的焦点[10]。

TQM的典型定义[11]包括以下短语:

- 以客户为中心;
- 全体员工参与;
- 持续改进;
- 整合整个组织的质量管理。

虽然所有的定义很相似,但存在一些混淆。尚不明确按照TQM的定义,哪些实践、政策以及活动需要被实施。目前,组织中实施TQM可以使用如下工具:

- 质量屋或质量功能展开(quality function deployment,QFD)[12];
- 田口技术[13];
- 帕累托图[14];
- 流程图;
- 因果分析图,也称为根本原因分析;
- 统计过程控制(statistical process control,SPC)[15];
- 计划、执行、检查、处理(plan-do-check-act,PDCA)循环[16];
- EFQM业务卓越模型[17]或者简单地卓越经营。

在20世纪80年代早期和中期,B. Galvin担任摩托罗拉董事长期间,摩托罗拉工程师认为传统质量水平评价方法采取的通过测量数千个实例获得缺陷的方式并不准确,相反,他们需要衡量每百万个实例的缺陷才能获得真实的标准。1986

thousands of instances—did not provide enough granularity. Instead, they wanted to measure the defects per million instances. Six Sigma[18] was then theorized by Bill Smith in 1986, upon formulating the strategic concepts of Motorola.[19] Motorola developed this new standard and created the methodology and the needed cultural change associated with it. Since then, hundreds of companies around the world have adopted Six Sigma as a way of doing business. This is a direct result of many of America's leaders openly praising the benefits of Six Sigma, leaders such as Larry Bossidy of Allied Signal (now Honeywell) and Jack Welch of General Electric.

PROF. DR. H.C. MULT. AUGUST-WILHELM SCHEER
Introduction

One of the key figures in leading the new ways of thinking, working and modeling with business processes was the German Professor and Entrepreneur August-Wilhelm Scheer. He developed the concept of using the business process as the anchor for his "architecture of integrated information systems" (ARIS)[20] covering design, implementation, execution and control of business processes. His approach is considered innovative and groundbreaking. Scheer founded his first company, IDS Scheer AG, based on his ARIS approach. IDS Scheer is an international software and consulting company that develops, markets, and supports Business Process Management (BPM) solutions around the world. The software tool "ARIS Toolset"[21] became the global market leader for process modeling and repository tools. Amongst many other things, the following can be mentioned that Prof. Scheer has developed or left his mark upon by pushing the bar with his research, entrepreneurship as well as his focus on integrating business and IT:

- The link between process management and information technology. The ability to link processes and IT changed everything around process concepts, including the way of thinking, working, modeling and automating processes.
- His contribution to the academic world includes: in 1975 Scheer took over one of the first chairs for information systems and founded the Institute for Information Systems (IWI) at the Saarland University, which he led until 2005.
- His contribution to the Software world includes: In 1984 he founded IDS Scheer—a Business Process Management (BPM) software and consulting company. Its flagship product, the ARIS Toolset, is still today the market leader.
- His contribution to the Process Modeling[22] community includes: He developed among others the 2 main methods for Business Process Modeling:
  1) Value-added Chain Diagrams (VCD).
  2) Event-driven Process Chains (EPC).
- His contribution to Enterprise Modelling and Architecture includes: The ARIS House which is one of the first Enterprise Architecture and Modelling frameworks that combines and organizes information of an organization in five interrelated views: data, function, organization, output, and control (Figure 1).

年,B. Smith在制定摩托罗拉的战略概念时提出了Six Sigma(六西格玛)[18]的概念[19]。摩托罗拉开发了这一新标准,并创建了相关的方法和所需的文化变革。从那时起,通过联合信号公司(现在是霍尼韦尔公司)的L. Bossidy和通用电气公司的J. Welch等许多美国公司领导人公开称赞六西格玛的好处,全世界数百家公司都将六西格玛作为一种经营方式。

1.3.2　PROF. DR. H. C. MULT. AUGUST-WILHELM SCHEER

1. 介绍

德国教授兼企业家A. W. Scheer是引领业务流程新思维、工作和建模方式的关键人物之一。他提出了将业务流程用作集成信息系统架构(architecture of integrated information systems, ARIS)[20]的理念,该体系结构涵盖了业务流程的设计、实现、执行和控制。他的方法被认为极具创造性和开创性。A. W. Scheer基于他的ARIS方法建立了他的第一家公司——IDS Scheer。IDS Scheer是一家国际软件和咨询公司,在世界各地开发、推销和服务BPM解决方案。软件工具ARIS工具集[21]成为流程建模和存储库工具的全球市场领导者。除此之外,A. W. Scheer教授的研究、创业精神以及对商业与IT整合的关注推动了他的发展并留下了自己的印记。

- 构建了流程管理与工厂之间的联系。将流程和IT关联的能力改变了流程的思维方式、工作方式、建模方式和流程自动化。
- A. W. Scheer对学术界的贡献包括:1975年,在萨尔兰大学成立了信息系统研究所(Institute for Information Systems, IWI)并担任主席,直到2005年。
- 他对软件领域的贡献包括:1984年,创建了IDS Scheer——一家BPM软件和咨询公司,它的旗舰产品ARIS工具集如今仍然是市场领导者。
- 他对流程建模[22]领域的贡献包括:开发了两种主要的业务流程建模方法:

(1)价值增值链图(value-added chain diagram, VCD);

(2)事件驱动的流程链(EPC)。

- 他对企业建模和架构的贡献包括:ARIS屋,它是早期将组织信息以五个相互关联的视图(数据、功能、组织、输出和控制)进行组合和组织的企业体系结构和建模框架之一(图1)。

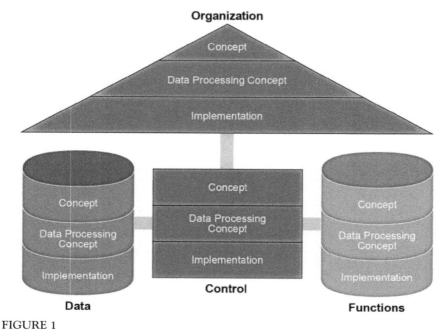

FIGURE 1

The Architecture of Integrated Information Systems (ARIS) model.

Prof. Scheer is widely regarded as the founder of the BPM industry. In this chapter we briefly explain why.

A New Way of Thinking: Linking Process and Information Concepts

It was in 1984 when two important events took place; Scheer's thinking about "IT oriented Business Administration"[23] was published, and in the same year he founded IDS Scheer which can be regarded as the frontrunner and foundation of the BPM industry. The notion of "business process" had been front and center of IDS Scheer from the beginning. Other important book publications followed, for example "Business Engineering"[24], "CIM—The Factory of the Future"[25] and, of course, the two books on ARIS. IDS Scheer quickly became a sought after advisor for large companies like Bosch and Daimler. The major business processes of the "CIM Organization" has been represented in the popular "Y Model" - a symbol that is a distinguishing feature of IDS Scheer's company logo.

Parallel to the set-up of his company IDS Scheer, Prof. Scheer researched methods of business user-oriented description of information systems[26] and the underlying technologies. Contrary to previous traditional thinking, the business process was now being treated as the critical link between people and IT. The ability to link process and information technology changed everything around process management

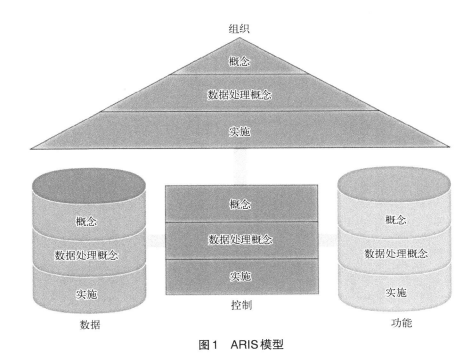

图1　ARIS模型

A. W. Scheer教授被广泛认为是BPM行业的创始人。在本节中,我们简要解释了原因。

2. 一种新的思维方式:链接流程与信息概念

1984年,有两件重大的事件发生:① A. W. Scheer关于"以IT为导向的企业管理"[23]的思想诞生;② 同年他创立了IDS Scheer,该公司被认为是BPM行业的领导者。业务流程的概念从一开始就是IDS Scheer的前沿和核心。随后A. W. Scheer出版了一些重要的书籍,如《业务工程》[24]《CIM——未来工厂》[25],当然,还有关于ARIS的两本书。IDS Scheer很快得到博世(Bosch)和戴姆勒(Daimler)等大公司的认可和追捧。"CIM组织"的主要业务流程以流行的"Y模型"表示,这也是IDS Scheer公司徽标的来源。

与他所建的IDS Scheer公司相提并论的是,A. W. Scheer教授还研究了面向业务用户的信息系统描述方法[26]和底层技术。与以前的传统思维相反,业务流程现在被视为人与IT之间联系的关键。将流程和信息技术相关联的能力改变了流

approaches because Scheer's approach met the demand for simplicity for the business user and—at the same time—delivers the methodological stringency for the transfer of process design into information technology-based execution. This approach is also represented in the ARIS Framework and was already being rolled out globally in 1991.

Architecture of Integrated Information Systems (ARIS)

Architecture of Integrated Information Systems (ARIS) is an approach to enterprise modeling and enterprise architecture. It offers methods for managing the entire life cycle of processes while taking a holistic view of the organization, data, functions, deliverables, and last but not least, the control flows (Figure 1). Based on the ARIS Framework[27], the ARIS software tool was developed—the ARIS Software tool is a modelling and repository software system.

The ARIS Framework facilitates the use of all available modeling methods, including the mentioned previously Event-driven Process Chains (EPC) and the Business Process Modeling Notation (BPMN). The ARIS Toolset supports over 250 modeling methods. These modeling methods address the different views on a business process as it is illustrated in ARIS (see Figure 2).

- Organizational view
- Data view
- Control view
- Functional view

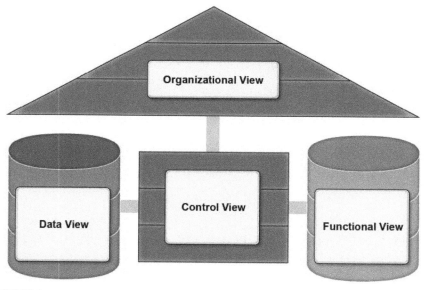

FIGURE 2

The Architecture of Integrated Information Systems (ARIS) view model.

程管理方法的一切,因为A. W. Scheer的方法满足了对业务用户简单性的要求,同时为将流程设计转换为基于信息技术的执行提供了严格的方法论。这种方法在ARIS框架中也有体现,并已于1991年在全球推广。

3. ARIS

ARIS是构建企业模型和结构的一种方法。它提供了管理整个流程生命周期的方法,同时对组织、数据、功能、可交付成果以及最后但同样重要的控制流程有着全面而整体的认知(图1)。基于ARIS框架[27],发展出一个用于建模和存储的ARIS软件系统。

ARIS框架有助于使用所有可用的建模方法,包括前面提到的EPC和BPMN。ARIS工具集支持超过250种建模方法,这些建模方法处理业务流程的不同视图,如ARIS所示(图2)。

- 组织视图;
- 数据视图;
- 控制视图;
- 功能视图。

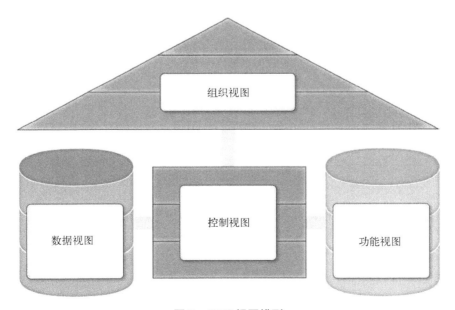

图2　ARIS视图模型

Each view of ARIS can be described in different levels of abstraction, starting with a pure business layer, ending in an implementation-oriented IT description layer. This allows the systematic link of business approaches to the underlying enabling IT.

Additional Important Aspects

In 1997, Scheer also founded a company focused on innovative learning and education technologies and approaches, called IMC AG. This company addresses the people-side of a process implementation. Other company start ups quickly followed—all of them focusing on topics around the integration of business processes and IT.

In 2003, Scheer was awarded the Philip Morris Research Prize and the Ernst & Young Entrepreneur of the Year Award. In December 2005, he received the Erich Gutenberg price and in the same month the Federal Cross of Merit first class of Germany was awarded to him. In 2005 he was elected a fellow of the "Gesellschaft für Informatik", a renowned German research organization. Since 2006, Scheer has been a member of the council for innovation and growth of the Federal Government of Germany. In 2007, he was honored as a "HPI-Fellow" by the Hasso-Plattner Institut (HPI) für Softwaresystemtechnik" and was elected President of the German Association for Information Technology, Telecommunications and New Media.

In 2010, Scheer was awarded the Design Science Lifetime Achievement Award at the University of St. Gallen. He received the award as a recognition of his contribution to design science research.

In 2014, the Global University Alliance (a collaborative, academic alliance consisting of more than 300 universities) and the LEADing Practice Community (a professional community consisting of more than 3200 certified practitioners) decided to honor August-Wilhelm Scheer with the LEADing Practice Lifetime Achievement Award in recognition of his long-term impact and contribution to evolve both the business and academic world and organizations structural way of thinking and way of working around enterprise modelling concepts.

Conclusion

Scheer has revolutionized the way we manage processes and link business to information technology. Frameworks like ARIS and the Y-CIM model have fundamentally transformed academia and practice regarding business process management.

JOHN A. ZACHMAN—THE ARRIVAL OF ENTERPRISE ARCHITECTURE

Introduction

John A. Zachman is the originator of the first "Concept of the Framework for Enterprise Architecture," which is widely regarded as the birth of enterprise architecture. With the arrival of the first enterprise architecture descriptions, concepts and frameworks, there was a clear link being established and built between the already

ARIS的每个视图都可以在不同的层级中进行描述,从纯业务层开始,到以实施为导向的IT实施层结束,这允许在IT方法的支持下将业务方法系统地链接到底层技术。

4. 其他重要方面

1997年,A. W. Scheer还成立了一家专注于创新学习和教育技术与方法的公司,名为IMC AG。该公司从事流程实施人员的培养。在这种背景下,其他公司很快跟进,都专注于业务流程和IT集成。

2003年,A. W. Scheer获得"菲利普莫里斯研究奖"和"安永青年企业家奖"。2005年12月,他获得了埃里希·古登堡奖,同月,他又被授予了"德国联邦功勋一等奖"。2005年,他被选为德国著名研究机构"计算机科学公司"的研究员。自2006年以来,A. W. Scheer一直是德国联邦政府创新与增长委员会的成员。2007年,他被哈索-普拉特纳研究所(Hasso-Plattner Institute,HPI)授予"HPI研究员"称号,并当选为德国信息技术、电信和新媒体协会主席。

2010年,为了表彰他对设计科学研究的贡献,A. W. Scheer获得圣加仑大学设计科学终身成就奖。

2014年,全球大学联盟(一个由300多所大学组成的合作性学术联盟)和领导实践团体(一个由3200多名认证从业者组成的专业社区)决定授予A. W. Scheer "领导实践终身成就奖",以表彰他长期以来对商业、学术界以及组织结构思维方式和围绕企业建模概念工作方式的影响和贡献。

5. 结论

A. W. Scheer彻底改变了我们管理流程和将业务与工厂进行联系的方式,尤其是像ARIS和Y-CIM模型这样的框架从根本上改变了学术界和BPM实践。

1.3.3   J. A. Zachman——企业架构的到来

1. 介绍

J. A. Zachman是第一个提出企业架构框架概念的人,本概念的提出被认为是企业架构的诞生。随着更多企业架构描述、概念和框架的出现,它为各业务流程之

established—although young and immature—business process landscape that had evolved prior to the introduction of enterprise architecture. This link, in particular, was mostly focusing on establishing a connection between ownerships (who is accountable), interrogatives (why are we doing this, and how should we do it), information systems[28] (connecting data to functionality and information flows), and the need for detailed representation of all of these functions and activities.

Zachman is not only known for his work on enterprise architecture, but is also known for his early contributions to IBM's Information Strategy methodology (Business Systems Planning)[29] as well as to their executive team planning techniques (Intensive Planning). Mr. Zachman retired from IBM in 1990, having served them for 26 years. He is the founder and chairman of Zachman International, an enterprise architecture research and education company. He is also the Executive Director of FEAC Inc, the Federated Enterprise Architecture Certification Institute, that issues certifications for DoDAF (the Department of Defense Architecture Framework), FEAF (the Federal Enterprise Architecture Framework), and Zachman (the Zachman Ontological Framework).[30]

Mr. Zachman serves on the Executive Council for Information Management and Technology (ECIMT) of the United States Government Accountability Office (GAO) and on the Advisory Board of the Data Administration Management Association International (DAMA-I) from whom he was awarded the 2002 Lifetime Achievement Award. In November 2013, he was acknowledged for Achievement and Excellence for Distinguished Innovative Academic Contribution by the IEEE Systems, Man and Cybernetics Society Technical Committees on Enterprise Information Systems and on Enterprise Architecture and Engineering. In August 2011, he was awarded the General Colin Powell Public Sector Image Award by the Armed Services Alliance Program. He was awarded the 2009 Enterprise Architecture Professional Lifetime Achievement Award from the Center for Advancement of the Enterprise Architecture Profession as well as the 2004 Oakland University Applied Technology in Business (ATIB) Award for IS Excellence and Innovation.

Mr. Zachman has been focusing on enterprise architecture since the 1980s and has written extensively on the subject, and he is also the author of the book *The Zachman Framework for Enterprise Architecture™: A Primer on Enterprise Engineering and Manufacturing*.[31]

The Zachman Framework for Enterprise Architecture

"A framework as it applies to enterprises is simply a logical structure for classifying and organizing the descriptive representations of an enterprise that are significant to the management of the enterprise as well as to the development of the enterprise's system (with the aim of) rationalizing the various concepts and specifications in order to provide for clarity of professional communication, to allow for improving and integrating development methodologies and tools, and to establish credibility and confidence in the investment of systems and resources."

John Zachman, 2003

间建立了清晰的联系。在引入企业架构之前，年轻且不成熟的业务流程布局就已经形成了。这一联系主要集中体现在建立所有权（谁负责）、问题（为什么要这样做，以及我们应该如何做）、信息系统[28]（将数据连接到功能和信息流）及所有这些功能和活动的详细表述的必要性之间的联系。

J. A. Zachman不仅因其在企业体系结构方面的工作而闻名，还因其对IBM信息战略方法论（业务系统规划）[29]以及其执行团队规划技术（集约化规划）的早期贡献而闻名。J. A. Zachman先生于1990年从服务了26年的IBM退休。他是从事企业架构研究与教育的Zachman International公司的创始人和董事长。他还是联邦企业架构认证协会（Federated Enterprise Architecture Certification Institute, FEAC）的执行董事，该协会为国防部架构框架（the Department of Defense architecture framework, DoDAF）、联邦企业架构框架（FEAF）和Zachman本体论框架（the Zachman ontological framework）提供认证服务并颁发证书[30]。

J. A. Zachman先生在美国政府问责办公室（United States government accountability office, GAO）的信息管理和技术执行理事会（Executive Council for Information Management and Technology, ECIMT）和国际数据管理协会（Advisory Board of the Data Administration Management Association International, DAMA-I）咨询委员会任职，并获得2002年终身成就奖。2013年11月，他因杰出的学术创新贡献获得了IEEE系统、人与控制论协会企业信息系统技术委员会以及企业架构与工程技术委员会的表彰。2011年8月，他被军队服务联盟项目授予"科林·鲍威尔将军公共部门形象奖"。他获得了2009年企业架构专业发展中心颁发的"企业架构专业终身成就奖"，以及2004年奥克兰大学颁发的"商业应用技术卓越与创新奖"。

J. A. Zachman先生自20世纪80年代以来一直专注于企业架构，并就这一主题进行了广泛的写作，他还是《Zachman企业架构框架：企业工程和制造入门》一书的作者[31]。

2. 企业架构的Zachman框架

"应用于企业的框架只是一种逻辑结构，用于对一个企业描述性的表述进行分类和组织，这对企业的管理和企业系统的开发（目的）提供了合理化的各种概念和规范，对实现专业沟通的清晰性、改进和集成开发方法和工具，以及建立对系统和资源投资的可信度和信心具有重要意义。"

J. A. Zachman, 2003

The Zachman Framework for Enterprise Architecture[32] is a two-dimensional classification scheme for descriptive representations of an enterprise. It was derived through observation of descriptive representations (design artifacts) of various physical objects like airplanes, buildings, ships, computers, etc., in which it was empirically observed that the design artifacts (the descriptive representations, the product descriptions, the engineering documentation) of complex products can be classified by the audience for which the artifact was constructed (the Perspective) as well as classified by the content or subject focus of the artifact (the Abstraction) (Figure 3).

Perspectives of the Framework

Different perspectives are being represented over the process of engineering and manufacturing complex products. The descriptive representations of the product that are prepared over this process are designed to express concepts and constraints relevant to the various perspectives. That is, not only do the design artifacts depict the necessary engineering information, but they depict it in such a fashion that it is intelligible to the perspective (audience) for which they were created.

The principal perspectives are easily identifiable including:

1. **The Owner's Perspective (Row 2)**[33]—the recipient (customer, user) of the end product (e.g., airplane, house, enterprise, etc.). These descriptive representations reflect the usage characteristics of the end product, what the owner(s) are going to do with the end product, or how they will use it once they get it in their possession. This is the conceptual view of the end product, whatever the owner can think about relative to its use.
2. **The Designer's Perspective (Row 3)**—the engineer, the architect, the intermediary between what is desirable (Row 2) and what is physically and technically possible (Row 4). These descriptive representations reflect the laws of nature, the system, or logical constraints for the design of the product. This is the logical view of the end product. For enterprises, this is the logical representation of the enterprise which forms the basis for the white collar system, the record-keeping system of the enterprise as well as the basis for the design of the blue-collar system, the material manipulation system for manipulating the tangible aspects of the enterprise.
3. **The Builder's Perspective (Row 4)**—the manufacturing engineer, the general contractor, the employer of some technical capacity for producing the end product. These descriptive representations reflect the physical constraints of applying the technology in the construction of the product.

Empirically, there are two identifiable additional perspectives which include:

1. **A Scope Perspective (Row 1)**[34]—the context that establishes the universe of discourse, the inner and outer limits, the list of relevant constituents that must be accounted for in the descriptive representations (models) for the remaining perspectives.

企业架构的Zachman框架[32]是一种用于描述企业的二维分类方案。它源于对各种物理对象(如飞机、建筑物、船舶、计算机等)的描述性表示(产品设计工作)的观察,复杂事物的制品(描述性表示、产品描述、工程文档)可以根据创建制品的观察者视角进行分类,也可以根据制品的内容焦点(摘要)进行分类。

3. 框架的概念

由于人们在工程建设和制造复杂产品的过程中表现出不同的观点,在此过程中准备的产品描述性内容旨在表达与各种观点相关的概念和约束。也就是说,设计制品不仅描绘了必要的工程信息,而且它们以这样一种方式描绘它,使得它们对于创建它们的视角(观众)是可理解的。

主要视角很容易识别,包括以下三种。

(1)拥有者的视角(图3第2行)[33]——最终产品(如飞机、住宅、企业等)的使用者(客户、用户)。这些描述性信息反映了最终产品的使用特性,拥有者将如何操作最终产品,或者一旦拥有最终产品,他们将如何使用它。无论拥有者对其使用有什么想法,这都是产品的概念视角。

(2)设计师的视角(图3第3行)——工程师、架构师是思想上(图3第2行)、制造上和技术上可行性(图3第4行)之间的中介。这些描述性表示反映了产品设计需要遵循的设计法则、系统或逻辑约束,这是最终产品的逻辑视图。对企业而言,这是企业的逻辑表现,是企业管理、生产制造、工艺设计的基础,是企业有形方面的体现。

(3)制造商的视角(图3第4行)——制造工程师、总承包商、具有生产最终产品技术能力的雇主。这些描述性信息反映了在产品制造中应用技术的物理约束。

根据经验,有两个可识别的附加视角,如下所述。

(1)范围视角(图3第1行)[34]——建立语义的内容、内在和外在的界限、相关成分的清单,必须在其余视角的描述性表征(模型)中加以考虑。

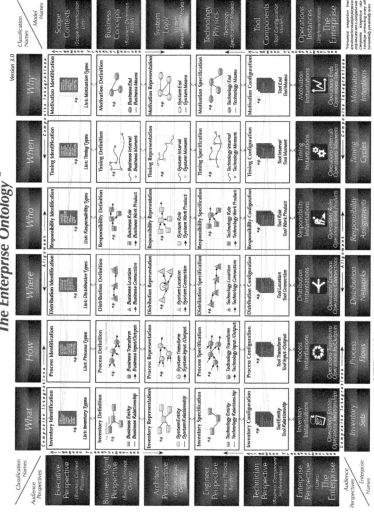

FIGURE 3

The most up-to-date (in 2014) version of the Zachman Framework for Enterprise Architecture.

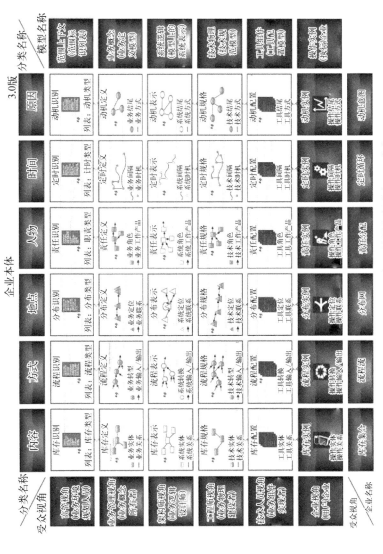

图 3 企业架构 Zachman 框架的最新版本（2014 年）

2. **An Out-of-Context Perspective (Row 5)**—a detailed description that disassociates the parts or pieces of the complex object for manufacturing purposes. These out-of-context representations play a part in the transformation from the media of the design of the product to the media of the end product itself. For example, in physical products, like airplanes, the medium of the design is typically paper and ink (or more recently, electronic). Whereas, the media of the end product itself is aluminum, titanium, composites, etc. The out-of-context artifacts are employed in this media transformation between the media of the design and the media of the end product. For enterprises, these are the product specifications relating the technology constraints of row 4 to the vendor products in which the technology constraints are materialized.

Conclusion

As a part of the historic development of process concepts, we have covered the influential and key people and concepts impacting the way of thinking, working, and modeling of processes. In this context, we have focused in this section on what John A. Zachman, Sr. has developed and published and how it changed enterprise modeling and enterprise architecture around the world. His concept of adding different views as the anchor for answering relevant questions of what, how, where, who, when, and why was innovative and groundbreaking in every way. His contribution to enterprise modeling, enterprise engineering, and enterprise architecture[35] has left its mark on this generation and generations to come.

LEAN THINKING, LEAN PRACTICE AND LEAN CONSUMPTION
The evolution of Toyota's practices

Lean thinking and Lean practices are generic versions of the Toyota Production System (TPS) and the Toyota Way of Management system.[36] Lean did not derive from theory, but through observing practices at Toyota that were delivering superior performance in terms of product quality, efficiency (hours per car) and time-to-market for new products, leading Toyota to eventually become the largest car maker in the world.

The problem the founder of TPS, Taiichi Ohno, was trying to solve in the 1950s was how to build several different products on the limited equipment that Toyota could afford at that time. Instead of resorting to producing in batches, he carried out many pioneering experiments to build an integrated production system that was able to make a variety of products in single-piece flow in line with demand. This challenged the assumptions that there is a trade-off between quality and productivity, and that bigger batches lead to economies of scale and lower costs. His experiments also led to the development of many new tools, such as Kanban pull systems and SMED quick changeovers.[37] Ohno also built on Toyota's Jidoka system for making abnormalities visible immediately, and was invented in the 1930s.

（2）脱离背景的视角（图3第5行）——一种详细的描述，它将部件从复杂对象中分离出来，以便于制造。这些背景的表述在从产品设计的媒介到最终产品本身的媒介的转换中起到了一定的作用。例如，在实物产品中，如飞机，设计的媒介通常是纸和墨水（或者最近的电子产品），然而，最终产品本身的介质是铝、钛、复合材料等。在设计介质和最终产品介质之间的这种介质转换中使用了语境外的制品。对于企业来说，这些是将图3第4行的技术约束与实现技术约束的供应商产品相关联的产品规范。

4. 结论

作为流程概念历史发展的一部分，我们研究了影响流程思维、工作和建模方式的有影响力的关键人物和概念。在此背景下，我们在本节中重点介绍了J. A. Zachman的研究成果，以及它如何改变世界各地的企业建模和企业架构。他提出了添加不同观点作为锚回答相关问题的概念，这些问题包括：内容、方式、地点、人物、时间以及原因，在各个方面都具有创新性和开创性。他对企业建模、企业工程和企业体系结构的贡献[35]在这一代人并且将在未来几代人身上留下印记。

1.3.4　精益思维、精益实践、精益消费

1. 丰田实践的演变

精益思维和精益实践是TPS和丰田管理系统的集中体现[36]。精益并非源于理论，而是通过观察丰田在产品质量、效率（每辆车的小时数）和新产品上市时间方面的卓越表现，最终使丰田成为世界上最大的汽车制造商。

TPS的创始人Taiichi Ohno在20世纪50年代试图解决如何在有限设备上生产几种不同的产品的难题。他没有采取批量生产的方式，而是进行了很多开创性的试验，包括建立了一个能够按照需求单件生产多类产品的完整生产体系，这给质量和生产率之间的均衡带来挑战，但也实现了批量越大越能够实现规模经济和成本越低的目标。他的试验还导致了许多新工具的开发，比如看板系统和SMED（快速换模）[37]。Taiichi Ohno在20世纪30年代基于丰田的自働化系统发明的这一系列方式能够使异常情况立即显现。

However, the distinguishing feature of Ohno's approach was to challenge and teach front line and support staff how to design their own work, using the Training Within Industry system pioneered during WWII in the USA.[38] This enabled the front line to establish a standard way of doing their work as a baseline for improvement, which in turn enabled them to see and respond to any deviations from this standard immediately. In analyzing the root causes of the many issues that interrupted their work, he also taught them how to use the scientific approach to solving problems, using Deming's PDCA method.

Indeed it is the repeated daily practice of PDCA that develops the capabilities of individuals and teams to continually improve their work and improve the performance of the system as a whole. Toyota is often quoted as saying it "makes people in order to make cars". These enhanced problem solving capabilities enabled Ohno to link activities together, remove all kinds of buffers and delays and with much shorter lead times, and to use simpler planning systems driven by demand rather than by forecasts. This accelerating continuous improvement system is called Kaizen.[39]

Similar logic was used to develop very different approaches in other areas of the business including managing product development projects, production engineering of right-sized tooling, supplier coordination and sales and marketing. Eiji Toyoda, the long-time Chairman of Toyota, also used these principles to build a management system to support Kaizen and to focus and align activities towards key corporate objectives, called the Toyota Way.[40] Again the key to doing so is building common capabilities at every level of management to plan and solve business problems using another version of PDCA called A3 thinking.[41]

The evolving understanding of Lean

TPS was developed in the 1950s and was taught to its suppliers in the 1970s and brought together as a management system in the Toyota Way in 2001. It continues to evolve as Toyota faces new challenges today. Likewise, our understanding of Lean has deepened over time. We initially bench marked their superior performance and coined the term Lean to describe this system in *The Machine that Changed the World*.[42] This caused quite a stir across the global auto industry and beyond.

But it quickly became apparent that simply collecting and training all the lean tools was not enough for others to follow Toyota's example. The authors (James P. Womack and Daniel T. Jones) set out to observe Toyota's leading practices in more detail and those of other pioneering leading practice organizations who had learnt directly from Toyota's practices. From this, they were able to distil a set of five principles behind a Lean system and a common action path to realize them in *Lean Thinking*.[43]

One of the insights from this research is that no one can see or is responsible for the horizontal sequence of activities that creates the value customers pay for, from concept to launch, from raw material to finished product and from purchase to disposal. Vertically organised departments instead focus solely on optimising their activities and assets to make their numbers.

Taiichi Ohno 的方法效果显著,二战期间美国利用该体系进行培训,教授并支持前线与后勤人员如何安排自己的工作[38],这使得前线能够建立起一种标准的方式来执行他们的任务,并将其作为工作改进的标准,该标准反过来又使他们能够看到和响应任何偏离本标准的情况。在分析影响他们工作的许多问题的根本原因时,他还教他们如何使用 W. E. Deming 的 PDCA 等科学的方法来解决问题。

实际上,正是 PDCA 不断重复的日常实践,培养了个人和团队不断改进工作和提高系统整体性能的能力。丰田经常被引用说它是"制造汽车前先制造人"。Taiichi Ohno 将活动连接在一起以达到解决问题能力的提升、消除各种缓冲区和延迟、缩短提前期,并使用由需求驱动而不是由预测驱动的更简单的计划系统。这种加速的持续改进系统称为改善[39]。

类似的逻辑常被用于其他领域,包括管理产品开发项目、合适尺寸工具的生产项目、供应商协调以及销售和营销。丰田董事长 Eiji Toyoda 也利用这些原则建立了一个管理系统,并集中资源在企业关键目标上进行持续改善,这就是所谓的"丰田之道"[40]。同样,做到这一点的关键是在管理的各个层面上用另一个叫做"A3思维"的 PDCA 版本来计划和解决业务问题[41]。

2. 对精益的理解

TPS 是在20世纪50年代发展起来的,在20世纪70年代输出给供应商,并在2001年作为一个管理系统与丰田模式合在一起。如今它仍随着丰田面临新的挑战而不断进化。同样,我们对精益的理解也随着时间的推移而加深。我们最初对它们的效果进行了评估,并创造了精益一词来体现"通过机器改变世界"的理念[42]。这在全球汽车工业及其他领域引起了相当大的轰动。

但大家很快就看清仅仅收集和培训所有的精益工具还不足以让其他人效仿丰田。作者(J. P. Womack 和 D. T. Jones)开始更详细地观察丰田领先的实践方式,以及其他直接从丰田实践中学习的组织的实践。从这一点上,他们能够在精益系统和共同的行动路径背后提炼出一套"五项原则",从而在精益思维中实现这些原则[43]。

这项研究的一个深刻见解是:从概念到发布、从原材料到成品、从购买到处置,垂直组织的部门只专注于优化他们的活动和资产以实现他们的数量,但是没有人能够看到或负责创造客户所支付的价值的横向活动序列。

To help teams to see the end-to-end processes or value streams they are involved in we used another Toyota tool, which we called Value Stream Mapping.[44] As teams map their value streams they realize the problem is not the people, but a broken process, and having stabilized their own work they now see new opportunities for collaboration to improve the flow of work and align it with the pull from real customer demand.

In industry after industry, they saw value streams that used to take many months from beginning to end now take a matter of days, with far fewer defects and more reliable delivery. This is only possible because front line staff know how to react quickly and tackle the root causes of problems that will *arise* in any tightly synchronised and interdependent system. It is also much easier to adapt to changing circumstances. Over time, these emergent capabilities achieve superior performance than systems designed and supported solely by experts. This is the main difference between Value Stream Analysis and Business Process Reengineering.

The other insight is that the traditional approach to managing by the numbers and through functional politics at HQ wastes a lot of management time, fails to align activities with corporate objectives, hides problems and takes management away from front line value creating activities. Relying on expensive enterprise systems to force compliance with the command and control instructions from the top has in many cases made things worse and much harder to adapt to changing circumstances.

Toyota uses a very different strategy formulation process, called Hoshin Kanri, to define the overall direction of the organization and to conduct a dialogue up and down the organization on proposed actions to achieve it, again based on PDCA.[45] As a result, resources and energies are prioritized and aligned through a visual process that reaches right down to the front line. This also lays the basis for collaboration across functional silos. Management in turn spends a lot more time at the front line, understanding their issues, eliminating obstacles and coaching problem solving. In this way management learns by helping colleagues to learn and does this by asking questions rather than telling them what to do. This builds very different behaviors and an environment where employees are challenged to fulfil their potential.[46]

The spread of Lean and Lean Consumption

Lean thinking and Lean practice has spread across almost every sector of activity, from retailing and distribution, discrete and process manufacturing, service and repair, financial services and administration, construction, software development and IT, healthcare and service delivery in governments. It has even created a framework for improving the viability of digital startups.[47] While the focus on value creation, value streams and learning is common, the sequence of improvement steps varies for different types of activity. However, Lean practices seem to work equally well in different cultures.

为了帮助团队看到他们参与的端到端流程或价值流,我们使用了另一个被称为价值流图的丰田工具[44]。当团队利用价值流图设计一个价值流时,他们意识到出现问题的原因不是人而是一个破碎的流程,并且在稳定自己的工作之后,他们能够看到新的协作机会,以改进工作流程并使其与来自实际客户需求的方向相一致。

通过在一个又一个行业中的实践,过去需要数月才能完成的价值流现在只需要几天时间,而且出现的缺陷更少、交付质量也更可靠,达到这种结果的原因是一线员工知道如何快速反应和解决任何系统中出现的问题。同时,对环境变化的适应也容易得多。随着时间的推移,这些应变能力实现了比专门由专家设计和支持的系统更好的性能,这是"价值流分析"和BPR之间的主要区别。

另一个领悟是:传统的按人数管理和通过总部职能进行管理的方法浪费了大量的管理时间,管理层远离一线价值创造活动也使得问题难以及时发现,不能使活动与企业目标保持一致。因此,依靠昂贵的企业系统来强制遵从高层的命令和控制指令的方式,在许多情况下使事情变得更糟,更难适应不断变化的环境。

丰田使用一个非常不同的被称为方针管理(Hoshin Kanri)的方法进行战略制定,以确定组织的整体方向,并为实现这一目标而再次基于PDCA在组织内部进行循环改进[45]。因此,资源通过一个可以直接到达一线的可视化流程被优先分配和调整,这也为跨职能条块分割的协作奠定了基础。反过来,管理层需要在一线花费更多的时间了解他们的问题、消除障碍和指导解决问题。通过这种方式,管理层通过"问题反馈"帮助员工学习并做到而不是告诉他们该做什么,这就建立了非同寻常的行为体系和激励员工发挥潜能的环境[46]。

3. 精益和精益消费的发展

精益思维和精益实践几乎遍布零售和分销、零散和流程制造、服务和维修、金融服务和管理、建筑、软件开发和IT、政府的医疗保健和服务交付等各个领域,它甚至为提高技术初创企业的生存能力创造了一个框架[47]。虽然价值创造、价值流和学习很常见,但不同类型的活动的改进步骤、顺序并不相同。然而,精益实践在不同的文化中似乎同样有效。

The full potential of Lean is realized when it is embraced by the whole supply chain. Toyota's aftermarket parts distribution system is still the global benchmark supply chain, delivering near perfect availability of the basket of parts at the point of use with only a tenth of the lead time and inventory in the pipeline from the point of production.[48] Not surprisingly this inspired retailers like Tesco and Amazon to develop their own rapid response distribution systems that are essential for convenience retailing and home shopping. Manufacturers like GKN have also moved away from concentrating activities in focused factories in distant low cost locations to creating rapid response supply chains to serve customers in each region.[49] GE is also using Lean to design a new product range and production system for household appliances in North America, bringing this activity back from China.

While most of the attention has been focused on the upstream supply chain, Lean actually begins with the customer's use of the product or service. In Lean Solutions, James Womack and Daniel Jones developed a framework for using Lean to define value from the user's perspective.[50] Consumption is in fact a series of processes that interact with the provider's processes. Mapping both processes reveals where they are broken and cause mutual frustration and unnecessary cost. This reveals opportunities for improving user experience at lower cost and even new business models. In the digital age it is now possible to track the customer's use of the product or service and enter into a two way dialogue with them. In a very real sense, customers and users are becoming an important part of the supply chain delivering today's products and services and co-developing tomorrow's solutions.

Conclusion

From the above it should be clear that Lean is not just another improvement methodology, but a very different set of behaviours and management system. It is not just a set of tools for production operations in the auto industry, but a much broader framework for creating more productive value creation systems in all kinds of sectors and activities. Readers should be aware of the confusion that is caused by partial descriptions of Lean that often miss the key elements that make it work as a system.

Lean shares the same scientific approach to the analysis of work as many other improvement methodologies, like BPR, Six Sigma and TQM. But it differs from them in how it is used. Rather than experts using scientific methods to design better systems, Lean builds superior performance by developing the problem solving capabilities of the front line, supported by a hands-on management system.

Lean is therefore a path or journey of both individual and organizational learning, and leads to more challenging and fulfilling work for those involved. It is learnt by doing through repeated practice rather than by studying books or in the classroom. While it is driven by practice, and not theory, Lean raises many interesting new hypotheses about learning and collaborative working for different academic disciplines to think about and research.

当精益被整个供应链所接受时,它的全部潜力就实现了。丰田的售后零部件分销系统仍然是全球供应链的标准,在生产时提供近乎完美的零部件供应,交货和库存部分仅占生产所用的十分之一[48]。同样,这对于便利零售和家庭购物至关重要,这促使特易购(Tesco)和亚马逊(Amazon)这样的零售商开发了自己的快速反应分销系统,像吉凯恩(GKN)这样的制造商为服务各地的客户,也已经把精力从遥远的低成本工厂转移到建立快速反应的供应链方面[49]。通用电气公司还利用从中国带回来的精益思维设计了一个新的产品系列和北美家用电器生产系统。

虽然大多数注意力都集中在上游供应链上,但精益实际上是从客户使用产品或服务开始的。在精益解决方案中,J. P. Womack和D. T. Jones开发了一个框架,用于从用户的角度使用精益定义价值[50]。消费实际上是一系列与提供者交互的过程。绘制这两个流程可以发现它们在哪里被割裂、导致相互的挫折,以及不必要的成本。这揭示了以更低的成本甚至新的商业模式改善用户体验的机会。在数字时代,现在可以跟踪客户对产品或服务的使用,并与他们进行双向沟通。从本质意义上讲,客户和用户正成为提供今天的产品和服务的供应链的重要组成部分,并共同制订明天的解决方案。

4. 结论

从上面可以清楚地看到,精益不仅仅是一种改进方法,而且是一组非常不同寻常的行为和管理系统。它不仅是一套用于汽车行业生产运营的工具,而且是一个用于在各种行业和活动中创建更具生产力的价值创造的更广泛的框架。但读者应该意识到由对精益的片面描述常常忽略了使其作为一个系统工作的关键元素而导致的错觉。

精益与许多其他改进方法(如BPR、六西格玛和TQM)一样是工作分析中使用的科学方法。但它的使用方式不同于其他方法,主要体现在精益实践在管理系统的支持下,主要是通过解决一线的问题实现更卓越的绩效,而不是使用科学方法设计更好的系统。

因此,精益是个人和组织学习的一条道路或路径,并为相关人员带来更具挑战性和更充实的工作。它是通过反复练习而不是读书或在课堂上学来的。虽然精益是由实践驱动而不是理论驱动的,但它在学习、工作协作方面提出了许多有趣的新假设,都需要进行思考和研究。

BUSINESS PROCESS REENGINEERING
Dr Michael Hammer

Michael Hammer[51] (1948–2008) was an American engineer, management author, and a former professor of computer science at the MIT, and is most commonly known as the founder of the management theory of BPR. BPR was widely perceived as one of the biggest business ideas of the 1990s—which Hammer defined as "the fundamental rethinking and radical redesign of business processes to achieve dramatic improvements in critical measures of performance." The terms *process improvement*, *process excellence* and *process innovation* all came from him.

The idea, first propounded in an article in Harvard Business Review[52], was later expanded into a book that Hammer wrote with James Champy, the founder of CSC Index, a consulting firm. The book[53] sold several million copies. So popular was reengineering that one survey in the 1990s showed it to have been adopted by almost 80% of Fortune 500 companies. It was often blamed for the widespread lay-offs that became part of almost every company's radical redesign at that time.

As an engineer by training, Hammer was the proponent of a process-oriented view of business management. He earned BS and MS degrees and a Ph.D. in EECS from the MIT in 1968, 1970, and 1973, respectively. He was a professor at the MIT in the department of computer science and lecturer in the MIT Sloan School of Management. Articles written by Hammer were published in business periodicals, such as the Harvard Business Review and The Economist. TIME named him as one of America's 25 most influential individuals[54] in its first such list. *Reengineering the Corporation*[55] was ranked among the "three most important business books of the past 20 years" by Forbes magazine.

Introducing Business Process Reengineering

Business Process Reengineering focuses exclusively on redesigning the core business processes of an organizational unit—or the enterprise as a whole. The aim is to increase productivity, improve process cycle times and the resulting quality thereof. The basic idea of BPR is to completely rethink an organization's core processes from scratch by turning to focus on processes that deliver the most value to the consumer (whether external or internal). This approach often leads to the adoption of a brand new value delivery system with a dramatically increased attention and focus upon the needs of the customer. The overall goal is to simply—and entirely—eliminate non-productive activities in two key areas; first, it includes the redesign of functional organizations into cross-functional teams to increase productivity and efficiency across organizational boundaries; second, the usage of technology to achieve a much higher degree of data dissemination in order to greatly improve decision making.

Many times before, companies simply haven't received the expected performance boost from the use of usual methods such as process rationalization and process automation. Heavy investments in information technology, in particular, has shown to deliver increasingly disappointing results because many

1.3.5　BPR

1. M. Hammer

M. Hammer[51]（1948—2008年）是美国工程师、管理学作家、麻省理工学院前计算机科学教授,通常被认为是BPR管理理论的创始人,术语流程改进、流程优化和流程创新都来自他。BPR被广泛认为是20世纪90年代最伟大的商业理念之一,M. Hammer将其定义为"对业务流程进行根本性的重新思考和彻底的重新设计,以实现关键绩效指标的显著改进"。

这一想法最初在《哈佛商业评论》[52]的一篇文章中提出,后来被扩展成M. Hammer与咨询公司CSC指数的创始人J. A. Champy合著的一本书,这本书[53]卖了几百万册。在20世纪90年代的一项调查中显示,几乎80%的《财富》500强企业都采用了这种方法。它经常被指责为大规模裁员的罪魁祸首,而裁员几乎成为当时每家公司彻底重新设计流程的一部分。

作为一名经过培训的工程师,M. Hammer是"以流程为导向进行业务管理"观点的支持者。他分别于1968年、1970年和1973年从麻省理工学院获得理学学士、理学硕士学位以及欧洲经济共同体博士学位。他是麻省理工学院计算机科学系的教授,也是麻省理工学院斯隆管理学院的讲师。M. Hammer写的文章发表在如《哈佛商业评论》和《经济学人》商业期刊上,《时代周刊》将他列为美国最具影响力的25个人之一[54]。《企业再造》[55]被《福布斯》杂志评为"过去20年中三本最重要的商业书籍"之一。

2. BPR介绍

BPR专门致力于为组织单位或整个企业重新设计核心业务流程,其目的是提高生产率、缩短生产时间及改进产品质量。BPR的基本思想是基于为消费者提供最大价值(无论是外部的还是内部的)的思想,从零开始全面地重新思考组织的核心流程。这种方法通常会导致采用全新的价值交付系统,并显著增加对客户需求的关注。总体目标是简单而完全地消除两个关键领域的非生产性活动:第一,它包括将组织的职能部门重新设计为跨职能团队,以提高跨组织边界的生产力和效率;第二,利用技术实现更高程度的数据传输从而大大提高决策水平。

在此之前的大多时候,在IT方面的大量投资面前,公司根本没有从流程合理化和流程自动化等常用方法的使用中获得预期的绩效提升。而且由于许多公司倾向于使用旧的或在某些方面冗余的方式完成机械化和自动化,投入已经显示出越来越令人失望的结果。正如A. Einstein(阿尔伯特·爱因斯坦)曾经说过的:"疯狂

companies tend to use technology to mechanize and automate old or, in some ways, redundant ways of doing business. As Albert Einstein once said: "Insanity: doing the same thing over and over again and expecting different results." It is important to note that technology is but a tool that acts on behalf of its user, so if you do not rethink and redesign how your processes should perform as well as educate employees on the process changes, you'll almost always end up with disappointing results when it comes to meeting the performance and value expectations from the original goals of business process redesign. The result is that technology is used only to speed up process execution while leaving existing processes intact.

Using technology to speed up process execution—in exchange of focusing on process redesign—simply will not adhere to the importance of addressing fundamental performance deficiencies. Job designs, business work flows, information flows, control, monitoring and governance mechanisms as well as organizational structures often come of age continuously across different competitive environments throughout history. Most of these areas tend to be geared towards effectiveness, efficiency and control, however, but in today's global business ecosystem, the terms to really look out for and pay attention to are around *process innovation, process speed* and *process quality.*

Width before Depth

Hammer realized early on that the processes really proved useful when you worked together across organizational borders—it may even be something as simple as cooperation among departments. Compare this with the fact that most process initiatives take place within one's own sphere of power. Of course, it is quite okay to find your sea legs in-house at first, but you meet the real possibilities and problems when you venture out of your own sphere of power. It is so much easier to get the internal areas to behave, both because of your direct authority, but also because the individual subprocesses are well known. You have to work in a more factual manner if you are to cooperate with/persuade an external partner, whom you have no authority over.

The reason being that the real potential for savings lies in the cooperation between two organizations, since you need a customer need to fulfill in order to make such a cooperation viable. And there must be a need for interorganizational contact in order to fulfill this customer need. The relevant optimizations at this level will have a lot of potential since we are optimization on a larger scale.

This touched on yet another point; *processes need width before depth,* which means that we identify and establish the overriding processes end-to-end, *before we go deeper into them.* We do this when the coherence and the external requirements have been established. This point of view matches the concepts of Mura—a Japanese term meaning "unevenness; irregularity; lack of uniformity; nonuniformity; inequality." Mura is a key concept in the TPS as one of the three types of waste (muda, mura, muri).[56]

就是一遍又一遍地做同样的事情，并期望得到不同的结果。"重要的是要认识到技术只是一种代表用户行动的工具，所以如果您不重新思考和重新设计您的流程应该如何执行，也不训练员工适应流程的变化，当您在设定业务流程重新设计的最初目标和期望时，您几乎总是会得到令人失望的结果。其结果是，技术仅用于在保持现有流程完好无损的情况下加速流程执行。

通常来说，使用技术来促进流程执行、变更流程关注点等行为，均不能解决流程存在的缺陷。工作职能设计、业务工作流程、信息流、控制、监控和治理机制以及组织结构在历史上的不同竞争环境中经常不断出现，然而，这些领域大多倾向于有效性、效率和控制，但在当今的全球商业生态系统中，真正需要注意的是流程创新、流程提速和流程质量。

3. 宽度优于深度

M. Hammer 很早就意识到，当从事跨组织边界工作——可能只是一些简单的工作，如部门之间的合作时，如果将这一点与发生在自己权力范围内的大多数流程进行比较的话，流程的重要性会更加突显。当然，在公司内部平衡关系很容易，但是当您走出自己的权力范围时，您会遇到真正的问题。这主要是因为在内部更容易表现出来，这不仅是因为您的直接权威，也因为每个子流程都是众所周知的。如果您要与一个您没有权力控制的外部合作伙伴进行合作，您必须以更实际的方式工作。

原因在于两个组织之间的合作是为了实现共赢。因为您需要一个客户来完成这种合作，所以，为了通过合作满足客户的需求，组织间必须有沟通渠道。通过进行大规模的流程优化，这将具有很大的可行性。

这又涉及另一个问题：我们在深入到流程之前，流程需要宽度优于深度，这意味着识别并建立覆盖端到端流程。当一致性和外部需求已经建立时，我们就这样做了。这一观点与 mura 的概念相吻合，mura 是一个日文术语，意思是"不均匀；不规则；缺乏一致性；不一致；不平等"。mura 是 TPS 三种浪费（muda、mura、muri）中的一个关键概念[56]。

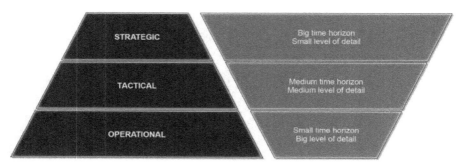

FIGURE 4

The three levels of management.

The Three Levels of Management

As employees become more and more specialized and as leadership no longer knows the most about the various disciplines (see phases 1 and 2), a need arises for making projects across the board in companies, and thus the project culture, emerged. Most organizations today are process-driven and have a flat structure. Decisions are still being made on a strategic (planning), tactical (management), or operational (executional) level, and depending on the leadership level, the decisions have a longer or shorter time horizon, and a greater or lesser level of detail. The process flow and the tasks that this solves will only benefit the company, however, if they are connected to the company's overriding goals.[57]

The figure 4 illustrates the three levels of management combined with the time horizon and level of detail. The levels of management are interconnected via information streams. From the leadership, who are the ones to decide the strategic part, and downward in the hierarchy, the level of agency is limited more and more the further you go down.

This is done through plans and budgets. From the lower level, control information, i.e., reports, are sent upward and become more and more detailed the higher you get in the hierarchy. The authors behind BPR were Michael Hammer and James A. Champy. The concept looks at the organization's business processes from a "clean slate" point of view and then decides how to best redesign and reconstruct these processes so as to improve the way they run the company.[58] In spite of the fact that BPR is considered a concept from the early 1990s, you could argue that there is nothing new in reengineering.[59]

The components of BPR existed before 1990, when the first articles about BPR were written. But it was not until the 1990s that these components were brought together to form a coherent management concept.

The Three Important Ingredients

Business Process Reengineering (BPR) is all about creating change by identifying and focusing on business processes and then make them more efficient by simplifying

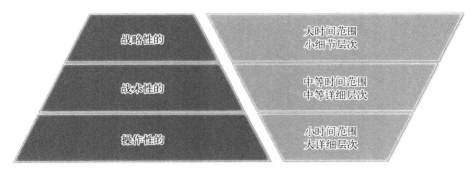

图4　三级管理

4. 三级管理

随着员工变得越来越专业化，领导层对各种举措的了解也越来越少（见阶段1和阶段2），因此需要在公司中形成流程文化。今天的大多数组织都是流程驱动的，并且是扁平结构，决策仍然是在战略（规划）、战术（管理）或运营（执行）层面上做出，根据领导层的不同，决策的时间范围更长或更短，细节层次更高或更低。但是，如果将流程、任务与公司的首要目标相关联，那么它们将对公司有价值[57]。

图4说明了结合时间范围和详细程度的三个管理级别。管理层通过信息流相互连接。决定战略的人的领导力最高，在管理层级结构中，被管理的人的级别越低越受到约束。

这是通过计划和预算来实现的。通过报表的方式控制信息，将报表从较低的层次逐级向上报送，并且越高的层级控制的信息就越详细。BPR的作者是M. Hammer和J. A. Champy。这个概念从"从头再来"的角度来看待组织的业务流程，然后决定如何最好地重新设计和重构这些流程，以改进他们管理公司的方式[58]。尽管BPR被认为是20世纪90年代初的一个概念，但您可以说在再造工程中没有什么新的东西[59]。

BPR的组成部分早在1990年前就已经存在了，关于BPR的第一篇文章就是当时写的。但直到20世纪90年代，这些组成部分才结合在一起，形成了一个连贯的管理概念。

5. 三个重要成分

BPR就是通过识别和关注业务流程来创建变更，然后通过使用现代技术和整个组织的参与来简化这些流程，从而提高效率。这意味着您不仅仅是在研究单个

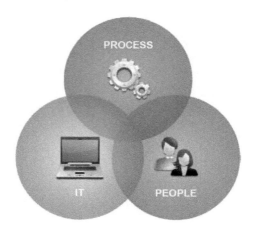

FIGURE 5

Illustration of the vital integration of process, information technology, and people.

them with the use of modern technology and the involvement of the entire organization. This means that you are not just looking at parts of the organization of individual technologies. When BPR is launched, the business process, technology, and the organization have to be developed simultaneously (Figure 5).

In BPR, the three circles should not slide away from one another. Even though these three parts of BPR cannot and must not be separated, IT is considered the key element in BPR (Hammer, 1990).

The Business Process Reengineering Cycle

A Business Process Reengineering(BPR) cycle[60] can be illustrated in the following way (Figure 6):

First you identify the process, and then an analysis of the existing process is carried out. Then, you design the new process, which is ultimately implemented. A clarification of the cycle follows below. Before launching a BPR project, the company should typically start with a BPR pre-analysis to identify the state of the present business process and establish the level of interaction between the organization, the technology, and the processes. Ambeck and Beyer[61] claim that the changes brought on by BPR happen when the development of processes, technology, and organization take place simultaneously and in accordance with the company's strategy and vision. Furthermore, they write that this exact approach is very different from the traditional methods where technological, organizational, and strategic issues are kept separate, which can result in a loss of overview and synergy effects.

A company's BPR project portfolio can span from the minor here-and-now processes of a shorter duration to longer lasting, complex processes, which require greater improvements and typically lasts from 1 to 6 months. Last, but not least, there can be long, continual process improvements, which can last from a few months to several years.

图5　流程、IT和人员集成的重要性

技术的组织部分,当BPR启动时,必须同时开发业务流程、技术和组织(图5)。

在BPR中,这三个圆不应该彼此分离。尽管BPR的这三个部分不能也必须不分离,工厂也被认为是BPR的关键要素(M. Hammer,1990)。

6. BPR周期

BPR周期[60]可以用以下方式进行说明(图6)。

首先确定流程;其次对现有流程进行分析;再次,设计新的流程;最后实施。循环说明如下。在启动BPR项目之前,公司通常应该从BPR预分析开始,以确定当前业务流程的状态,并建立组织、技术和流程之间的交互级别。Ambeck和Beyer[61]认为,当流程、技术和组织的发展同时发生并且符合公司的战略和愿景时,BPR带来的变化才会发生。此外,他们还写道,这种精确的方法与传统方法非常不同,传统方法将技术、组织和战略问题分开,这可能会导致综合和协同效应的缺失。

一个公司的BPR项目组合可以从现在的耗时较短的简单流程扩展到耗时较长的复杂流程,这通常需要较大的改进,因此通常持续1 ～ 6个月。最后同样重要的是,长期的、持续的流程改进可能要持续几个月到几年。

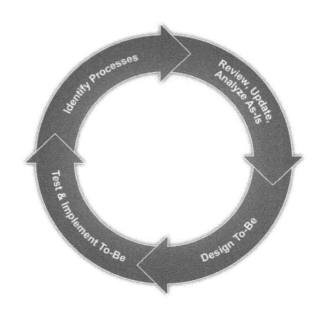

FIGURE 6

The Business Process Reengineering cycle.[62]

Methods and Approaches to Business Process Reengineering

In the literature written about BPR, several methodical approaches are mentioned. A BPR method can, in many ways, be compared to a road map, which helps the driver get his bearings before, during, and after the road trip. The same goes for BPR. You need to know where you are and where you are going and at the same time find the right way. Then you can start the process (the road trip), and then you can continuously check how far you have come. In the following, we will look at three different methods, which one thing is in common, the fact that their approach to the BPR project is divided into steps.

One such approach to BPR is process reengineering life cycle (PRLC).[63] The method was developed by Dr Subashish Guha. PRLC is based on analysis of different BPR methods in use today. Guha (et al.) has studied the different BPR methods, and even though the studies did not result in a standard BPR method, they showed that all the different BPR methods follow the same principle—the aforementioned *step* principle, which can be divided into three overriding phases (Figure 7).

By identifying the company's visions and goals, you can decide where to concentrate your efforts. Then, by identifying which processes to work with, you can ensure that those processes support the company's visions and goals. In that way, you can be sure that the effort is worthwhile. If you have not identified visions and goals, you risk spending time and resources on a process that should have been cut off.

You have to understand the existing processes—what they do for the company and how they fit together. This is important if you are to create an optimum

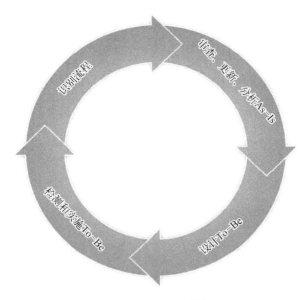

图6　BPR周期[62]

7. BPR 的方法与途径

在有关BPR的文献中，提到了几种方法。例如，路线图有助于驾驶员在道路行驶之前、期间和之后获得方向，BPR方法可以在许多方面与之进行类比。您需要知道您在哪里、您要去哪里，同时找到正确的方法。然后您可以开始这个流程（公路旅行），随后您可以不断地评估您已经走了多远。在下面的内容中，我们将讨论三种不同的方法（其中一种方法是常见的），它们都将实现BPR项目的方法划分为多个步骤。

其中一种方法是流程再造生命周期（process reengineering life cycle，PRLC）[63]。该方法是由Guha博士研发的，Subashish Guha等通过对不同BPR方法进行研究，尽管没有得出标准的BPR方法，但他们表明所有不同的BPR方法都遵循相同的原则，即上述可分为三个主要阶段的步骤原则（图7）。

通过确定公司的愿景和目标，您可以决定把精力集中在哪里。然后，通过确定要使用的流程，您可以确保这些流程支持公司的愿景和目标。这样，您就可以确保所做的努力是值得的。如果您没有确定愿景和目标，您就有可能把时间和资源花在一个应该被切断的流程上。

如果要创建最佳的重新设计，您必须了解现有的流程它们为公司做了什么，以及它们如何结合在一起，这一点很重要。否则，您将面临最终设计无法达到预期

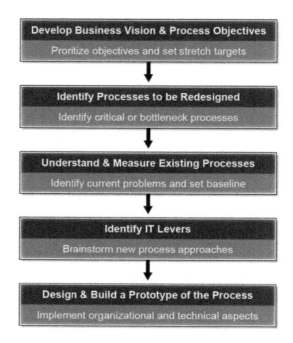

FIGURE 7

The five steps of process redesign.[64]

redesign. Otherwise, you risk ending up with a design that does not deliver the desired result. If you are aware of the IT level in your company, it can provide you with ideas for new approaches to processes. Where are we, and where are we going? When all of this is in place, you can test the new process. The process must be implemented, and you have to take organizational changes and technical aspects into consideration.

Hammer and Champy postulate that the article "Business Process Reengineering: Analysis and Recommendations" is not very detailed concerning the BPR techniques,[65] which has probably resulted in the wide range of methods.

Business Process Reengineering Project and Success Criteria

In his account of BPR, Nicolaisen reaches the criteria, stated below, that need to be realized in order to say that BPR has been a success since the criticism in the report from 1994, "State of re-engineering rapport" states that two-thirds of the initiatives had either unsatisfactory or mediocre results. This is in line with the estimate made by several of the BPR founders that between 50 and 70% of all BPR project results are unsatisfactory:

- Whether or not the project fulfilled the requirements set up for it, for instance, if you have utilized the right innovative technology or if you were able to reach an agreement on the necessary changes.

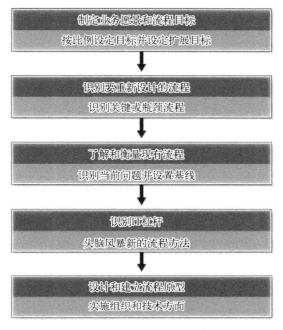

图7　流程重设的五个步骤[64]

结果的风险。如果您了解公司的IT水平,它可以为您提供新的流程方法。我们在哪里? 我们要去哪里? 当所有这些都就位时,您可以测试新流程。您必须实现该过程,并且必须考虑组织和技术方面的更改。

　　M. Hammer和J. A. Champy认为《业务流程再造: 分析和建议》一文对于BPR技术没有很详细的介绍[65],这可能导致出现了更多演绎的方法。

8. BPR项目及成功标准

Nicolaisen在他对BPR的叙述中提出了下列标准,即需要实现这些标准才能说BPR自1994年提出并在质疑的声音中一直是成功的。"重新设计融洽关系的状态"指出,BPR三分之二的做法要么结果不令人满意,要么结果平庸。这与一些BPR创始人的估计一致,即所有BPR项目的结果中有50% ~ 70%是不令人满意的。

- 项目是否满足为其设定的要求。例如,您是否使用了正确的创新技术,或者您是否能够就必要的变更达成一致。

- Whether or not the project created competitive advantages for the company, like have you increased customer satisfaction or have you boosted morale and productivity among the employees.
- Whether or not the project has improved the results on the bottom line, like if you have increased your market shares, increased your profits, or improved ROI.

Even though the company follows the PRLC principles or some other method and/or uses the five steps, in process change, this does not imply that BPR has been successful. The company has to be ready for change. Smith divides change up in two, a nonradical change, which is low risk and, therefore, has a greater chance of success, and it does not require a lot of resources.[66]

Apart from this, the leadership needs to have enough clout and the ability to communicate the radical changes, so as to minimize opposition to change among the employees. You need to employ a process "owner," who is responsible for keeping the reengineering team together and ensure progress. The reengineering team often consists of 5 to 10 individuals who will examine the existing processes and then redesign and implement the processes. Aside from the 5 to 10 individuals, the team should have consultants or coworkers from other departments.[67]

There are a lot of project criteria for a successful BPR; it is, therefore, important that the company has the right approach and an understanding of how demanding the process can be.

The Pros of Business Process Reengineering

When companies work with BPR, they should be able to identify company processes and analyze and redesign if need be. Kaplan and Murdock feel that this way of working provides the company with certain advantages when it comes to their core processes,[68] and at the same time, the company is encouraged to focus on the results of the entire company and not just the results of a single department.

The advantages of working with business process reengineering can be summed up as follows[69]:

- Optimizing business processes
- Improving efficiency and higher quality
- Reducing costs
- Better working environment and understanding among employees
- Increasing flexibility and the ability to handle change
- Creating a foundation for growth
- Increasing cooperation between departments

To get a BPR project to succeed, it is important, aside from having the full support of the leadership, to have the right people aboard. Senior management needs to be represented in management committees, and there is a need for project managers and BPR consultants.

Davenport states that even though BPR is demanding, it is a process that the company should consider as it ultimately creates a bonus for the company both

- 项目是否为公司创造了竞争优势。例如,您是否提高了客户满意度,或者您是否提高了员工的士气和生产力。
- 项目是否提高了收益。比如提高了市场份额、增加了利润或提高了投资回报率(ROI)。

即使公司遵循PRLC原则或其他方法或使用流程重设的五个步骤,在流程变更中,这并不意味着BPR已经成功。公司必须做好变革的准备,Smith把变革分为两种,其中一种是非根本性的变革,风险很低,因此成功的机会更大,而且不需要很多资源[66]。

除此之外,领导层还需要有足够的影响力和沟通能力,以尽量减少员工对彻底变革的反对。您需要任命一个流程所有人,他负责保持重组团队在一起并确保进展。重组团队通常由5 ～ 10个人组成,他们将检查现有流程,然后重新设计和实施流程。除了这5 ～ 10个人,团队还应该有来自其他部门的顾问或同事[67]。

对于一个成功的BPR,有很多项目标准。因此,公司有正确的方法和理解流程的要求是很重要的。

9. BPR 的优点

当公司使用BPR时,他们应该能够识别公司流程,并根据需要进行分析和重新设计。R. B. Kaplan 和 L. Murdock 认为,这种工作方式在核心流程方面为公司提供了一定的优势[68],同时鼓励公司关注整个公司的业绩,而不仅仅是单个部门的业绩。

BPR 的价值可以概括为以下几点[69]:

- 优化业务流程;
- 提高效率和质量;
- 降低成本;
- 营造更好的工作环境和促进员工之间的理解;
- 提高灵活性和处理变化的能力;
- 为增长创造基础;
- 加强部门间的合作。

要使BPR项目成功,除了需要得到领导层的全力支持,还要有合适的人员。此外,管理委员会中需要有高级管理层的代表,并且需要项目经理和BPR顾问。

T. H. Davenport 指出,即使BPR要求很高,这也是一个公司应该考虑的流程,因为它最终会为公司创造财务和竞争方面的优势。通过流程工作,公司最终可能

financially and competitively. Through the process work, the company might eventually expose processes that need to be eliminated or redesigned.[70]

The Cons of Business Process Reengineering

We do not feel that there are any disadvantages of a successful BPR, and we have not been able to find any documentation that indicates this. But there are some issues that the company needs to take into consideration when planning to implement BPR. These issues have to do with the company's resources to meet the project criteria. The leadership could feel that the organization is not ready for change, and then they should gradually make changes in the culture of the organization before implementing BPR.

Another issue could be whether or not the organization can allocate the needed resources to provide the leadership with the necessary clout.

And finally, it can be difficult to implement a BPR if the organization is not International Organization for Standardization (ISO) certified, i.e., all processes have been documented. If you attempt to implement BPR in an organization in which all the processes have not been documented or established, it can require a lot of resources. This could mean that a BPR in itself will not create financial profit, but combined with an ISO certification and cheaper insurances, the overall financial benefit of the BPR can be increased. An added bonus of BPR could be that all procedures are identified and written down. This could then be used for an implementation of activity-based costing (ABC) or time-driven activity-based costing (TDABC), in which the project group normally requires a lot of resources to identify processes, speak to employees, and maybe even measure resource consumption themselves. To implement an ABC or TDABC under this condition would require fewer resources and can create an added profit.

Conclusion

Setting up a business process re-engineering initiative and designing the many different projects involved in the execution of such an initiative is a daunting task for any organization. Not only does the process redesign itself pose a number of technical challenges, but one of the most prominent issues comes from the requirement of organizational change management. LEGO[71] has portrayed this issue in detail when they set out to document all of their business processes across their organizational boundaries. As redesigning the core business processes of any organization requires a different point of view and take on both performance and value delivery, there will be a fundamental need for the employees involved with such projects to change not only their way of thinking and their way of working, but also their way of modelling with processes. It is therefore vital for any leadership to incite and govern employee adaptation and education so that the organization has acquired the correct skills to carry out such projects successfully, and also to enable them to govern and monitor the processes and their performance. This will ultimately enable the organization as a whole to continuously improve and adapt processes to a constantly changing business environment.

会公开需要消除或重新设计的流程[70]。

10. BPR 的弊端

我们通常认为成功的BPR没有任何缺点,尽管我们还没有找到任何支持这一观点的证据。但是,在计划实施BPR时,公司需要考虑一些问题,这些问题与满足项目标准的公司的资源有关。领导层可以感觉到组织还没有准备好变更,因此他们应该在实施BPR之前,逐步改变组织的文化。

另一个问题可能是,组织是否能够分配所需的资源,为领导层提供必要的影响力。

最后,一个企业如果未经过ISO认证,也就是说,所有的流程如果没有文档记录,那么就很难实施BPR。如果您试图在一个组织中实现BPR,在这个组织中,所有流程都没有被记录或建立,那么它可能需要大量的资源。这可能意味着BPR本身不会产生财务利润,但如果与ISO认证和更便宜的保险相结合,BPR的整体财务效益可以提高。BPR的另一个好处可能是识别并记录所有程序。然后,这可以用于作业成本法(activity-based costing,ABC)或时间驱动作业成本法(time-driven activity-based costing,TDABC)的实施,在这种方法中,项目组通常需要大量资源来识别流程、与员工交谈,甚至可以自己衡量资源消耗。在这种情况下实施ABC或TDABC需要的资源更少,并且可以创造额外的利润。

11. 结论

对于任何组织来说,建立一个业务流程重新设计计划和设计许多不同的项目都是一项艰巨的任务。不仅流程重新设计本身带来许多技术挑战,而且最突出的问题之一来自组织变更管理的需求。乐高[71]已经详细描述了这个问题,当他们开始记录跨组织边界的所有业务流程时,由于对任何组织的核心业务流程进行重新设计都需要不同的观点,并同时承担绩效和价值责任,参与此类项目的员工不仅需要改变他们的思维方式和工作方式,还需要改变他们的流程建模方式。因此,任何领导都必须激励和管理员工适应当前阶段环境和参加培训,使组织获得成功实施此类项目的正确技能,并使他们能够管理和监控流程及其绩效。这最终将使组织形成一个整体,通过不断改进和调整流程,以适应不断变化的业务环境。

TOTAL QUALITY MANAGEMENT

Introduction

Total Quality Management (TQM) as defined by the International Organization for Standardization (ISO) is:

> *"…a management model for an organization, centered on quality, based on the participation of all its members, and achieves long-term success through client satisfaction, and creates advantages for all members of the organization, and the society as a whole."*

source: ISO 8402:1994.

In Japan, TQM consists of four process steps:

1. ***Kaizen***: Focuses on "constant process improvement", on making the processes visible, measurable and repeatable.
2. ***Kansei***: An examination of the way the consumer uses the product, which leads to improvement of the product itself.
3. ***Atarimae Hinshitsu***: The idea and the way that "things function as they are supposed to."
4. ***Miryokuteki Hinshitsu***: The idea that "things need an aesthetic quality"

TQM is therefore not a new concept, even though that in the shape and form we know it now, it is said to have been formed and influenced by W. Edwards Deming, Philip B. Crosby, J. M. Juran, and Kaoru Ishikawa. Deming was a consultant for Toyota, and TQM and Toyota Production System have many common traits. There are other related concepts and techniques like Six Sigma and Lean that are very closely related to TQM and its concepts of "constant process improvement," customer feedback loops, on making the processes visible, measurable and repeatable. Six Sigma was created by Motorola, which also uses TQM. Some of the companies that have implemented TQM include Ford Motor Company , Motorola, NXP Semiconductors (formerly Philips Semiconductor, Inc), SGL Carbon and Toyota Motor Company.[72]

Throughout the 1980s, American companies adopted TQM concepts and standardized their processes with Baldrige standards and Six Sigma since 1990. Companies today need to be flexible and innovative to be competitive. This requires the companies to have the methods and the software to distance themselves from their competitors. Future generations of companies are based on best practice with an integrated, holistic perspective on process leadership. A renowned expert in this field, W. Edwards Deming, stated 14 steps to the recognized implementation of TQM[38]:

1. Create a common purpose and goal.
2. The leadership decides on transformation and it is they that should promote change.
3. Integrate quality into the product; do not depend on inspection to expose problems.
4. Establish long-term relations based on performance rather than price.

1.3.6　TQM

1. 介绍

ISO定义的TQM为：

　　"……一个以质量为中心，以全体成员参与为基础，通过达成客户满意，实现长期成功的组织管理模式，为组织全体成员和整个社会创造优势。"

来源：ISO 8402：1994。

在日本，TQM包括四个过程步骤：

（1）Kaizen（改善）：注重持续的流程改进，使流程可见、可测量和可重复。

（2）Kansei（感性）：对消费者使用产品的方式的检验，这会导致产品本身的改进。

（3）Atarimae Hinshitsu：事物按预期运行的理念和方式。

（4）Miryokuteki Hinshitsu：事物需要完美品质的观点。

　　TQM并不是一个新概念，据说它是由W. E. Deming（质量管理大师）、P. B.Crosby（零缺陷之父、世界质量先生、伟大的管理思想家）、J. M. Juran（约瑟夫·M.朱兰：现代质量管理的领军人物）和Kaoru Ishikawa（石川馨，鱼骨图发明者、日本质量控制兼统计专家）共同创立和推广的。W. E. Deming是丰田的顾问，TQM和TPS有许多共同的特点。还有其他使流程可见、可测量和可重复并与TQM及其持续的流程改进概念、客户反馈循环密切相关的概念和技术，如六西格玛和精益。六西格玛由摩托罗拉公司创建，该公司也使用TQM。实施TQM的一些公司包括福特汽车公司、摩托罗拉公司、NXP（恩智浦）半导体公司（前身为飞利浦半导体公司）、SGL（西格里）碳素集团和丰田[72]。

　　整个20世纪80年代，美国公司采用了TQM的概念，并自1990年以来用波多里奇质量模型标准（Baldrige standards）和六西格玛标准化了他们的流程。现在的公司需要灵活和创新才能具有竞争力。这就要求公司需要用方法和软件来与竞争对手保持距离。未来几代公司都是以最佳实践为基础，用综合、全面的视角看待流程领导。该领域的一位著名专家W. E. Deming介绍了公认的TQM实施的14个步骤[38]：

（1）创造一个共同的目的和目标；

（2）领导层决定并推动变革；

（3）整合质量与产品，不要依靠检验来发现问题；

（4）以业绩而非价格建立长期关系；

5. Constant improvement of product, quality, and service.
6. Start training people.
7. Focus on leadership.
8. Dispel fear
9. Tear down barriers and silos between departments.
10. Abstain from preaching to the employees.
11. Support, help, and create improvements.
12. Remove barriers for pride, arrogance in a job well done.
13. Implement an energetic program for learning and self-improvement.
14. Have everyone in the organization work on the transformation.

These 14 steps best describe what the concept of TQM encompasses. The concept can also be summed up in five simple main points:

1. Continuous improvement
2. Employee empowerment
3. Analysis, comparing and benchmarking
4. Just-In-Time (JIT) principles
5. Knowledge of the TQM tools

This main point provides an opportunity to explain the use of TQM in process work, see "The Use of TQM in Process Work".

Implementing Total Quality Management in the Organization

According to Hashmi,[73] the first task is to consider if the organization is ready for the change that the implementation of TQM will bring. The right prerequisites for the implementation of TQM deal with the company history, its present needs, the implementation of initiatives leading to TQM, and the quality of the existing staff. An organization with a culture of flexibility will be better prepared to implement TQM. If this flexibility does not exist, it is important that the leadership makes changes so that they can become culture bearers in the implementation of jTQM in order to perform the required change management. In other words, a holistic approach to change is needed.

The next task previous to the implementation of a new system like TQM is identification of the tasks to be solved, establishing the necessary leadership structures, preparation of strategies for support of involvement, design of communication tools in accordance with the change, and the allocation of resources for the implementation.[74]

Key Elements for Change

When implementing TQM, one attempts to establish a company culture in the entire organization, where focus is on constant improvement of quality. Dr S. Kumar[75] describes eight key elements that an organization should focus on to ensure a successful implementation of TQM.[75]

（5）不断改进产品、质量和服务；

（6）开始培训员工；

（7）专注于领导力；

（8）消除恐惧；

（9）打破部门之间的壁垒和隔阂；

（10）禁止向员工说教；

（11）支持、帮助和创建改进；

（12）在出色的工作中消除自负和傲慢的障碍；

（13）实施积极的学习和自我提高计划；

（14）让组织中的每个人都致力于变革。

这14个步骤最能说明TQM的概念是什么。这个概念也可以概括为5个简单的要点：

（1）持续改进；

（2）员工授权；

（3）分析、比较和基准；

（4）准时制原则；

（5）了解TQM工具。

这些要点为解释TQM在流程工作中的应用提供了机会，详见"TQM在流程工作中的应用"。

2. 在组织中实施TQM

根据Hashmi[73]的说法，第一个任务是考虑组织是否准备好接受TQM实现带来的改变。实施TQM的先决条件包括：公司的历史、目前的需求、实施TQM的措施以及现有员工的素质。具有灵活性文化的组织可以更好地准备实施TQM。如果不存在这种灵活性，那么领导层必须做出改变，使他们能够成为TQM实施过程中的文化承载者，以便执行所需的变革管理。换句话说，需要一种整体的变革方法。

在实施TQM等新体系之前，确定要解决的任务、建立必要的领导结构、制定支持参与的战略、根据变化设计沟通工具以及分配实施资源[74]。

3. 变更的关键要素

在实施TQM时，我们试图在整个组织中建立一种公司文化，其重点是持续改进质量。S. Kumar博士[75]描述了组织应重点关注的8个关键要素，以确保TQM的成功实施[75]。

The key elements are *Ethics, Integrity, Trust, Training, Cooperation, Leadership, Recognition*, and *Communication*. These can be illustrated in the following way (Figure 8)[76]:

The foundation consists of ethics and integrity: The basis for ethics is an ethical code in the organization. This organization must include all persons, and it serves as a guideline for their performance in their jobs. Integrity includes traits like honesty, morals, values etc., which the customers would expect of the organization. Trust in connection with the customers is essential to all persons in the organization since customer satisfaction is of the essence.

The bricks are made up of training, cooperation, and leadership: Training is extremely important in connection with productivity. For the leadership, the task is to ensure the skills of the employees, teamwork, problem-solving, decision-making etc. Cooperation ensures a faster and probably better solution to a problem. At the same time, there is a constant improvement in the processes and the work, and the employees feel safe in solving problems amongst themselves. Leadership is seen as the most important element in TQM. The visions, strategies, values, etc., that form the basis for the culture of the organization must come from the leadership. It is essential to have a leadership that is dedicated to leading, and so effectively. This goes for the low-level leader as well as for senior management. The leadership must support the entire process and have an extensive knowledge of the process and be personally involved in the initiation of methods, systems, and measurement to be able to realize the company goals.

FIGURE 8

Components and key elements in connection with the implementation of Total Quality Management.[77]

关键要素分别是：道德、正直、信任、培训、合作、领导、认可和沟通。这些可以用以下方式说明（图8）[76]。

"地基"是道德和正直：道德的基础是组织中的道德规范。该组织必须包括所有人员，并以其道德规范作为他们在工作中表现的指导方针。正直包括诚实、道德、价值观等客户对组织的期望。与客户相关的信任对于组织中的所有人都是至关重要的，因为客户满意度是至关重要的。

"砖块"由培训、合作和领导组成：培训对于提高生产力至关重要。对于领导层来说，任务用来确保员工的技能、团队合作、解决问题、决策等。合作确保更快更好地解决问题。同时，流程和工作也在不断改进，员工在解决问题时感到安心。在TQM中，领导力被视为最重要的因素。构成组织文化基础的愿景、战略、价值观等必须来自领导层。必须有一个致力于管理的领导层，这样才能有效。这不仅适用于低层领导，也适用于高级管理层。领导层必须支持整个过程，对流程有广泛的了解，并亲自参与方法、系统和测量的启动，以实现公司的目标。

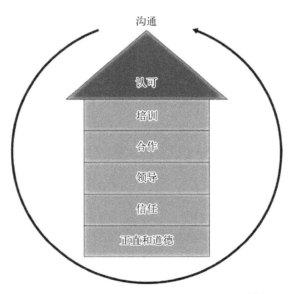

图8　TQM的组成部分和实施的关键要素[77]

The mortar is communication and that is what binds the entire organization together: The communication must ensure a common and frequent frame of reference and must ensure a common way of understanding of the concepts and ideas that are being spread among the individuals in the organization. This also involves contractors and customers. The leadership needs to maintain an open communication with the employees, and the common frame of reference ensures that both the receiver and the sender understand the information that is communicated.

The roof is recognition: Recognition is a very important element in the system, and it should be granted immediately after the completion of a task. A task can be both suggestions for improvements and actual improvements based on suggestions. Recognition can lead to increased self-worth, higher productivity, better quality, and a faster performance of the task. You could argue that there is a relatively close connection between Deming's 14 steps and Kumar's eight key elements as focus areas.

It is obvious that leadership is important to a successful implementation of TQM since the management needs to be in the front on several levels, and in many ways drive the process, or at the very least, support the employees in the process. The implementation of TQM is of importance to management itself, too. Lakhe and Mohanty (1994) argued that not only did the implementation of TQM mean a radical change in the culture of the organization, but it also meant that the leadership actually had to start leading the employees, and at the same time, there needs to be a constant process for quality improvement.

This puts a lot of pressure on the individual manager's flexibility and his or her role as a culture bearer and motive power. You cannot help reaching the conclusion that the implementation of TQM will not be a success until everyone in the organization has come to an understanding of these eight key elements and acts accordingly.

The Use of Total Quality Management in Process Work

Why work on quality—why is it important? According to Figure 9,[78] quality provides market shares and reduces costs.

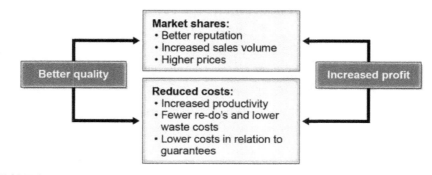

FIGURE 9

Total Quality Management's figure for quality.

"黏合剂"是沟通,这将整个组织联系在一起:沟通必须确保一个共同和频繁的参考框架,并且必须确保一个共同的标准来理解正在组织中的个人之间传播的概念和想法,这也涉及承包商和客户。领导层需要与员工保持开放的沟通,共同的参考框架确保接收者和发送者都了解所传达的信息。

"屋顶"是认可:赏识认可是系统中一个非常重要的元素,应该在任务完成后立即授予。任务可以是改进的建议,也可以是基于建议的实际改进。认可可以提高自我价值、提高生产力、提高质量和更快地完成任务。您可以说,其在 W. E. Deming 的 14 个步骤和 S. Kumar 的 8 个关键要素作为重点领域之间有着相对密切的联系。

很明显,领导力对于 TQM 的成功实施很重要,因为管理层需要在多个层面处于领先地位,并且在许多方面推动过程,或者至少在流程中支持员工。TQM 的实施对管理本身也具有重要意义。Lakhe 和 Mohanty(1994)认为,TQM 的实施不仅意味着组织文化的根本改变,而且意味着领导层实际上必须开始领导员工,同时,需要有一个持续的质量改进流程。

这给人员的灵活性和作为文化承载者及动力的角色带来了很大的压力。只有在组织中的每个人都认识到这 8 个关键要素并采取相应的行动之后,您才可能得出这样的结论:TQM 实施成功。

4. TQM 在流程工作中的应用

为什么要提高质量?为什么这很重要?根据图 9[78]可知,质量能够扩展市场份额并降低成本。

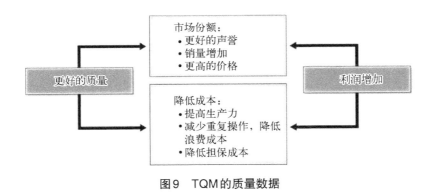

图 9　TQM 的质量数据

Besides, quality has an influence on the company's reputation and legal responsibilities. The involvement of TQM in process work can best be communicated with a basis in the aforementioned five main points for TQM.

1. **Constant improvement**: In Japan, they use the term "Kaizen," and in the United States and Europe, "Six Sigma" is used to describe efforts in this field. The point is to set higher goals all the time and constantly improve all processes.
2. **Empowerment of employees**: The idea is that the people who work with the process, like the ones running the machines in a production plant, are also

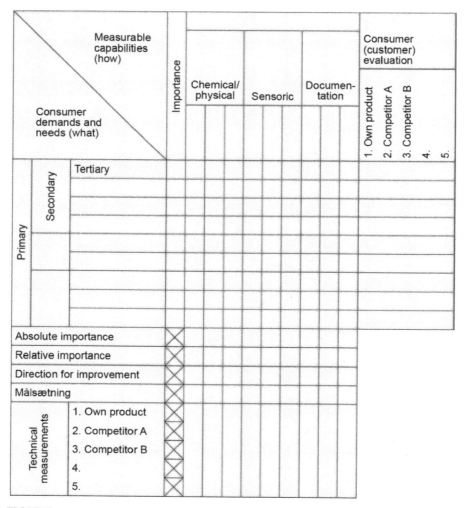

FIGURE 10

House of Quality.[79]

此外,质量对公司的声誉和法律责任也有影响。TQM在流程工作中的参与可以最好地与上述5个TQM要点的基础进行沟通。

(1)持续改进:在日本,他们使用"改善"一词,在美国和欧洲则用"六西格玛"一词描述这一领域的努力,尽管称谓不同,但关键是要始终设定更高的目标,不断改进所有流程。

(2)员工授权:其理念是,与流程合作的人,就像在生产车间运行机器人一样,

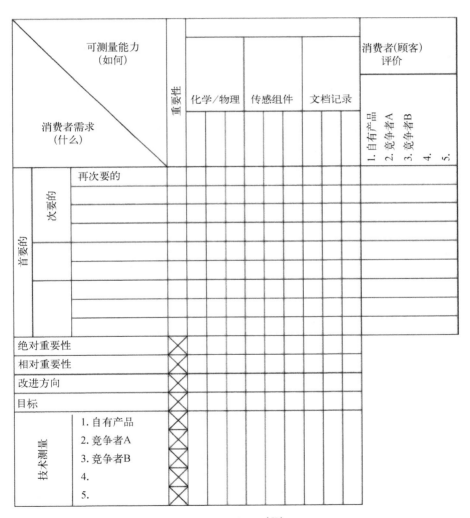

图10 质量屋[79]

those who know the process best, and thus, are best equipped to improve upon it. We speak of a "quality circle", which consists of a group of employees who meet and discuss work-related problems. These quality circles have proven to provide results in the form of improved productivity and quality.

3. **Benchmarking**: Benchmarking involves best practice in the process work, so that a demonstrated standard of performance represents the highest possible goal for the company within a certain type of process. You make a serious effort to reach that goal. You benchmark within every field and across all departments. TQM requires a measurable benchmarking.

4. **Just-in-Time (JIT)**: The philosophy from JIT stands for constant improvement and proactive problem solving. On a more tangible level, it means minimizing storage via control measures.

5. **A knowledge of the TQM tools**: The management should continually train all employees in the use of TQM, and for this purpose, you utilize tools from Quality Function Deployment (QFD) and "House of Quality" (Figure 10)," Taguchi-techniques, Pareto diagrams, process diagrams, cause and effect diagrams, and Statistical Process Control (SPC). The tools are described in greater detail in the upcoming section of this chapter.

TOOLS FOR TQM

House of Quality or Quality Function Deployment

Breyfogle (1999) has listed the following 10 items for the "translation" of the input coming from the organization in the house of quality, which will then be used as useful information for the entire organization:[80]

1. Make a list of the characteristics of your customers. This list can be made through interviews with customers and/or research and studies.
2. Identify the importance of each of these characteristics. This information can be gleaned from the customer studies.
3. Get the customer's input on the existing design and the competitors' designs.
4. Designers need to make up a list of the technical features to meet the customer's characteristics.
5. Relations should be identified in relation to the matrix and assigned a qualitative value (weak, medium, strong).
6. Technical tests must be done on existing designs and competitors' designs to measure objective differences.
7. The importance of all technical features should be calculated as either absolute values or sorted by their relative importance.
8. The difficulties of maneuvering each technical feature must be evaluated.
9. Correlation matrices are to be worked out.
10. Goal values for each technical feature should be established. This could be based on the customers' evaluation in step 3. Based on the calculations of the technical importance under step 7 and the evaluation of the technical difficulties under step 8, choose which technical features to focus on.

也是最了解流程的人,因此,他们最有能力改进流程。我们说的是一个"质量圈",它由一群员工组成,他们会面并讨论与工作有关的问题。事实证明,这些"质量圈"为提高生产率和质量提供了保障。

(3)标杆学习:标杆学习是流程工作中的最佳实践方式,因此,一个已证明的绩效标准代表了公司在特定流程类型中可能达到的最高目标。您为达到这个目标需要付出巨大的努力。您在每个领域和所有部门进行标准检验,因此TQM需要一个可测量的标准。

(4)准时化生产:准时化生产的理念是持续改进和主动解决问题。在更具体的层面上,这意味着通过控制措施最小化存储。

(5)了解TQM工具:管理层应持续培训所有员工使用TQM,为此,您应使用QFD和质量屋(图10)、田口技术、帕累托图、流程图、因果图和SPC等工具。在本节的下一小节中将更详细地描述这些工具。

1.3.7　TQM工具

1. 质量屋或QFD

F. W. Breyfogle(1999)列出了以下10项来自质量屋组织的解释,然后将其用作整个组织的有用信息[80]。

(1)列出客户的特征列表,此列表可以通过与客户的面谈或研究探讨制定。

(2)确定每个特征的重要性,这些信息可以从对客户的研究中获得。

(3)获取客户对现有设计和竞争对手设计的意见。

(4)设计人员需要编制一份个性化的技术清单,以满足客户的个性化要求。

(5)应根据矩阵确定关系,并分配一个定性值(弱、中、强)。

(6)必须对现有设计和竞争对手的设计进行技术评估,以测量客观差异。

(7)所有技术特征的重要性应以绝对值的方式计算或按其相对重要性排序。

(8)必须评估实现每个技术特征的难度。

(9)要计算出相关矩阵。

(10)应确定每个技术特征的目标值,这可以基于步骤(3)中客户的评估。根据第(7)步下的技术重要性计算和第(8)步下的技术难点评估,选择要关注的技术特征。

QFD is used early in the production process to determine where to launch a special effort in the quality work. Basically, you need to translate customer needs into specific features, which can be designed and produced. The House of Quality is a graphical tool that can provide a visual picture of how well the company meets the (prioritized) customer needs.

Taguchi Techniques

Taking the fact that most problems originate from product or process design, there are three concepts, of which we will mention two:

1. **Sturdiness of quality**: The product must be produced in uniform quality, regardless of changing, external factors. The variation in input materials must not influence the quality.
2. **Quality loss function**: This function is a simple mathematical function that describes how costs increase the more the product deviates from the targeted customer needs. This function can help you set the limits for acceptable and unacceptable product output, which could be utilized in the company's financial management (Figure 11).

Pareto Diagrams

In 1906, the economist Vilfredo Pareto illustrated in his research a stunting discovery: 80% of the land in Italy was owned by just 20% of the people. This principle is today known as the Pareto principle— or 80–20 rule—and has been widely adopted

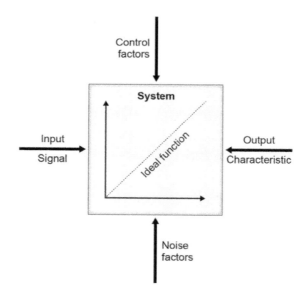

FIGURE 11

An example Taguchi diagram.

在生产过程的早期,通常使用QFD来确定在哪里启动质量改进工作。基本上,您需要将客户的需求转换为可以设计和生产的特定功能。质量屋是一个图形化的工具,可以提供公司如何满足(优先级)客户需求的可视化图像。

2.田口技术

鉴于大多数问题都源于产品或流程设计,我们将提到其三个概念中的两个。

(1)质量的可靠性:产品必须以统一的质量生产,不受外界因素的影响。输入材料的变化不得影响质量。

(2)质量损失函数:该函数是一个简单的数学函数,描述产品偏离目标客户需求越多时成本如何增加。此功能可以帮助您设置可接受和不可接受产品输出的限制,这些限制可用于公司的财务管理(图11)。

3.帕累托图

1906年,经济学家V. Pareto(维弗雷多·帕累托)在他的研究中说明了一个惊人的发现:意大利20%的人拥有80%的土地。这一原则今天被称为帕累托原则(Pareto principle),即80-20规则。它已被广泛采用并应用于商业、经济、数学和流程的各个方面[81]。帕累托分析作为一种简单的技术用来确定根本原因或问题解

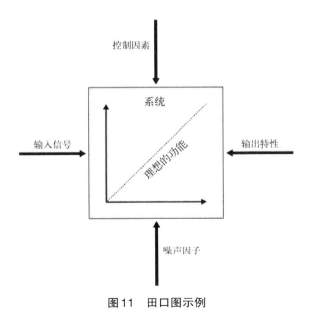

图11　田口图示例

and used across all aspects of business, economics, mathematics, and processes - just to name a few.[81] The Pareto analysis is applied in a straightforward technique to prioritize the root-cause and/or problem solving, subsequently that the first part resolves the greatest number of problems.[82] It is based on the idea that 80 percent of problems may be caused by as few as 20 percent of causes.

Identify a list of problems, followed by scoring and mapping each of the problems by root-cause and subsequently summarizing the scores of each group. At this point, the effort of finding the source of the problem should be clear, and work is then focused on finding an answer or a solution to the root-cause of the problems for the highest scoring group. The Pareto analysis thereby not only identifies the most important problem to solve, it also provides the use of a rating system as to how difficult the problem will be to solve Figure 12.[83]

Process Diagrams

Process diagrams describe the sequence of events in a process, for instance, around the manufacture or sales of a product or a service to a customer. You can glean vital information from these diagrams, such as data accumulation,

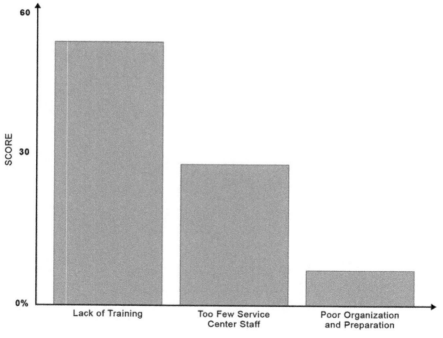

FIGURE 12

An example of a Pareto analysis chart showing that 51 complaints are due to employee lack of training, 27 complaints are due to too few service center staff, and seven complaints are about poor organization and preparation.

决的优先级,接下来先解决比重最大的问题[82]。它是基于这样一个观点:80%的
问题可能是由20%的原因引起的。

使用帕累托方法的步骤是首先确定问题列表,其次按每个问题产生的原因对
各组进行评分和描述,并汇总每组的得分,为得分最高的问题小组寻找答案或解决
方案。在这一点上,寻找问题根源是必须要做的。因此,帕累托分析不仅确定了需
要解决的最重要的问题,还提供了评级系统来确定解决问题的难度(图12)[83]。

4. 流程图

流程图描述流程中事件的顺序,如围绕提供给客户产品或服务的制造或销售。
您可以从这些图表中收集重要信息,如数据积累、隔离、积累的问题区域、审计领

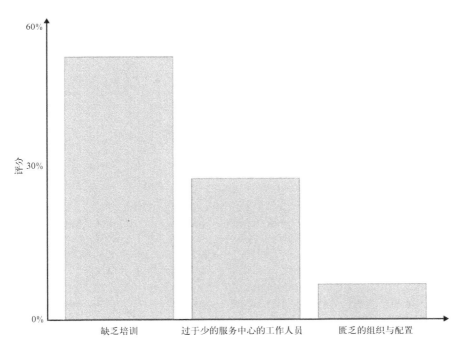

图12　帕累托分析图的一个示例,显示51起投诉是由于员工缺乏培训,27起投诉是
因为服务中心员工太少,7起投诉是关于组织和准备不足

FIGURE 13

An example of a simple sales process diagram.[84]

isolation, accumulation of problem areas, ideal areas for audits, opportunities for reduction of transport distances, etc. In the process diagram of Figure 13, we see an illustration of a very simple and generic sales process—from customer engagement to closing the deal.

Cause & Effect Analysis Diagrams

In 1960, Professor Dr. Kaoru Ishikawa established the cause and effect analysis discipline. Often referred to as a fishbone diagram as the technique maps the possible causes of a problem in a diagram that looks a lot like an entire fishbone stripped of meat. This method visualizes the challenges and problems in a situation in relation to possible causes and complications. The four steps to use the cause and effect analysis are:[85]

1. Identify the challenge and problem (situation).
2. Identify the complication and main factors involved
3. Identify potential root causes.
4. Analyze and examine your diagram.

These diagrams are used for identification of possible quality problems and for locating appropriate areas for inspection (Figure 14).

Statistical Process Control

Statistical Process Control (SPC) is a technique of quality control which uses statistical methods. SPC is applied in order to monitor and control a process. Monitoring and controlling the process ensures that it operates at its full potential. At its full potential, the process can make as much conforming product as possible with a minimum (if not an elimination) of waste (rework or scrap). SPC can be applied to any process where the "conforming product" (product meeting specifications) output can be measured. Key tools used in SPC include control charts, a focus on continuous improvement, and the design of experiments. An example of a process where SPC is applied is manufacturing lines[86] (Figure 15).

The PDCA Cycle

Deming developed a model for use for constant improvement.[87] The model is known as the PDCA Cycle (Figure 16). Throughout this cycle you look at the following:

1. **Plan**—you have to establish the goals, and then the necessary processes to meet the goals of the organization.

图13 简单销售流程图示例[84]

域、设计最佳运输距离等。在如图13所示的流程图中,我们可以看到从客户参与到完成交易的非常简单和通用的销售流程的示意。

5. 因果分析图

1960年,Kaoru Ishikawa教授创立了因果分析技术。这项技术通常被称为鱼骨图,它在一张图中描绘了问题的可能原因,这张图看起来很像一整根被剥去肉的鱼骨。该方法将问题产生的可能原因与解决问题的难度和问题可视化。使用因果分析的四个步骤是[85]:

(1)确定难度和问题(情况);

(2)识别复杂问题及主要影响因素;

(3)确定潜在的根本原因;

(4)分析并检查图表。

这些图表用于识别可能出现的质量问题,并定位适当的检查区域(图14)。

6. SPC

SPC是一种采用统计方法进行质量控制的技术。应用SPC可以监控和控制流程以确保其充分发挥能力。在其全部能力下,该流程可以以最少(如果不是消除)浪费(返工或报废)来制造尽可能多的合格产品。SPC可应用于任何可测量合格产品(符合规范的产品)输出的流程。SPC中使用的关键工具包括控制图、关注持续改进和实验设计。应用SPC的流程示例是生产线[86](图15)。

7. PDCA 循环

W. E. Deming开发了一个用于持续改进的模型[87],该模型被称为PDCA循环(图16),在整个循环中,您将看到以下内容:

(1)计划——您必须建立目标,然后制定必要的流程来实现组织的目标。

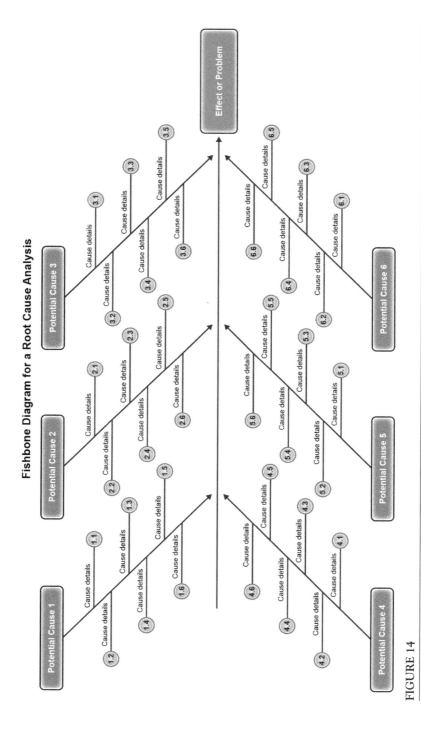

FIGURE 14

A root cause analysis (cause and effect) fishbone diagram.

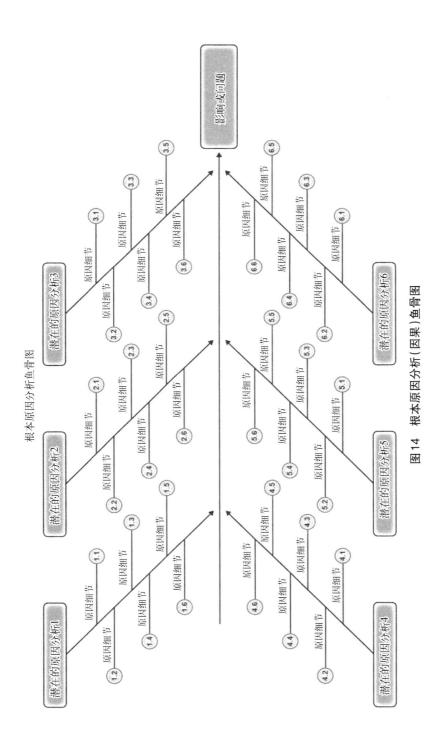

根本原因分析鱼骨图

图 14 根本原因分析（因果）鱼骨图

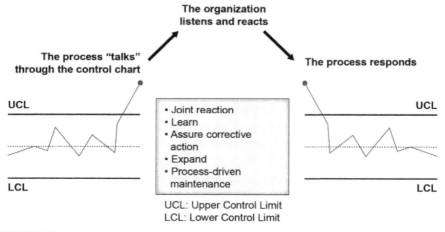

FIGURE 15

A Statistical process control chart.

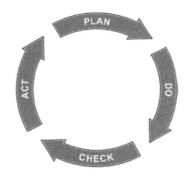

FIGURE 16

The Plan-Do-Check-Act (PDCA) cycle.

2. **Implement** the processes.
3. **Monitor and evaluate** the processes and the results in relation to the goals and report the results.
4. **Act and make decisions** on possible areas of improvement. Evaluate the entire PDCA process and, if necessary, modify it before the next implementation phase.

The model presupposes that the organization is flexible or that the right culture and the support of the management is a given. It is interesting to note that Deming in the PDCA cycle argues for a monitoring of results and for prompt action on this monitoring. In his 14 steps to the implementation of TQM, he argues that the improvement of quality should not be a consequence of inspection and measuring,

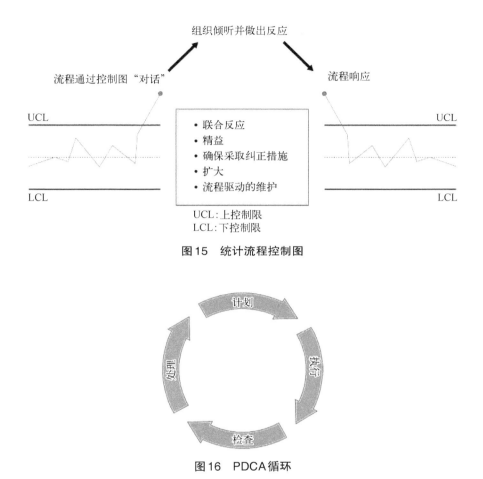

图15　统计流程控制图

图16　PDCA循环

（2）执行流程。

（3）监控和评估与目标相关的流程和结果,并报告结果。

（4）对可能的改进领域采取行动并做出决定。评估整个PDCA流程,如有必要,在下一个实施阶段之前对其进行修改。

该模型假定组织是灵活的,或者给定了正确的文化和管理层的支持。值得注意的是,W. E. Deming主张在PDCA循环中对结果进行监控,并对此监控即时操控。在实施TQM的14个步骤中,他认为质量的提高不应该是检验和测量的结果,

but should rather be integrated with the product. To improve on the quality of a certain product or service, the starting point must be a frame of reference to measure the improvement against.

Business Excellence and EFQM

The principles and tools from total quality management (TQM) are also known as "business excellence" when they are applied to management and general improvements of the company. The stakeholder value, customer focus, and process management are the parameters the company needs to improve on through a systematic approach.[88] (Figure 17).

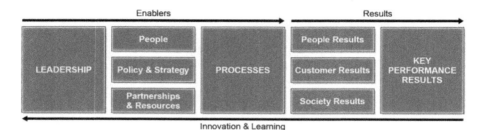

FIGURE 17

A Systematic approach to business excellence.

Conclusion

Quality management as a foundational discipline has been around for many years now, and it has gained a significant focus and attention of, in particular, the manufacturing industry. Total Quality Management (TQM) has since its early development phases in the 1940s laid the groundwork for an organizational management discipline that focuses on quality across organizational boundaries and disciplines. The involvement of people in this discipline is all-encompassing, and as such, implementing TQM in any organization requires a heavy investment of both time and resources. Training is also required for the personnel, but for many companies, it is generally considered – if implemented right and adequately – to be worth the effort.

Some of the many benefits that TQM has to offer are:
- The ability to produce high quality products for the end consumer
- Elimination of waste, and reduction of the resources needed to produce goods
- Continuous process improvement and product optimization
- Achieve labor savings due to the streamlining of processes
- Increased product quality awareness across the company

而是应该与产品相结合。为了提高某一产品或服务的质量,起点必须是衡量改进的参照系。

8. 卓越经营和EFQM

TQM的原则和工具在应用于公司的管理和总体改进时,也被称为卓越经营。利益相关者价值、以客户为中心和流程管理都是公司需要通过系统方法改进的参数[88](图17)。

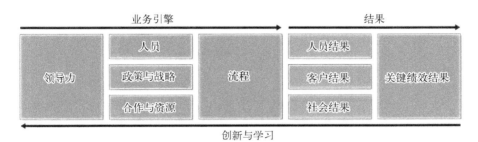

图17 卓越经营的系统方法

9. 结论

质量管理作为一门基础性的学科已经存在了很多年,它得到了制造业等行业显著的关注。TQM自20世纪40年代早期的发展阶段以来,为组织管理学科奠定了基础,该学科侧重于跨组织边界和跨学科的质量。人们参与这一学科是包罗万象的,因此,在任何组织中实施TQM都需要大量的时间和资源投入。员工也需要接受培训,对于许多公司来说,通常认为培训的实施是正确的,并且是值得努力的。

TQM提供了许多好处,包括:

- 为最终消费者生产高质量产品的能力;
- 消除浪费,减少生产商品所需的资源;
- 持续的流程改进和产品优化;
- 通过简化流程实现劳动力节约;
- 提高整个公司的产品质量意识。

SIX SIGMA

Introduction

Six Sigma, or 6σ as it is also written, is a collection of methods, techniques and tools for process improvement. The basic thinking of Six Sigma can be traced back to Carl Friedrich Gauss (1777–1855) as a measurement standard by introducing the concept of the normal curve. In the 1920, Walter Shewhart demonstrated that sigma imply where a process requires improvement. In 1986, Bill Smith and Mikel Harry, two engineers at Motorola were accredited to having developed "Six Sigma", and in 1995, Jack Welch made it the central business strategy of General Electric.

Today, the principles are widely adopted across industry sectors, and in recent years the Six Sigma ideas have merged with the Lean manufacturing methodology, naming it Lean Six Sigma. The Lean Six Sigma methodology aim to support the business and operational excellence by focusing on variation, design, waste issues and process flows.[89] Companies such as Motorola, General Electric, Verizon, and IBM uses Lean Six Sigma as a growth strategy to rethink and transform themselves through efficiency—from organizational setup to manufacturing, software development to sales and distribution, and finally for service delivery functions.

What Six Sigma Means

Sigma is used in statistics to indicate a deviation from the standard of 99,99999980268%.[90] This means that almost no errors are produced. In reality a process that is designed to be in accordance with Six Sigma may not continue to do so. It has been empirically proven that, in time, it decreases from 6 to 4.5 sigma. It is debated whether or not the 1.5 is universal, but nobody doubts that there is a decrease.[91]

The Purpose of Six Sigma

The purpose of Six Sigma is to improve the quality of process outputs by identifying and removing the causes of defects (errors) and minimizing variability in manufacturing and business processes. It uses a set of quality management methods, including statistical methods, and creates a special team of people within the organization (Champions, Black Belts, Green Belts, Yellow Belts, etc.) who are experts in these methods. Each Six Sigma project carried out within an organization follows a defined sequence of steps and has quantified value targets, for example:[92]

- Reduce process cycle time.
- Reduce pollution.
- Reduce costs.
- Increase customer satisfaction.
- Increase profits.

1.3.8　六西格玛

1. 介绍

六西格玛,也被写作 6σ,是流程改进的方法、技术和工具的集合。六西格玛的基本思想可以追溯到 C. F. Gauss(1777—1855 年,数学家、物理学家和天文学家),他引入正态曲线的概念作为测量标准。1920 年,W. A. Shewhart(沃特·阿曼德·休哈特,现代质量管理的奠基者,美国工程师、统计学家、管理咨询顾问、被人们尊称为"统计质量控制之父")论证了西格玛意味着流程需要改进的地方。1986 年,摩托罗拉的两位工程师 B. Smith 和 M. Harry 发明了六西格玛,1995 年,J. Welch 将其作为通用电气的核心业务战略。

如今,这些原则被广泛应用于各个行业。近年来,六西格玛理念与精益制造方法相融合,并将其命名为精益六西格玛。精益六西格玛方法旨在通过关注变化、设计、浪费问题和流程来支持业务和运营的卓越性[89]。摩托罗拉、通用电气、Verizon 和 IBM 等公司将精益六西格玛作为一种增长战略,通过从组织设置到生产、从软件开发到销售和分销,最后再到服务交付功能的效率来重新思考和转变自己。

2. 六西格玛的意思

在统计中使用西格玛表示与标准的偏差为 99.99999980268%[90],这意味着几乎没有错误产生。实际上,按照六西格玛设计的流程可能不会继续如此。经验证明,随着时间的推移,它从 6 西格玛下降到 4.5 西格玛。1.5 西格玛是否具有普遍性还存在争议,但没有人怀疑它会有所减少[91]。

3. 六西格玛的目的

六西格玛的目的是通过识别和消除造成缺陷(错误)的原因以及最小化制造和业务流程的可变性来提高流程输出的质量。它使用一套质量管理方法,包括统计方法,并在组织内创建一个专门的团队(冠军、黑带、绿带、黄带等),他们是这些方法的专家。在一个组织内执行的每一个六西格玛项目都遵循一系列确定的步骤,并量化价值目标,例如[92]:

- 缩短流程循环时间;
- 减少污染;
- 降低成本;
- 提高客户满意度;
- 增加利润。

Six Sigma makes a distinction between improving existing processes and designing new ones. For this purpose, Deming's Plan-Do-Check–Act cycle has been developed into the following two methodologies (DMAIC of DMADV).

DMAIC

DMAIC (define, measure, analyze, improve, and control) is used for projects aimed at improving an existing business process. The DMAIC project methodology has five phases (Figure 18):

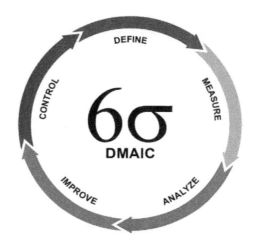

FIGURE 18

The Six Sigma DMAIC (define, measure, analyze, improve, and control) model.

1. **Define** the system, the voice of the customer and their requirements, and the project goals, specifically.
2. **Measure** key aspects of the current process and collect relevant data.
3. **Analyze** the data to investigate and verify cause-and-effect relationships. Determine what the relationships are, and attempt to ensure that all factors have been considered. Seek out the root cause of the defect under investigation.
4. **Improve** or optimize the current process based upon data analysis using techniques such as design of experiments, poka yoke or mistake proofing, and standard work to create a new, future-state process. Set up pilot runs to establish process capability.
5. **Control** the future-state process to ensure that any deviations from the target are corrected before they result in defects. Implement control systems such as SPC, production boards, visual workplaces, and continuously monitor the process.

Some organizations add a "recognize" step at the beginning, which is to recognize the right problem to work on, thus yielding an RDMAIC methodology.[93]

六西格玛区分了改进现有流程和设计新流程。为此，W. E. Deming的PDCA循环已发展成以下两种方法（DMAIC和DMADV）。

4. DMAIC

DMAIC（定义、测量、分析、改进和控制）用于旨在改进现有业务流程的项目。DMAIC项目方法有五个阶段（图18）。

图18　六西格玛DMAIC模型

（1）明确定义系统、客户的声音及其需求，以及项目目标。

（2）测量当前流程的关键方面并收集相关数据。

（3）分析数据以调查和验证因果关系。确定关系是什么，并尝试确保已考虑所有因素。找出被调查缺陷的根本原因。

（4）根据数据分析，利用实验设计、防误防错等技术改进或优化当前流程，并通过标准工作创建新的未来状态流程。建立试运行以构建流程能力。

（5）控制未来的流程状态，确保任何偏离目标的情况在导致缺陷之前得到纠正。实施SPC、生产板、可视工作场所等控制系统，并持续监控流程。

一些组织在开始时添加了识别步骤，即识别要处理的正确问题，从而生成RDMAIC方法[93]。

DMADV

DMADV (define, measure, analyze, design, and verify) is used for projects aimed at creating new product or process designs. The DMADV project methodology, also known as DFSS (Design for Six Sigma[94]), also features five phases (Figure 19):

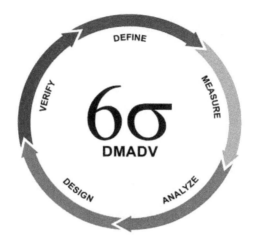

FIGURE 19

The Six Sigma DMADV (define, measure, analyze, design, and verify) model.

1. **Define** design goals that are consistent with customer demands and the enterprise strategy.
2. **Measure** and identify CTQs (characteristics that are Critical To Quality), product capabilities, production process capability, and risks.
3. **Analyze** to develop and design alternatives.
4. **Design** an improved alternative, best suited per analysis in the previous step.
5. **Verify** the design, set up pilot runs, implement the production process, and hand it over to the process owner(s).

First you define the goal in accordance with the demand and the strategy. Then, you measure and identify the quality criteria for the product and process and the risks involved in both. The analysis will result in several designs, which are compared and analyzed to find the very best design. Then, you verify if the chosen design meets the goal, and then you launch it.

Conclusion

We have in the above section elaborated on what Six Sigma is and the way of thinking and working around the concept. One of the things to consider is why to choose Six Sigma in the first place? You could, of course, consider other process methodologies that require a higher (Zero Defect) or lower level of accuracy. In the end, it's a question of cost-benefit in every situation. Both improvement methodologies

5. DMADV

DMADV（定义、测量、分析、设计和验证）用于旨在创建新产品或流程设计的项目。DMADV项目方法论，也称为DFSS（六西格玛设计[94]），也具有五个阶段（图19）。

图19　六西格玛DMADV模型

（1）定义与客户需求和企业战略一致的目标。

（2）测量和识别CTQ（critical to quality，对质量至关重要的特性）、产品能力、生产流程能力和风险。

（3）对开发和设计替代方案进行分析。

（4）为上一步中的每个分析设计一个改进的替代方案。

（5）验证和试运行，最后实施流程，并移交给流程所有人。

首先，根据需求和策略定义目标。其次，度量、识别产品和流程的质量标准以及这两者所涉及的风险。再次，对若干个子方案进行比较和分析，以找到最佳方案。最后，验证所选方案是否满足目标，然后启动它。

6. 结论

在上面的章节中，我们已经详细阐述了六西格玛是什么，以及围绕这个概念的思考和工作方式。要考虑的一件事是为什么首先选择六西格玛。当然，您可以考虑其他更高（零缺陷）或更低精确度的流程方法。最后，这是一个在任何情况下都

(Six Sigma and Zero Defect) are based on the analysis of data. It is therefore important to know that Six Sigma requires very precise data to yield any benefit. If the performance and measuring tools are not accurate enough, you cannot calculate the process improvements accurately. The accuracy of the measuring is proportional with the cost of measuring, and this can be a hindrance that will ultimately make you choose other optimization or improvement techniques, such as BPR or Lean, rather than Six Sigma.

Regardless, Six Sigma offers a substantial range of adequate benefits and rewards once it has been successfully implemented in an organization. Some of the most sought for benefits are for instance:

- The goal is to improve the process quality to a level of less than 3.4 errors per one million possibilities.
- Six Sigma distinguishes between designing a new process and improving an existing one.
- Six Sigma is as a concept aware of the roles and qualifications of the involved parties in the process management.
- Six Sigma is designed to correct an existing process, but is not suited to create new products or technologies.
- It provides a clear focus for achieving measurable and quantifiable financial advantages.
- It increases focus on strong leadership and support.

It provides a clear duty to make decisions based on verifiable data instead of merely guessing.

End Notes

1. Business Process Management: Profiting From Process, Roger Burlton, Sams Publishing, 2001.
2. An Introduction To Enterprise Architecture: Third Edition, Scott A. Bernard, AuthorHouse, 2012.
3. A framework for information systems architecture, John A. Zachman, IBM Systems Journal, Vol 26. No 3, 1987.
4. Toyota Production System: Beyond large-scale production, Taiichi Ohno, Productivity Press, 1988.
5. Ibid.
6. Lean Consumption, J.P. Womack and D.T. Jones, *Harvard Business Review*, 2005.
7. Reengineering Work: Don't Automate, Obliterate, Michael Hammer, *Harvard Business Review*, 1990.
8. Total Quality Management: Origins and evolution of the term, Martínez Lorente, Ángel Rafael, Emerald Group. Publishing Limited, 1998.
9. Quality Management: Strategies, Methods, Techniques, Tilo Pfeifer, Hanser, 2002.
10. A History of Managing for Quality: The Evolution, Trends, and Future Directions of Managing for Quality, Joseph M. Juran, ASQC Quality Press, 1995.
11. What is total quality control? The Japanese way, Kaoru Ishikawa, Prentice Hall, 1985.
12. The House of Quality, John R. Hauser, Don Clausing, Harvard Business Review, 1988.

要考虑成本效益的问题。两种改进方法(六西格玛和零缺陷)都基于数据分析。因此,重要的是要知道六西格玛需要非常精确的数据才能产生效益。如果性能和测量工具不够精确,则无法准确计算流程改进。测量的准确度与测量成本成正比,这可能是一个障碍,最终会使您选择其他优化或改进技术,如BPR或精益,而不是六西格玛。

不管怎样,一旦六西格玛在一个组织中成功实施,它将提供大量的、适当的好处和奖励。例如,一些最受欢迎的收益是

- 将流程质量提高到每100万误差小于3.4可能性的水平;
- 六西格玛能够区分设计新流程和改进现有流程;
- 六西格玛能够帮助了解流程管理中相关方的角色和资格;
- 六西格玛旨在纠正现有流程,但不适合创建新的产品或技术;
- 它为实现可测得和可量化的财务优势提供了明确的重点;
- 加强对强有力领导和支持的关注。

它提供了一个明确的职责,以可验证的数据为基础做出决策,而不仅仅是猜测。

参考文献

[1] Business Process Management: Profiting From Process, Roger Burlton, Sams Publishing, 2001.
[2] An Introduction To Enterprise Architecture: Third Edition, Scott A. Bernard, AuthorHouse, 2012.
[3] A framework for information systems architecture, John A. Zachman, IBM Systems Journal, Vol 26. No 3, 1987.
[4] Toyota Production System: Beyond large-scale production, Taiichi Ohno, Productivity Press, 1988.
[5] Ibid.
[6] Lean Consumption, J.P. Womack and D.T. Jones, Harvard Business Review, 2005.
[7] Reengineering Work: Don't Automate, Obliterate, Michael Hammer, Harvard Business Review, 1990.
[8] Total Quality Management: Origins and evolution of the term, Martínez Lorente, Ángel Rafael, Emerald Group. Publishing Limited, 1998.
[9] Quality Management: Strategies, Methods, Techniques, Tilo Pfeifer, Hanser, 2002.
[10] A History of Managing for Quality: The Evolution, Trends, and Future Directions of Managing for Quality, Joseph M. Juran, ASQC Quality Press, 1995.
[11] What is total quality control? The Japanese way, Kaoru Ishikawa, Prentice Hall, 1985.
[12] The House of Quality, John R. Hauser, Don Clausing, Harvard Business Review, 1988.

13. Application of Taguchi Method for Optimizing Turning Process by the effects of Machining Parameters, Krishankant, Jatin Taneja, Mohit Bector, Rajesh Kumar, International Journal of Engineering and Advanced. Technology, 2012.
14. Quantitative methods for quality improvement, Hart, K. M., & Hart, R. F., ASQC Quality Press, 1989.
15. Foundations of statistical quality control, Barlow, Richard E.; Irony, Telba Z., Institute of Mathematical Statistics, Hayward, CA, 1992.
16. Statistical Method from the Viewpoint of Quality Control, Walter A. Shewhart, W. Edwards Deming, Dover. Books on Mathematics, 2011.
17. EFQM Excellence Model, AKYAY UYGUR, SEVGI SÜMERLI, International Review of Management and Business Research, 2013.
18. The Six Sigma Handbook, Fourth Edition, Thomas Pyzdek, Paul Keller, McGraw-Hill Education, 2014.
19. *The Six Sigma Way: How GE, Motorola, and Other Top Companies are Honing Their Performance*, Pande, P. S., Robert P, Neuman and Roland R., Cavanagh, McGraw-Hill Professional, 2001.
20. Architecture of Integrated Information Systems: Foundations of Enterprise Modelling, August-Wilhelm Scheer, Springer, 1992.
21. Start-ups are easy, but..., August-Wilhelm Scheer, Springer, 2001.
22. ARIS — business process modeling, August-Wilhelm Scheer, Springer, 1999.
23. Computer: A Challenge for Business Administration, August-Wilhelm Scheer, Springer, 1985.
24. Business Process Engineering: Aris-Navigator for Reference Models for Industrial Engineers/Book, August-Wilhelm Scheer, Springer, 1995.
25. CIM Towards the Factory of the Future, August-Wilhelm Scheer, Springer, 1994.
26. Principles of efficient information management, August-Wilhelm Scheer, Springer, 1991.
27. ARIS – Business Process Frameworks, August-Wilhelm Scheer, Springer, 1998.
28. A framework for information systems architecture, John A. Zachman, IBM Systems Journal, 1987.
29. Business systems planning and business Information control study: a comparison, John A. Zachman, IBM Systems Journal, 1982.
30. John Zachman's Concise Definition of The Zachman Framework™, John A. Zachman, Zachman International, Inc., 2008.
31. *The Zachman Framework for Enterprise Architecture™: A Primer For Enterprise Engineering and Manufacturing*, John Zachman, Zachman International, 2003.
32. The Zachman Framework For Enterprise Architecture: Primer for Enterprise Engineering and Manufacturing, John A. Zachman, Zachman International, Inc., 2003.
33. Conceptual, Logical, Physical: It Is Simple, John A. Zachman, Zachman International, Inc., 2000-2011.
34. Architecture, John A. Zachman, Zachman International, Inc., 2007–2011.
35. Cloud Computing and Enterprise Architecture, John A. Zachman, Zachman International, Inc., 2011.
36. The original texts are Taiichi Ohno, The Toyota Production System, Productivity Press, Oregon, 1988 and The Toyota Way, Toyota Motor Corporation, Tokyo, 2001.
37. The Lean Lexicon, Lean Enterprise Institute (LEI), Cambridge, 2003, The use of these tools is described in a series of workbooks published by LEI, including Creating Continuous Flow, 2001, Making Materials Flow, 2003 and Creating Level Pull, 2009.

［13］Application of Taguchi Method for Optimizing Turning Process by the effects of Machining Parameters, Krishankant, Jatin Taneja, Mohit Bector, Rajesh Kumar, International Journal of Engineering and Advanced. Technology, 2012.

［14］Quantitative methods for quality improvement, Hart, K. M., & Hart, R. F., ASQC Quality Press, 1989.

［15］Foundations of statistical quality control, Barlow, Richard E.; Irony, Telba Z., Institute of Mathematical Statistics, Hayward, CA, 1992.

［16］Statistical Method from the Viewpoint of Quality Control, Walter A. Shewhart, W. Edwards Deming, Dover. Books on Mathematics, 2011.

［17］EFQM Excellence Model, AKYAY UYGUR, SEVGI SÜMERLI, International Review of Management and Business Research, 2013.

［18］The Six Sigma Handbook, Fourth Edition, Thomas Pyzdek, Paul Keller, McGraw-Hill Education, 2014.

［19］The Six Sigma Way: How GE, Motorola, and Other Top Companies are Honing Their Performance, Pande, P. S., Robert P, Neuman and Roland R., Cavanagh, McGraw-Hill Professional, 2001.

［20］Architecture of Integrated Information Systems: Foundations of Enterprise Modelling, August-Wilhelm Scheer, Springer, 1992.

［21］Start-ups are easy, but..., August-Wilhelm Scheer, Springer, 2001.

［22］ARIS—business process modeling, August-Wilhelm Scheer, Springer, 1999.

［23］Computer: A Challenge for Business Administration, August-Wilhelm Scheer, Springer, 1985.

［24］Business Process Engineering: Aris-Navigator for Reference Models for Industrial Engineers/Book, August-Wilhelm Scheer, Springer, 1995.

［25］CIM Towards the Factory of the Future, August-Wilhelm Scheer, Springer, 1994.

［26］Principles of efficient information management, August-Wilhelm Scheer, Springer, 1991.

［27］ARIS−Business Process Frameworks, August-Wilhelm Scheer, Springer, 1998.

［28］A framework for information systems architecture, John A. Zachman, IBM Systems Journal, 1987.

［29］Business systems planning and business Information control study: a comparison, John A. Zachman, IBM Systems Journal, 1982.

［30］John Zachman's Concise Definition of The Zachman Framework^{TM}, John A. Zachman, Zachman International, Inc., 2008.

［31］The Zachman Framework for Enterprise Architecture^{TM}: A Primer For Enterprise Engineering and Manufacturing, John Zachman, Zachman International, 2003.

［32］The Zachman Framework For Enterprise Architecture: Primer for Enterprise Engineering and Manufacturing, John A. Zachman, Zachman International, Inc., 2003.

［33］Conceptual, Logical, Physical: It Is Simple, John A. Zachman, Zachman International, Inc., 2000−2011.

［34］Architecture, John A. Zachman, Zachman International, Inc., 2007−2011.

［35］Cloud Computing and Enterprise Architecture, John A. Zachman, Zachman International, Inc., 2011.

［36］The original texts are Taiichi Ohno, The Toyota Production System, Productivity Press, Oregon, 1988 and The Toyota Way, Toyota Motor Corporation, Tokyo, 2001.

［37］The Lean Lexicon, Lean Enterprise Institute (LEI), Cambridge, 2003, The use of these tools is described in a series of workbooks published by LEI, including Creating Continuous Flow, 2001, Making Materials Flow, 2003 and Creating Level Pull, 2009.

38. For a description of TWI see Donald A. Dinero, *Training Within Industry*, Productivity Press, New York, 2005.
39. Masaki Imai, Kaizen, McGraw Hill, New York, 1991.
40. Jeff Liker, The Toyota Way, McGraw Hill, New York, 2004.
41. John Shook, Managing to Learn, Lean Enterprise Institute, Cambridge, 2009.
42. James P Womack, Daniel T Jones & Daniel Roos, The Machine that Changed the World, Rawson Macmillan, New York, 1991.
43. James P Womack & Daniel T Jones, Lean Thinking, Simon & Schuster, New York, 1996 and 2003.
44. John Shook and Mike Rother, Learning to See, Lean Enterprise Institute, Cambridge, 1998.
45. Pascal Dennis, Getting the Right Things Done, Lean Enterprise Institute, Cambridge, 2009.
46. This management system is described in Jeff Liker and Gary Convis, The Toyota Way to Lean Leadership, McGraw Hill, New York, 2012, Mike Rother, Toyota Kata, McGraw Hill, New York, 2010 and Michael and Freddy Balle, Lead with Respect, Lean Enterprise Institute, Cambridge, 2014.
47. Eric Ries, The Lean Startup, Penguin, New York, 2013.
48. Described in detail in Chapter 4 of Lean Thinking, ibid.
49. The Tesco and GKN examples are outlined in Daniel Jones and James Womack, Seeing the Whole Value Stream, Lean Enterprise Institute, Cambridge, 2011.
50. James Womack and Daniel Jones, Lean Solutions, Simon & Schuster, New York, 2005.
51. The Economist Guide to Management Ideas and Gurus, Tim Hindle, 2012.
52. Reengineering Work: Don't Automate, Obliterate, Michael Hammer, Harvard Business Review, 1990.
53. Reengineering the Corporation: A Manifesto for Business Revolution (Collins Business Essentials), Michael Hammer, James Champy, Harper Business Essentials, 2006.
54. Time 25: They Range In Age From 31 To 67, Time Magazine, 1996.
55. Michael Hammer and James Champy. *Reengineering the Corporation: A Manifesto for Business Revolution*, (Harper Business, 1993).
56. Identifying and Eliminating The Seven Wastes or Muda, Rene T. Domingo, Asian Institute of Management.
57. von Rosing, M., Hagemann Snabe, J. Rosenberg, A. Møller, C. Scavillo, M. *Business Process Management – the SAP® Roadmap*, 2009.
58. Hammer, M., "Reengineering Work: Don't Automate, Obliterate", *Harvard Business Review*, July/August,1990, 104–112.
59. The New Industrial Engineering: Information Technology and Business Process Redesign, Davenport, Thomas &. Short, J., Sloan Management Review, 1990.
60. A Practical Guide to Business Process Re-engineering, Mike Robson, Philip Ullah, Gower Publishing, Ltd., 1996.
61. Kristian D. Ambeck and Peter Beyer, *The Road to Renewal* (Danish publication).
62. Hammer, Michael and Champy, James, Reengineering the Corporation: A Manifesto for Business Revolution, (Harper Business, 1993), Chapter 1 excerpt.
63. Business Process Reengineering: Building a Comprehensive Methodology, Subashish Guha, William J. Kettinger & James T.C. Teng, Taylor & Francis Group, 1993.
64. Davenport, Thomas, *Process Innovation: Reengineering work through information technology*, (Harvard Business School Press, Boston, 1993).
65. Maureen Weicher, William W. Chu, Wan Ching Lin, Van Le and Dominic Yu, *Business Process Reengineering: Analysis and Recommendations*.

［38］ For a description of TWI see Donald A. Dinero, Training Within Industry, Productivity Press, New York, 2005.

［39］ Masaki Imai, Kaizen, McGraw Hill, New York, 1991.

［40］ Jeff Liker, The Toyota Way, McGraw Hill, New York, 2004.

［41］ John Shook, Managing to Learn, Lean Enterprise Institute, Cambridge, 2009.

［42］ James P Womack, Daniel T Jones & Daniel Roos, The Machine that Changed the World, Rawson Macmillan, New York, 1991.

［43］ James P Womack & Daniel T Jones, Lean Thinking, Simon & Schuster, New York, 1996 and 2003.

［44］ John Shook and Mike Rother, Learning to See, Lean Enterprise Institute, Cambridge, 1998.

［45］ Pascal Dennis, Getting the Right Things Done, Lean Enterprise Institute, Cambridge, 2009.

［46］ This management system is described in Jeff Liker and Gary Convis, The Toyota Way to Lean Leadership, McGraw Hill, New York, 2012, Mike Rother, Toyota Kata, McGraw Hill, New York, 2010 and Michael and Freddy Balle, Lead with Respect, Lean Enterprise Institute, Cambridge, 2014.

［47］ Eric Ries, The Lean Startup, Penguin, New York, 2013.

［48］ Described in detail in Chapter 4 of Lean Thinking, ibid.

［49］ The Tesco and GKN examples are outlined in Daniel Jones and James Womack, Seeing the Whole Value Stream, Lean Enterprise Institute, Cambridge, 2011.

［50］ James Womack and Daniel Jones, Lean Solutions, Simon & Schuster, New York, 2005.

［51］ The Economist Guide to Management Ideas and Gurus, Tim Hindle, 2012.

［52］ Reengineering Work: Don't Automate, Obliterate, Michael Hammer, Harvard Business Review, 1990.

［53］ Reengineering the Corporation: A Manifesto for Business Revolution (Collins Business Essentials), Michael Hammer, James Champy, Harper Business Essentials, 2006.

［54］ Time 25: They Range In Age From 31 To 67, Time Magazine, 1996.

［55］ Michael Hammer and James Champy. Reengineering the Corporation: A Manifesto for Business Revolution, (Harper Business, 1993).

［56］ Identifying and Eliminating The Seven Wastes or Muda, Rene T. Domingo, Asian Institute of Management.

［57］ von Rosing, M., Hagemann Snabe, J. Rosenberg, A. Møller, C. Scavillo, M. Business Process Management— the SAP® Roadmap, 2009.

［58］ Hammer, M., "Reengineering Work: Don't Automate, Obliterate", Harvard Business Review, July/ August,1990, 104−112.

［59］ The New Industrial Engineering: Information Technology and Business Process Redesign, Davenport, Thomas &. Short, J., Sloan Management Review, 1990.

［60］ A Practical Guide to Business Process Re-engineering, Mike Robson, Philip Ullah, Gower Publishing, Ltd., 1996.

［61］ Kristian D. Ambeck and Peter Beyer, The Road to Renewal (Danish publication).

［62］ Hammer, Michael and Champy, James, Reengineering the Corporation: A Manifesto for Business Revolution, (Harper Business, 1993), Chapter 1 excerpt.

［63］ Business Process Reengineering: Building a Comprehensive Methodology, Subashish Guha, William J. Kettinger & James T.C. Teng, Taylor & Francis Group, 1993.

［64］ Davenport, Thomas, Process Innovation: Reengineering work through information technology, (Harvard Business School Press, Boston, 1993).

66. Smith, K.K., "Philosophical Problems in Thinking about Organisational Change", in Change in Organisations, (Jossey-Bass, Oxford, 1982).

67. Michael Hammer; James Champy, *Reengineering the Corporation*.

68. Kaplan, R.B. and Murdock, L., "Rethinking the Corporation: Core Process Redesign", *The McKinsey Quarterly*, No. 2, 1991.

69. There's no Better Time to Do Business Process Improvement than the Bad Time, Hutex™ Management Consulting.

70. Thomas H. Davenport, Laurence Prusak and H. James Wilson "Article from Computerworld: Reengineering revisited: What went wrong with the business-process reengineering fad. And will it come back?"

71. LEADing BPM Practice Case Story, Mark von Rosing, Henrik von Scheel, Anette Falk Bøgebjerg, LEADing Practice, 2013.

72. T. Priyavrat, (2007),'Teams, Traits And Tasks [QT3] For Total Quality', Quality World, Vol V, April 2007.

73. Hashmi, K. "Introduction and Implementation of Total Quality management (TQM)".

74. Beckhard, R., Pritchard W. *Changing the Essence* (San Francisco: Jossey-Bass, 1992).

75. Kumar, Dr. S., 2006 *Total Quality Management*, 13–17. ISBN-10: 8131805689.

76. Nayantara Padhi. Total Quality Management of Distance Education. The Eight Elements of TQM. ISBN-10: 0415961602.

77. Thareja, P., "Each One is Capable (Part 16 of A Total Quality Organisation Thru' People). FOUNDRY", *Journal For Progressive Metal Casters*, Vol. 20, No. 4, July/Aug, 2008.

78. Thareja, P., Each One is Capable (Part 16 of A Total Quality Organisation Thru' People). Journal: Foundry, July/August 2008, Vol. 20, No. 4, P61-69. ISBN20083325.

79. Breyfogle, F.W., *Implementing Six Sigma: Smarter Solutions Using Statistical Methods*, (John Wiley & Sons Inc., New York, NY, 1999).

80. Breyfogle, F.W. 1999, Implementing Six Sigma: Smarter Solutions Using Statistical Methods, John Wiley & Sons Inc., New York, NY. ISBN-10: 0471265721.

81. Mandelbrot, Benoit; Richard L Hudson (2004). The (mis)behavior of markets: A fractal view of risk, ruin, and reward. New York: Basic Books. p. 153.

82. Vijay K. Mathur, "How Well Do We Know Pareto Optimality?" Journal of Economic Education 22#2 (1991) p. 172–78.

83. The Economics of Vilfredo Pareto, Renato Cirillo, Routledge, 1978.

84. Sales Process Diagram, LEADing Practice Business Process Reference Content #LEAD-ES20005BP.

85. Ishikawa, Kaoru (1985) [First published in Japanese 1981]. What is Total Quality Control? The Japanese Way. Prentice Hall. ISBN 0-13-952433-9.

86. Barlow, R. E. & Irony, T. Z. (1992) "Foundations of statistical quality control" in Ghosh, M. & Pathak, P.K. (eds.) Current Issues in Statistical Inference: Essays in Honor of D. Basu, Hayward, CA: Institute of Mathematical Statistics, p. 99-112.

87. Deming, W.E., "Out of The Crises" MIT Center for Advanced Engineering Study, 1986. ISBN-10: 8176710377.

88. Priyavrat, Thareja. GEMI (Global Environmental Management Aman Mohamed Initiative: Total Quality Environmental Management. The Primer. Washington D.C.: GEMI, 1993.

89. Pande, P. S.; Neuman, Robert P.; Cavanagh, Roland R. (2001). The Six Sigma Way: How GE, Motorola, and Other Top Companies are Honing Their Performance. New York: McGraw-Hill Professional.

［65］ Maureen Weicher, William W. Chu, Wan Ching Lin, Van Le and Dominic Yu, Business Process Reengineering: Analysis and Recommendations.

［66］ Smith, K.K., "Philosophical Problems in Thinking about Organisational Change", in Change in Organisations, (Jossey-Bass, Oxford, 1982).

［67］ Michael Hammer; James Champy, Reengineering the Corporation.

［68］ Kaplan, R.B. and Murdock, L., "Rethinking the Corporation: Core Process Redesign", The McKinsey Quarterly, No. 2, 1991.

［69］ There's no Better Time to Do Business Process Improvement than the Bad Time, Hutex™ Management Consulting.

［70］ Thomas H. Davenport, Laurence Prusak and H. James Wilson "Article from Computerworld: Reengineering revisited: What went wrong with the business-process reengineering fad. And will it come back?"

［71］ LEADing BPM Practice Case Story, Mark von Rosing, Henrik von Scheel, Anette Falk Bøgebjerg, LEADing Practice, 2013.

［72］ T. Priyavrat, (2007),'Teams, Traits And Tasks［QT3］For Total Quality', Quality World, Vol Ⅴ, April 2007.

［73］ Hashmi, K. "Introduction and Implementation of Total Quality management (TQM)".

［74］ Beckhard, R., Pritchard W. Changing the Essence (San Francisco: Jossey-Bass, 1992).

［75］ Kumar, Dr. S., 2006 Total Quality Management, 13−17. ISBN-10: 8131805689.

［76］ Nayantara Padhi. Total Quality Management of Distance Education. The Eight Elements of TQM. ISBN-10: 0415961602.

［77］ Thareja, P., "Each One is Capable (Part 16 of A Total Quality Organisation Thru' People). FOUNDRY", Journal For Progressive Metal Casters, Vol. 20, No. 4, July/Aug, 2008.

［78］ Thareja, P., Each One is Capable (Part 16 of A Total Quality Organisation Thru' People). Journal: Foundry, July/August 2008, Vol. 20, No. 4, P61−69. ISBN20083325.

［79］ Breyfogle, F.W., Implementing Six Sigma: Smarter Solutions Using Statistical Methods, (John Wiley & Sons Inc., New York, NY, 1999).

［80］ Breyfogle, F.W. 1999, Implementing Six Sigma: Smarter Solutions Using Statistical Methods, John Wiley & Sons Inc., New York, NY. ISBN-10: 0471265721.

［81］ Mandelbrot, Benoit; Richard L Hudson (2004). The (mis)behavior of markets: A fractal view of risk, ruin, and reward. New York: Basic Books. p. 153.

［82］ Vijay K. Mathur, "How Well Do We Know Pareto Optimality?" Journal of Economic Education 22#2 (1991) p. 172−78.

［83］ The Economics of Vilfredo Pareto, Renato Cirillo, Routledge, 1978.

［84］ Sales Process Diagram, LEADing Practice Business Process Reference Content #LEAD-ES20005BP.

［85］ Ishikawa, Kaoru (1985)［First published in Japanese 1981］. What is Total Quality Control? The Japanese Way. Prentice Hall. ISBN 0−13−952433−9.

［86］ Barlow, R. E. & Irony, T. Z. (1992) "Foundations of statistical quality control" in Ghosh, M. & Pathak, P.K. (eds.) Current Issues in Statistical Inference: Essays in Honor of D. Basu, Hayward, CA: Institute of Mathematical Statistics, p. 99−112.

［87］ Deming, W.E., "Out of The Crises" MIT Center for Advanced Engineering Study, 1986. ISBN−10: 8176710377.

［88］ Priyavrat, Thareja. GEMI (Global Environmental Management Aman Mohamed Initiative: Total Quality Environmental Management. The Primer. Washington D.C.: GEMI, 1993.

［89］ Pande, P. S.; Neuman, Robert P.; Cavanagh, Roland R. (2001). The Six Sigma Way: How GE, Motorola, and Other Top Companies are Honing Their Performance. New York: McGraw-Hill Professional.

90. Adams, C.W., Gupta, Praveen, *Six Sigma Deployment*. Butterworth-Heinemann, Burlington, MA, 2003.

91. Stamatis, D.H., *Six Sigma Fundamentals: A Complete Guide to the System, Methods, and Tools*, (Productivity Press, New York, 2004) p. 1.

92. Mikel J. Harry, The Vision of Six Sigma: Tools and Methods for Breakthrough, Volume II, Fifth Edition, January1, 1997. ASIN: B003X67OX4.

93. Webber, L., Wallace, M. *Quality Control for Dummies*, December 15, 2006.

94. De Feo, J.A., Barnard, W. Juran, *Institute's Six Sigma Breakthrough and Beyond – Quality Performance Breakthrough Method*, (Tata McGraw-Hill Publishing Company Limited, 2005).

［90］ Adams, C.W., Gupta, Praveen, Six Sigma Deployment. Butterworth-Heinemann, Burlington, MA, 2003.

［91］ Stamatis, D.H., Six Sigma Fundamentals: A Complete Guide to the System, Methods, and Tools, (Productivity Press, New York, 2004) p. 1.

［92］ Mikel J. Harry, The Vision of Six Sigma: Tools and Methods for Breakthrough, Volume Ⅱ, Fifth Edition, January1, 1997. ASIN: B003X67OX4.

［93］ Webber, L., Wallace, M. Quality Control for Dummies, December 15, 2006.

［94］ De Feo, J.A., Barnard, W. Juran, Institute's Six Sigma Breakthrough and Beyond-Quality Performance Breakthrough Method, (Tata McGraw-Hill Publishing Company Limited, 2005).

Phase 4: What Is Business Process Management?

Keith D. Swenson, Mark von Rosing

INTRODUCTION

As explored in an earlier chapter, the term *business process modeling* was coined in the field of systems engineering by S. Williams in his 1967 article entitled, "Business Process Modeling Improves Administrative Control."[1] However, it was not until the 1990s that the term became popular[2] and the term *process* became a new productivity paradigm.[3] Companies were encouraged to think in terms of processes instead of functions and procedures. Today, there are multiple books, white papers, articles, blogs, and even entire conferences on the subject of business process management (BPM). However, many people still struggle to find a precise definition of BPM. As we will show, viewpoints vary wildly, making people unclear whether BPM is a process, technology, or management discipline. The answer depends upon whom you ask. If the question is asked of a technology company, most likely the definition of BPM will be centered more on technology than business. Going into this analysis, we believed that there was a strong bias toward technology within the industry, as indicated by the way that software companies refer to BPM in the context of the capabilities of their particular technology, such as SAP to describe the BPM Netweaver engine or Oracle to describe their BPM collaboration platform. We are greatly concerned by the way that the BPM acronym is used loosely, with its meaning depending upon the context. The lack of a widely accepted definition has arguably had the single most harmful effect on the industry.

DEFINITION AND RESEARCH

During a concerted review and assessment of approximately 100 articles, we found dozens of differing definitions of BPM; what was offered as authoritative meanings varied significantly by publication, which creates a significant problem. Without a common understanding of BPM, one cannot make a conclusive statement about what BPM does or how one might or might not use it. The goal of this research effort is to find the definition that most closely represents the concept that people (expert and otherwise) generally have for the term BPM. To this end, we want to uncover a definition that will resonate with most people in a meaningful and useful way when we say "BPM." With the goal to find a common definition, we will cite well-known definitions of BPM, then provide our analysis of the usefulness in terms of various contexts, such as business, technology, length, etc.

The Complete Business Process Handbook. http://dx.doi.org/10.1016/B978-0-12-799959-3.00004-5

1.4　第4阶段：什么是BPM

Keith D. Swenson, Mark von Rosing

1.4.1　介绍

正如前面一章所探讨的,术语"业务流程建模"是由S. Williams于1967年在系统工程领域一篇题为《业务流程建模改进管理控制》的文章中提出的[1]。然而,直到20世纪90年代,术语"流程"才变得流行[2],并成为了一种新的生产力范式[3]。我们鼓励各公司从流程而不是职能和程序方面进行思考。今天,有许多关于BPM主题的书籍、白皮书报告、文章、博客,甚至以其为主题的会议。然而,许多人仍然很难理解BPM的精确定义,很多观点甚至千差万别,使得人们不清楚BPM是流程、技术还是管理规程。事实上,答案取决于您问谁,如果问的对象是一家技术公司,那么很可能BPM的定义更多地集中在技术上而不是业务上。在本章分析中,正如软件公司在其特定技术(如SAP描述BPM NetWeaver引擎或Oracle描述其BPM协作平台)的功能背景下引用BPM的方式所表明的那样,我们认为业界对技术存在着强烈的偏见。我们认为BPM这个缩略词的使用取决于语境。毫无疑问,缺乏广泛接受的定义可能会对该行业产生最有害的影响。

1.4.2　定义与研究

在对大约100篇文章进行回顾和研究后,我们发现了许多关于BPM的不同定义。此外,BPM的含义因出版物的不同而有很大的差异,这就造成了一个重大的问题,即如果没有对BPM的共同认知,人们就不能对BPM的作用、如何使用或不使用它做出针对性的结论。这项研究工作的目标是得出最能代表人们(专家和其他人)对于术语BPM理解的准确定义。为此,我们希望找到一个在我们提到BPM时能以有意义和有用的方式与大多数人认知达成一致的定义。为了找到这个共同的定义,我们将引用众所周知的BPM定义,然后根据不同的环境背景(如业务、技术和时间的长短等)对其有用性进行分析。

Cambridge Dictionary Online: "(BPM is) the development and control of processes used in a company, department, project, etc. to make sure they are effective."

This definition does not say anything about process efficiency or quality and ignores the idea of process improvement, all of which are necessary parts of BPM. With the focus on development and control, the definition might lead the reader to take on a very technology-centric view that suggests a focus on the automation aspects of processes, and it is the automation that is the focus and goal of BPM. The larger view is that processes exist outside of development, and they are subject to examination and improvement with or without technology.

Rummler and Brache, Improving Performance: How to Manage the White Space in the Organization Chart: "(BPM is the) management of the series of steps that a business executes to produce a product or service."

Management of the work steps is clearly an aspect of process. This definition potentially conflates the automation of the tasks (i.e., the process) with the continual improvement of the process over time. The reader could easily confuse the term "management" as used here with the idea of automation, but BPM is not necessarily or even particularly about automation.

Smith and Fingar, BPM: The Third Wave: "(BPM is) management of the complete and dynamically coordinated set of collaborative and transactional activities that deliver value to customers."

The idea of coordination imparts a sense that control and stewardship are aspects of the management of business process, but it does not talk about improvement. Management of the activities is different from the management of the process. This definition focuses on transactional work, ignoring the idea that work may be transformational or tacit within a process that delivers value. This definition lacks the specificity that would clearly show that process and process management can exist without automation.

Martyn Ould, BPM: A Rigorous Approach: "(BPM is the) management of a coherent set of activities carried out by a collaborating group to achieve a goal."

The management of activities of one instance of a process is different than the management of the flow of activities in a set of processes over time. The goal of BPM is not to successfully complete one process, but to control, steward, and continually improve all processes over time. This facet is not clear in this definition. Again we fear that management, in this setting, could be interpreted as automation.

Marlon Dumas, et al., Fundamentals of BPM: "BPM is the art and science of overseeing how work is performed in an organization to ensure consistent outcomes and to take advantage of improvement opportunities."

This is a very good definition. It is important that a definition of BPM clarify that it takes a process-oriented approach as opposed to a function-oriented approach. That being said, this definition, because it talks only of work and outcomes, could also be seen to apply to improvements aimed at a single step (unit of work) of a process in isolation of the entire process.

Wikipedia article on BPM, captured around November 28, 2013 *(for clarity, the text not directly related to defining BPM has been removed)*: "BPM has been referred

剑桥在线词典对BPM的定义是"（BPM是）被用于并确保公司、部门、项目等有效运转的流程的开发与控制方法。"

这个定义没有提到流程效率或质量，也忽略了流程改进的概念，但所有这些都是BPM的必要部分。随着对发展和控制的关注，该定义可能会诱导读者以技术为中心的视角关注流程，而非流程自动化视角，而自动化也是BPM的重点和目标。更广泛的观点是，流程存在于发展之外，无论有没有技术，它们都要接受检查和改进。

G. A. Rummler和A. P. Brache在《绩效改进：消除管理组织图中的空白地带》中说："（BPM是）企业为生产产品或服务而执行的一系列管理步骤。"

工作步骤的管理明显是流程的一个方面。这个定义可能会将任务（即流程）的自动化与流程持续改进相混淆。读者很容易将这里使用的术语"管理"与自动化的概念混淆，但是BPM不一定，或者说不特指自动化。

H. Smith和P. Fingar在《BPM：第三次浪潮》中认为："（BPM是）管理一组为客户提供价值的完整且动态的协作和活动。"

协作的概念给人一种感觉，即控制和管理是BPM的一个方面，但它并不讨论改进。活动的管理不同于流程的管理，这个定义侧重于事务性工作，但忽略了工作可能是转换性的，或在传递价值的过程中是隐性的。这一定义缺乏明确表明过程和过程管理不需要自动化就可以存在的特异性。

M. Ould的《BPM：一种严格的方法》中写道："（BPM是）协作小组为实现目标而执行的一组连贯的管理活动。"

一个流程实例中的活动管理不同于一组流程中随时间变化的活动流管理。BPM的目标不是成功地完成一个流程，而是控制、管理和不断地改进所有流程。这个方面在这个定义中不清楚。在这种情况下，我们再次担心管理知识被解释为自动化。

M. Dumas（马隆·达马斯）等在《BPM基础知识》中说："BPM是一门用于监督组织如何执行工作，以确保结果一致性并利用时机不断改进的艺术和科学。"

这是一个很好的定义。重要的是，该定义阐明了BPM应该采用面向流程的方法，而不是面向功能。尽管如此，这一定义由于只谈论了工作和成果，但也可被视为适用于整个流程中的一个步骤（工作单位）的改进。

截取维基百科2013年11月28日左右关于BPM的文章的部分（为了清晰起见，与定义BPM没有直接关系的文字已经删除）："BPM被称为一种整体管理方

to as a 'holistic management' approach to aligning an organization's business processes with the wants and needs of clients. BPM uses a systematic approach in an attempt to continuously improve business effectiveness and efficiency while striving for innovation, flexibility, and integration with technology. It can therefore be described as a 'process optimization process.' ... As a managerial approach, BPM sees processes as strategic assets of an organization that must be understood, managed, and improved to deliver value-added products and services to clients."

This is poorly written and clearly far too long to be a definition. The gratuitous use of "holistic" does not help in understanding, nor do the many uses of vague, conditional phrases (e.g., "has been," "could be," "attempt"). More to the point, there is no requirement for a reference to technology in the definition and an overemphasis on other areas.

IBM: "BPM is a discipline that leverages software and services to provide total visibility into an organization. Discover, document, automate, and continuously improve business processes to increase efficiency and reduce costs."

This narrowly defines BPM as software and services, conveniently matching the sort of thing that IBM can supply. There is wide agreement in the field that BPM is a management discipline, which could be done on paper and pencil if necessary; software is not a necessary ingredient.

Association for Information and Image Management: "BPM is a way of looking at and then controlling the processes that are present in an organization. It is an effective methodology to use in times of crisis to make certain that the processes are efficient and effective, as this will result in a better and more cost efficient organization."

This is a fairly reasonable definition, but positioning BPM as something to be used in a crisis is not consistent with what most people see as being an essential aspect.

BPM Institute, "What is BPM Anyway? BPM Explained": "(BPM is a) process of managing your business processes; a management discipline; a technology or set of technologies; a rapid application development framework. First and foremost, BPM is a process and a management discipline."

Here, BPM is defined as just about anything you want. Again, there is confusion with the technology perspective. BPM is certainly not a rapid application framework. This definition simply does not help people understand what BPM is.

Paul Harmon, 2005: "(BPM is) a management discipline focused on improving corporate performance by managing a company's business processes."

This is very good for a short definition. However, it is not clear about whether BPM is the management of the tasks within a single process (e.g., automation) or the means of how processes are repeatable, modified, and improved over time. It is important to say a few more words to make it clear that BPM is the latter and to emphasize the need to care for processes as part of the going concern of the enterprise.

Gartner: "BPM is the discipline of managing processes (rather than tasks) as the means for improving business performance outcomes and operational agility. Processes span organizational boundaries, linking together people, information

法,用于使组织的业务流程与客户的需求相一致。BPM使用一种系统化的方法,试图不断提高业务效益和效率,同时努力实现创新、灵活性和与技术的集成。因此,它可以被描述为'流程优化流程'。作为管理方法,BPM将流程视为组织的战略资产,必须对流程进行深入了解、管理和改进,才能向客户提供增值产品和服务。"

这个定义显然写得太长,无法作为一个定义。无端使用"整体"这个词语很难理解,也没有使用模糊的、有条件的短语(如已经、可能、尝试),更重要的是该定义过分强调了其他方面,而没有提及技术。

IBM给出的定义是:"BPM是一门利用软件和服务提供组织全面可见性的学科。通过发现、记录、自动化和持续改进业务流程,可以提高效率并降低成本。"

这狭隘地将BPM定义为软件和服务,旨在方便匹配IBM可以提供的产品和服务。事实上,大家有一个广泛的共识,即BPM是一门必要时可以用纸和笔来完成的管理学科,软件并不是一个必要的组成部分。

信息和图像管理协会认为:"BPM是一种查看和控制组织中存在的流程的方法。这是在危机时期使用的一种确保流程的效益性和有效性的方法,因为这将形成一个更好、更具成本效益的组织。"

这是一个相当合理的定义,但是将BPM定位为在危机中使用的东西,与大多数人认为的一个基本方面不一致。

BPM研究所认为:"什么是BPM？ BPM是一个管理业务流程的过程、一个管理规程、一种或一组技术、一个快速的应用程序开发框架。首先,BPM是一个流程和管理规程。"

在这里,BPM被定义为您想要的任何东西。同样,该定义也存在技术混淆观点,BPM当然不是一个快速的应用程序框架。所以,这个定义并不能帮助人们理解什么是BPM。

P. Harmon在2005年也给出了定义,他认为:"(BPM是)一种管理原则,专注于通过管理公司的业务流程来提高公司绩效。"

这是一个很好的简短定义。然而,目前还不清楚BPM是管理单个流程中的任务(如自动化),还是如何随着时间的推移对流程进行可重复、修改和改进的方法。重要的是多说几句话,以明确BPM是后者,并强调流程是企业持续经营需求的一部分。

高德纳(Gartner)公司认为:"BPM是管理流程(而不是任务)的准则,它是提高业务绩效成果和运营敏捷性的手段。流程跨越组织边界,将人员、信息流、系统

flows, systems and other assets to create and deliver value to customers and constituents."

This definition is not bad, although a little bit wordy. It does not fully capture the idea of stewardship or control of the process in the steady state.

Hammer, cited by Tim Weilkiens in Object Management Group's Certified Expert in BPM: "(BPM is a) structured approach to performance improvement that centers on the disciplined design and careful execution of company end to end processes."

This definition is exclusively about design and execution can easily be confused with automation of processes.

CIO Magazine: "BPM is a systematic approach to improving a company's business processes."

This definition needs to mention a process-oriented approach as opposed to a functional-oriented approach. It speaks solely to improvement, thus ignoring the idea of controlling and making decisions about the operating processes.

BusinessDictionary.com: "(BPM is an) activity undertaken by businesses to identify, evaluate, and improve business processes. With the advancement of technology, BPM can now be effectively managed with software that is customized based on the metrics and policies specified by a company. This type of action is essential to businesses seeking to improve process performance related issues so that they can better serve their clients."

This definition is not bad, although a bit wordy. The focus on technology is somewhat misplaced.

Techopedia: "BPM is a concept that focuses on aligning all organizational elements to improve operational performance. The BPM strategy is categorized with holistic management approaches which are used to develop better business efficiency, while channeling organizations toward more creative, flexible and technologically-integrated systems."

It is unclear what is meant by "aligning all organizational elements," "channeling organizations," or "holistic management." This is missing the aspect that BPM by nature is process oriented, and that performance must be measured across the entire process, as opposed to only a single step of the process. There is a problem using the term management as a way to describe what happens within the nature of BPM, so it would be helpful if this point of clarification was made.

Popkin Software, BPMN and BPM: "BPM is concerned with managing change to improve business processes."

Although this definition is nice and short, the use of business process within the definition makes it circular. The definition needs more explanation about how this is accomplished. It ignores the idea of controlling the process on an ongoing basis.

IBM Redbook, BPM Enabled by SOA: "BPM is most often associated with the life cycle of a business process. The process life cycle spans identifying and improving processes that deliver business capability to deploy and manage the process when it is operational. What is often forgotten is managing process performance after a process is operational."

和其他资产连接在一起,从而为客户和参与者创造和交付价值。"

这个定义不错,尽管有点冗长。它的缺点是并没有完全捕捉到在稳定状态下管理或控制流程的思想。

M. Hammer引用了T. Weilkiens作为OMG的认证BPM专家的看法:"(BPM是一种)结构化的性能改进方法,其核心是公司端到端流程的严格设计和精心执行。"

这个定义是专门关于设计和执行的,很容易与流程自动化混淆。

《首席信息官》杂志认为:"BPM是改善公司业务流程的系统方法。"

这个定义需要提到面向流程的方法,而不是面向功能的方法。它只提到了改进,因此忽略了控制和决策操作流程的方面。

BusinessDictionary.com认为:"(BPM是一种)企业为识别、评估和改进业务流程而进行的活动。随着技术的进步,公司可以通过基于指定的度量和策略定制的软件有效地管理BPM。这对于寻求改善流程性能等相关问题的企业至关重要,以便他们能够更好地为客户服务。"

这个定义虽然有点冗长,但也还可以,不过对技术的关注有些错位。

Techopedia公司认为:"BPM是一个专注于协调所有组织要素以提高运营绩效的概念。BPM策略是一个用于提高业务效率的整体管理方法,它是用于引导组织走向更具创造性、更灵活和技术集成的系统。"

目前还不清楚协调所有组织要素、"引导组织"或整体管理的含义。这忽略了BPM本质上是面向流程的这一方面,而且性能必须跨整个流程进行度量,而不是仅对流程的单个步骤进行度量。使用术语管理作为描述BPM性质内发生的事情的方法也存在问题,因此如果对这一点做出澄清将会对我们理解BPM有所帮助。

Popkin软件、BPMN和BPM认为:"BPM关注的是通过管理变革以改进业务流程。"

虽然这个定义很好而且简短,但是在该定义中涉及了业务流程的循环,因此,这个定义需要更多关于循环如何实现的解释。显然,它忽略了持续控制流程的思想。

IBM红皮书中面向服务的架构(SOA)支持的BPM认为:"BPM通常与业务流程的生命周期相关联。流程生命周期涵盖识别和流程改进,其目的是使业务能力能够在流程运行时进行持续的部署和优化。一个流程运行后,人们总是忘记持续优化流程的性能。"

The mention of "deploying a process" might be construed as a technological view centered on process automation, as might be the phrase "process is operational." There is no idea of control, or oversight, care for processes as an asset of value.

PC Magazine: "(BPM is a) structured approach that models an enterprise's human and machine tasks and the interactions between them as processes."

This is a very technological view, particularly the mention of "machine tasks". BPM might or might not use modeling, but this definition implies that BPM is the modeling of the process, and not the work that people do to steward or improve a process.

Business Analysis Body of Knowledge: "BPM covers a set of approaches that focus on how the organization performs work to deliver value across multiple functional areas to a customer. BPM aims for a view of value delivery that spans the entire organization and views the organization through a process-centric lens."

Although this definition avoids the technology bias, there is nothing that speaks to the nature of "management." It looks to be an analytic method rather than a management discipline.

INSIGHTS GAINED

We have discussed at length what many authorities see as representing BPM and observed what we believe to be the key shortfall of each case. In this section, we elaborate on what was learned about the nature of process to synthesize the best features of what we observed and to address the limitations we noted. The specific desire is to understand what works in terms of good patterns and to reflect on what does not work, thereby exposing useful antipatterns.

BPM Is Not About Automation of Processes

The implication is that BPM is not about automating business processes (in the "paving the cowpaths" meaning) but about improving them. It presumes that you view business as a set of processes, and BPM is the act of improving those processes. It is important to note that "skill" is different from "skill improvement." This can be confusing. For example, in competitive situations the two ideas are often intertwined: What is the act of playing tennis, if not also the act of trying to improve the way you play tennis? However, in other contexts, it is easier to distinguish: The activity of driving is different than taking a driving course to improve the way you drive. In the same way, reengineering a process is not simply about automating what is currently there. Some will say that automation by itself is an improvement over a manual process. BPM is the activity of discovering and designing the automated process, and it is done when the finished application is deployed to the organization. The running of the processes is not part of BPM. However, monitoring the process to find areas of improvement would still be an important part of BPM.

部署流程一词被认为是一种以流程自动化为中心的技术视角，或者"流程是可操作的"。事实上，这个概念没有把流程作为有价值的资产加以控制或监督。

《PC》杂志认为："（BPM是）一种结构化的方法，它将企业的人工、机器任务以及它们之间的交互作为流程进行建模。"

这是一个非常有技术性的观点，特别是提到了机器任务。BPM可能使用建模，也可能不使用建模，但这个定义意味着BPM是流程的建模，而不是人们为管理或改进流程所做的工作。

"业务分析知识体系"（*business analysis body of knowledge*）认为："BPM涵盖了一系列方法，这些方法专注于组织如何执行工作，从而在多个功能领域为客户提供价值。BPM的目标是一个价值交付视图，该视图跨越整个组织，并通过以流程为中心的视角来查看组织。"

虽然这一定义避免了技术偏差，但没有说明管理的本质。这个定义看起来是一种分析方法，而不是管理规程。

1.4.3　获得的见解

我们已经详细讨论了许多主体对BPM的定义，并观察了我们认为的每个定义的关键不足之处。在这一节中，我们详细介绍我们所了解的流程的性质，以综合得出我们所观察到的最佳特性，并解决我们注意到的局限性。具体的要求是在良好的模式下明白什么有用，反思什么无用，从而趋利避害。

1. BPM不是关于流程的自动化

这意味着BPM是关于改进业务流程，而不只是关于自动化。它假定您将业务视为一组流程，而BPM正是改进这些流程的行为。重要的是，技能不同于技能改进，这可能会令人困惑。例如，在竞争环境中，这两个想法经常交织在一起：打网球的动作是什么？如何努力改善您打网球的方式？如果这个例子不好理解，那么，用另外一个例子更容易区分，例如：日常的驾驶活动与通过参加驾驶课程以改善驾驶方式并不一样。同样，重新设计流程并不仅仅是将当前的流程自动化。有人会说，自动化本身就是对手动流程的改进。但BPM是发现和设计自动化流程的活动，它是将完成的应用程序部署到组织内部。尽管流程的运行不是BPM的一部分，但是，监控流程以找到改进方面仍然是BPM的重要组成部分。

BPM Is Done by People Concerned Primarily with Improvement of the Process

A business process will involve many people, but how many of them are concerned with improving it? Some will insist that improvement is everyone's job. That is, the receptionist should be thinking about how to improve the operations if possible. This interpretation is too broad to be useful. The in-house cook adding a new spice to a menu item—making it taste better, motivating more employees to eat in the building, cutting down on wasted time driving to an outside restaurant, and improving the amount of information interactions between workers, thus resulting in better performance—is not BPM by any account. Everybody in a business is working to do their best job, and every good job helps the business, but all of this is not BPM. BPM must be narrowly defined as the activity done by people who actively and primarily look specifically at the business processes and try to improve them. Clearly, those people must solicit input from as many others as possible, but those others are not doing BPM.

Misrepresentations of BPM

In this section, we suggest a variety of different ways that people abuse the BPM term and offer our thoughts on why the point is problematic.

BPM Is Not a Product

There is a product category called BPMS, which is a BPM suite or BPM system. Gartner has introduced a new product category called "intelligent BPMS." What is included depends very much on the vendor. Analysts have attempted to list features and capabilities that are necessary, but those features change from year to year. For example, in 2007, the suggestion was that BPM suites must have a Business Process Execution Language (BPEL) execution capability, but today this is entirely ignored or forgotten.

BPM Is Not a Market Segment

Again, there might be a market segment around products that support BPM or BPMS products, but BPM itself is a practice. Vendors may be labeled as a "BPMS vendor", which simply means they have some products that can support the activity of BPM, among other things.

An Application Does Not Do BPM

The application might be the result of BPM activity. Once finished, it either controls the business process or supports people engaged in doing the business process. It may, as a byproduct, capture metrics that help further improvement of the process.

2. BPM由主要关注流程改进的人员完成

一个业务流程会涉及许多人，但是有多少人关心改进它？有些人会坚持认为改进是每个人的工作，也就是说，如果可能，接待员都应该考虑如何改进操作。但这种解释太过宽泛、无用。例如，内部厨师在菜单项上添加了一种新的香料，使其味道更好，激励更多的员工在大楼里吃饭，减少开车到外面的餐厅所浪费的时间，并提高员工之间的信息交互量，从而导致更好的表现，这无论如何也不能称之为BPM。事实上，企业中的每个人都在努力做好他们最好的工作，每一份好的工作都有助于企业，但这一切都不是BPM。BPM必须被狭义地定义为"那些积极且主要关注业务流程并试图改进它们的人所做的活动"。显然，这些人必须尽可能多地征求没有做BPM的其他人的意见。

3. 对BPM的误解

在本节中，我们提出了人们滥用BPM术语的各种不同的方式，并提供了我们对问题原因的分析。

4. BPM不是一个产品

有一个产品类别称为BPMS，它可以被认为是BPM套件或BPM系统。Gartner公司推出了一个新的产品类别，叫做"智能BPMS"，其中包含的内容在很大程度上取决于供应商。分析人员试图列出必要的特性和功能，但这些特性会逐年变化。例如，在2007年的建议是BPM套件必须具有业务流程执行语言（BPEL）执行功能，但现在这一点完全被忽略或遗忘了。

5. BPM不是一个细分市场

同样，围绕支持BPM或BPMS产品的产品可能有一个细分市场，但BPM本身就是一种实践。供应商可能被贴上"BPMS供应商"的标签，这意味着他们只是拥有一些可以支持BPM活动的产品。

6. 应用程序不执行BPM

应用程序可能是BPM活动的结果。完成后，它要么控制业务流程，要么支持从事业务流程的人员。应用程序可能作为一个副产品，有助于进一步改进流程的指标。

In this sense, an application supports BPM in the same way that a receptionist may support BPM by coming up with good ideas; that is not enough to say that the application, or the receptionist, is "doing" BPM.

BPM As a Service Is Not Application Hosting

We use the term business process as a service to mean applications hosted outside the company that supports more than one function of a business process. Like the application above, it does the process, but it does not do BPM.

Entire Organizational Units Do Not Do BPM

To say that a company is doing BPM is simply a way of saying that there are some people in the company who are doing BPM. This kind of abstraction is normal. It should be obvious that when a company or division claims to be doing BPM, the majority of the people there are not actually doing BPM.

BPM Is Not Merely Anything that Improves Business

Some argue that every activity is part of a process, because a process is just a set of activities. Then, any action taken to improve any activity is BPM. We have argued against this interpretation because such a broad interpretation would make BPM meaningless: It would mean anything. There is broad acceptance that BPM is a practice of methodically improving a process that supports business, and that improvements in part of the process must be done only after the consideration of the entire end-to-end process.

BPM Is Not All Activities Supported by a BPMS

As mentioned earlier, a BPMS supports many things (e.g., application development) that are not BPM. A BPMS that only supported the exact activity of BPM would not be as useful as one that brought a lot of capabilities together. It is, however, a common mistake for people to say that because a BPMS supports something, it is then an aspect of BPM. While it is true that someone who does BPM needs to document a process, it is not true that anyone who documents a process is doing BPM. Also, while it is true that many BPMS support designing a screen form, it is not true that designing a screen form is BPM. The activity of BPM is fairly well defined, but a BPMS supports a much wider set of activities.

Just Because You Can Do Something with a BPMS Does Not Mean It Is BPM

A BPMS is designed to support the activity of BPM. However, there are many things a BPMS can do that are not BPM.

从这个意义上讲,应用程序支持BPM的方式与前台人员提出好的想法可能支持BPM的方式相同,但这不足以说明应用程序或前台人员正在执行BPM。

7. 作为服务的BPM不是应用程序的托管

我们使用术语"业务流程作为服务"意味着它托管在公司外部、支持业务流程中的多个功能。像上面的应用程序一样,它执行流程但不执行BPM。

8. 整个组织单位不执行BPM

一个公司在做BPM只是说公司里有一些人在做BPM,这种理解是正常的。很明显,当一个公司或部门声称正在做BPM时,大多数人实际上并没有做BPM。

9. BPM不仅仅是改善业务

因为一个流程只是一组活动,所以一些人认为每个活动都是一个流程的一部分,然后,为改进任何活动而采取的任何行动都是BPM。我们反对这种解释,因为如此泛泛的解释将使BPM毫无意义(它将意味着任何事情)。人们普遍认为,BPM是一种有条理地改进支持业务流程的实践,并且只有在考虑了整个端到端的流程之后,才能对流程的一部分进行改进。

10. BPM不是BPMS支持的所有活动

如前所述,BPMS支持许多不是BPM的东西(如应用程序开发)。一个只支持BPM的BPMS的功能远远比不上将很多功能结合在一起的BPM。然而,人们常犯的错误是:总是认为BPMS支持某些东西,所以是BPM的一部分。虽然执行BPM的人需要记录流程是真的,但记录流程的人执行BPM并不是真的。此外,虽然BPM人员确实需要为流程编制文档,但为流程编制文档的人员并不一定在执行BPM。另外,虽然许多BPM支持屏幕表单设计,但设计屏幕表单不是BPM。BPM的活动定义得相当好,但是BPM支持更广泛的活动集。

11. 仅仅因为可以使用BPMS做些什么并不意味着它就是BPM

BPMS设计用于支持BPM的活动。然而,BPMS可以做很多不是BPM的事情。

Participating in a Process Is Not Doing BPM

A manager approving a purchase order is not doing BPM, even though that approval is an activity in a process. A bank manager rejecting a loan application is not doing BPM, even though this activity is a step in a business process. These people are doing jobs that are part of a process, but they are not doing BPM.

Implementation (Coding) of the Process Application Is Not BPM

An application developer designing a form for data entry as a step in a process is not doing BPM at that moment. Once the "to-be" process has been adequately spelled out, the actual implementation of the application that supports it is no longer actively engaged in improving the process. A small caution should be noted here: Applications are often developed incrementally—show the customer, get feedback, improve, and iterate—and the process may be improved incrementally as well. Those incremental improvements should be included as the activity of BPM, but the activity of implementation of the application is not BPM. The criteria are clear: If you are actively and primarily engaged in improvement of the process, then it is BPM; otherwise, it is engineering.

Making a Suggestion for Process Improvement Is Not BPM

There is a distinction between many people who make suggestions and those who then actually do the BPM. When a process analyst is involved in BPM, it is expected that they will solicit lots of information about what is and is not working, as well as suggestions on how it might work. Those people who give the feedback are helping the BPM work but are not themselves doing BPM.

Improving a Single Step of a Process Is Not BPM

Some people have the mistaken idea that any possible action that improves a process is BPM, no matter how small. A person doing BPM needs to have some kind of big-picture view of the process; it has been described as an "end-to-end" view of the process. Optimizing one step in a process, without knowledge of the entire process, is exactly what Hammer and Champy were warning about: To understand the correct optimizations, we need to consider those optimizations within the context of a complete business process. A workman smoothing gravel on a road is improving all of the process that involves driving on that road, but it is not BPM because he does not have visibility of the whole process. The engineer finding a way to double the bandwidth of a fiber-optic cable is improving all the processes that require communications, but this is not BPM either. An office worker who finds that OpenOffice 4 helps to create documents faster than some other word processor is improving all the processes that involve writing documents; this is not BPM either. To have a discussion about BPM, we can consider only those activities by people who have a view to, and consider the effect on, the entire end-to-end process.

12. 参与流程并不是在做BPM

批准采购订单的经理不是在执行BPM,即使该批准属于流程中的活动;拒绝贷款申请的银行经理不是在执行BPM,即使此活动是业务流程中的一个步骤。这些人做的工作是流程的一部分,但他们不是在做BPM。

13. 流程应用程序的实现(编程)不是BPM

应用程序开发人员为数据输入设计表单仅仅是流程中的一个步骤,此时还没有执行BPM。一旦将来流程得到了充分的说明,支持它的应用程序的实际实现就不再积极地改进流程。这里应该注意一点:应用程序通常是以增量方式开发的——向客户展示、获取反馈、改进和迭代,并且流程也可能以增量方式改进。这些增量改进应该作为BPM的活动包括在内,但是应用程序的实现活动不是BPM。标准很明确:如果您积极并且主要致力于流程的改进,那么它就是BPM;否则,它只是一个工程。

14. 提出流程改进建议不是BPM

提出建议的人和实际执行BPM的人之间存在区别。当流程分析师参与到BPM中时,人们期望他们会收集大量关于什么是有效的和无效的信息,以及关于它如何工作的建议。所以,那些提供反馈的人属于帮助BPM运行,但他们自己却不属于做BPM的行列。

15. 改进流程的单个步骤不是BPM

有些人错误地认为,任何可能改善流程的行为都是BPM,不管它有多小。从事BPM的人需要对流程有某种宏观的认识,这种认识被描述为流程的端到端视图。在不了解整个流程的情况下优化流程中的某个步骤正是M. Hammer和J. A. Champy所警告的:我们只有在完整的业务流程环境中做的优化才是正确的优化。举个例子来说明:如果一个在道路上磨平碎石的工人改善道路上涉及所有行驶的流程,这不是BPM,因为他无法看到整个流程。再如,工程师找到了一种使光缆带宽加倍的方法,并正在改进所有需要通信的流程,这也不是BPM。一个办公室工作人员发现OpenOffice4帮助创建文档的速度比其他一些文字处理器快,于是他改进所有涉及文档编写的流程,这也不是BPM。要讨论BPM,我们只能考虑那些对整个端到端流程有看法并考虑其影响的人的活动。

CONCLUSION: ONE COMMON DEFINITION

From the above analysis and comparison of definitions, as well as a long research discussion lead by Keith Swenson, among others, on the LinkedIn BPM Guru group, bpm.com forum, Association of Business Process Management Professionals forum, and other places, brings us to propose this definition:

> Business process management (BPM) is a discipline involving any combination of modeling, automation, execution, control, measurement, and optimization of business activity flows in applicable combination to support enterprise goals, spanning organizational and system boundaries, and involving employees, customers, and partners within and beyond the enterprise boundaries.

This definition is designed to be short enough to use regularly in both business and technology context without gratuitous words. However, there is a trade-off: a longer definition might be clearer, perhaps at the cost of being more cumbersome. Here is clarification of what we mean by these words:

- BPM is a discipline: It is a practice; it is something you do. Predominant in the definitions is the idea that BPM is something you do, not a thing you own or buy. It is described in many definitions as a practice. There was wide agreement on this: Well over 90% of the participants expressed this view.
- Business stems from the state of being busy, and it implies commercially viable and profitable work. A business exists to provide value to customers in exchange for something else of value.
- Process means a flow of business activities and seeing those activities as connected toward the achievement of some business transaction. Flow is meant loosely here: The order may or may not be strictly defined.
- A person doing BPM must consider a process at the scope of interrelated business activities that holistically cooperate to fulfill a business objective. This is the key difference from a functional view of business, where the production of the different classes of output might be optimized independent of the other functions. In a complex system, such as a modern business, it is well known that local optimization of part of the system will rarely lead to good overall results. A BPM practitioner must consider the metrics of the entire system when evaluating a specific process.
- Modeling means that they would identify, define, and make a representation of the complete process to support communication about the process for the purpose of creating a shared understanding and to provide the tools to exercise control. There is no single standard way to model, but the model must encompass the process.
- Automation refers to the work that is done in advance to assure the smooth execution of the process instances. In many cases, this means writing software, but it might include building machinery or even creating signage to direct participants in the process.
- Execution means that instances of a process are performed or enacted, which may include automated aspects. Conceptually, the process instance executes

1.4.4　结论：一个共同定义

通过以上对定义的分析和比较，以及 K. D. Swenson 等在 LinkedIn BPM 专家组、bpm.com 论坛、BPM 专业人员协会论坛等地开展的一系列研究讨论，我们提出了这一定义：

BPM 是一门涉及建模、自动化、执行、控制、测量和优化业务活动流的任意组合的学科，这些组合适用于支持企业目标、跨越组织和系统边界，并涉及企业边界内外的员工、客户和合作伙伴。

这个定义设计得足够简短，可以在业务和技术环境中定期使用，而不必使用无谓的词语。然而，这是一种权衡：更长的定义可能更清晰，代价也许是更麻烦。以下是我们所说的这些词的含义。

- BPM 是一门学科：它是一种实践，它是您所做的事情。在定义中占主导地位的是这样一种观点，即 BPM 是您所做的事情，而不是您所拥有或购买的东西。它在许多定义中被描述为一种实践，90% 以上的参与者认同这一观点对此达成了广泛的共识；
- 业务源于忙碌的状态，这意味着在商业上可行并且能促进盈利。企业的存在是为了向客户提供价值，以换取其他有价值的东西；
- 流程是指业务活动的流程，并将这些活动视为与实现某些业务交易相关联的活动。"流"在这里是松散的：顺序可以严格定义，也可以不严格定义；
- 从事 BPM 的人员必须在相互关联的业务活动范围内考虑流程，这些业务活动整体协作以实现业务目标，这是与功能性业务视图的关键区别。在功能性业务视图中，不同类型产生的输出可能独立于其他功能。众所周知，在一个诸如现代企业的复杂系统中，对系统的局部优化很少会带来良好的总体结果。BPM 从业者在评估特定流程时必须考虑整个系统的指标；
- 建模：意味着它们将标识、定义并表示整个流程，以支持关于流程的交流，从而创建一致的理解，并提供用于实施控制的工具。建模没有单一的标准化的方法，但是流程必须包含在模型之内；
- 自动化：是指为确保流程实例的顺利执行而提前完成的工作。在许多情况下，这意味着需要进行编程，但也可能包括构建一套系统，甚至创建标识体系来指导流程中的参与者；
- 执行：意味着执行或实施流程的实例，其中可能包括自动化方面。从概念上讲，流程实例按照 BPM 执行者的模型实现独立于 BPM 执行者之外的自

itself, following the BPM practitioner's model, but unfolding independent of the BPM practitioner.

- Control means that the there is some aspect of making sure that the process follows the designed course. This can be strict control and enforcement, or it might be loose control in the form of guidelines, training, and manual practices.
- Measurement means that effort is taken to quantitatively determine how well the process is working in terms of serving the needs of customers.
- Optimization means that the discipline of BPM is an ongoing activity, which builds over time to steadily improve the measures of the process. Improvement is relative to the goals of the organization and ultimately in terms of meeting the needs of customers.
- Enterprise is used here simply to mean a business organization or organizations where people are working together to meet common goals; it does not need to be exceptionally large, and it does not need to be for profit.
- The mention of enterprise goals is included here to emphasize that BPM should be done in the context of the goals of the enterprise, and not some small part of it. This might seem a bit redundant in one sense: Any improvement of a process must be an improvement in terms of the enterprise goals; anything else would not be called an improvement.
- Within and beyond the enterprise boundaries recognizes that the enterprise is part of a larger system. Customers are part of the business process. Their interaction, along with those of employees, should be considered as part of the end-to-end interaction.

Our goal was to find a common definition. Our request is that you endorse this definition of BPM, so that there exists a common definition that we can use as a basis of creating shared understanding and advancing the discipline. We believe that a single common definition is of critical value to the entire marketplace; this wish started our research and analysis and led to the definition. In the aim and objective to isolate the single prominent definition, we did not wish to single out definitions. However, we realize that with such a publication, there will probably be people who will say, "I would prefer if the definition had X, or didn't have Y."

Everyone has an opinion. The question we asked ourselves—and we think that you should ask yourself—is this: Does this common definition cover the concepts that are inherent in the idea of BPM? To that end, our work is built on research and analysis, and single opinions were avoided to unite and build a common understanding of the subject.

End Notes

1. Williams S., "Business Process Modeling Improves Administrative Control," in *Automation* (December 1967), 44–50.
2. Michael Hammer, "Reengineering Work: Don't Automate, Obliterate," *Hardvard Business Review* (July 1990).
3. Asbjørn Rolstadås, "Business Process Modeling and Reengineering," in *Performance Management: A Business Process Benchmarking Approach* (1995), 148–150.

动执行；

- 控制：意味着确保流程在某些方面要遵循流程设计。这可以是严格的控制和执行，也可以是以指导方针、培训和手动实践的形式进行的松散控制；
- 度量：意味着要定量地确定流程在满足客户需求方面的效果情况；
- 优化：意味着BPM的规则是一个持续的活动，随着时间的推移它逐渐建立，以稳定地改进流程的度量。优化、改进与组织的目标有关，最终是为了满足客户的需求；
- 企业：在这里仅仅是指一个或多个企业组织，人们一起工作以实现共同目标，它不需要特别大，也不需要为了利润；
- 这里提到企业目标是为了强调BPM应该在企业目标的背景下进行，而不是在其中的一小部分。从某种意义上说，这似乎有点多余：流程的任何改进都必须是企业目标方面的改进，除此之外，其他任何事情都不能称为改进；
- 企业内部和外部都认识到企业是更大系统的一部分，客户是业务流程的一部分，他们的交互，以及员工的交互，应该被视为端到端交互的一部分。

我们的目标是找到一个共同的定义。我们的要求是，您认可BPM这个定义，这样就有了一个共同的定义，我们可以将其用作创建共享理解和推进规范的基础。我们相信，单一的共同定义对整个市场具有关键价值，这一愿望使我们开始了研究和分析，并得出了定义。我们的目的和目标是得出一个共同的定义，而不是逐一列出。然而，我们意识到，有了这样的出版物，可能会有人说："如果定义有X，或者没有Y，我更喜欢。"

每个人都有自己的观点。我们问自己的问题是：这个共同的定义是否涵盖了BPM理念中根本的思想？为此，我们的工作以研究和分析为基础，避免了单一的观点，以统一和建立对这一主题的共同理解。

参考文献

[1] Williams S., "Business Process Modeling Improves Administrative Control," in Automation (December 1967), 44–50.

[2] Michael Hammer, "Reengineering Work: Don't Automate, Obliterate," Hardvard Business Review (July 1990).

[3] Asbjørn Rolstadås, "Business Process Modeling and Reengineering," in Performance Management: A Business Process Benchmarking Approach (1995), 148–150.

The BPM Way of Thinking

Mark von Rosing, Henrik von Scheel, August-Wilhelm Scheer

INTRODUCTION

In Part II, we introduce the "way of thinking" around business process concepts. We focus on the value of an ontology and the business process management (BPM) ontology, which is the essential starting point that creates the guiding principles. We also provide structural concepts around strategic definitions, such as wants, needs, direction, issues, and problems.

Here, we exploring how BPM ontology can be applied within the areas of process modelling, process engineering, and process architecture. It provides the fundamental process concepts that can be used to document corporate knowledge and structure process knowledge by defining relation process concepts (e.g., the order of process steps).

We feel this way of thinking enables the right abstraction level and allows an understanding of the underlying thoughts, views, visions, and perspectives.

Today, many BPM and or process frameworks, methods and or approaches, such as LEAN, Six Sigma, BPR, TQM, Zero Defect, BPMN, and BPMS, have their own vocabularies. Each of these vocabularies has its own definition of terms, such as business process, process step, process activity, events, process role, process owner, process measure, and process rule.

Ontology is an essential discipline that can support understanding and structuring of BPM knowledge, create and define the fundamental concepts, and provide semantic relations and correlations between these concepts. Such definitions are not incorporated in contemporary BPM practice in an integrated and standardized way. Therefore, we present an ontology for BPM. Our focus will be on the value of an ontology and the BPM ontology itself.

This discussion builds the foundation for the rest of the book, as the BPM ontology includes our shared vocabulary (i.e., folksonomy) that structures knowledge in two ways. First, it allows practitioners to structure their business knowledge by adding meaningful relationships between the vocabulary terms. Second, it organizes concepts in hierarchic "is-a" relationships, which allow for a polymorphic inheritance of properties.

The Complete Business Process Handbook. http://dx.doi.org/10.1016/B978-0-12-799959-3.00005-7

第二部分

2.1 BPM的思维方式

Mark von Rosing, Henrik von Scheel, August-Wilhelm Scheer

介绍

在这一部分中,我们将围绕业务流程的概念介绍思维方式。我们关注本体和BPM本体的价值,基于价值来创建指导原则。我们还剖析围绕战略定义的结构概念,如目标、需求、方向、难点和问题。

在这里,我们将探讨如何将BPM本体应用于流程建模、流程工程和流程体系结构。并在本节提供通过定义流程概念间的关系(如流程步骤的顺序)来记录企业知识和构建流程知识的流程基本概念。

我们认为通过这种思维方式可以实现恰到好处的抽象概括,来帮助我们理解潜在的思想、见解、愿景和观点。

如今,许多BPM或流程框架、方式或方法都有自己的词汇表,如精益管理、六西格玛、BPR、TQM、零缺陷、BPMN和BPMS。每个词汇表都有自己的术语定义,如业务流程、流程步骤、流程活动、事件、流程角色、流程责任人、流程度量和流程规则。

本体论是一门重要的学科,它可以支持对BPM知识的理解和构造,创建和定义基本概念,并提供这些概念之间的语义关系和相关性。这些定义并没有以一种集成和标准化的方式被纳入到当代的BPM实践中。因此,我们为BPM提供一个本体,我们的重点将是本体的价值和BPM本体本身。

这一讨论为本书的其余部分奠定基础,因为BPM本体包括了这本书的共享词汇表(即大众分类),这个词汇表通过以下两种方式构造知识:其一是它允许从业者基于词汇表中的术语,添加有意义的关系来构造他们的业务知识;其二是它通过有层次结构的 "is-a" 关系组织概念,允许属性的多态继承。

The Value of Ontology

Mark von Rosing, Wim Laurier, Simon M. Polovina

INTRODUCTION

It is generally accepted that the creation of added value requires collaboration inside and between organizations.[1] Collaboration requires sharing knowledge (e.g., a shared understanding of business processes) between trading partners and between colleagues. It is on the (unique) knowledge that is shared between and created by colleagues that organizations build their competitive advantage.[2] To take full advantage of this knowledge, it should be disseminated as widely as possible within an organization. Nonaka distinguished *tacit* knowledge, which is personal, context specific, and not so easy to communicate (e.g., intuitions, unarticulated mental models, embodied technological skills), from *explicit* knowledge, which is meaningful information articulated in clear language, including numbers and diagrams.[3]

Tacit knowledge can be disseminated through *socialization* (e.g., face-to-face communication, sharing experiences), which implies a reduced dissemination speed, or can be *externalized*, which is the conversion of tacit into explicit knowledge. Although explicit knowledge can take many forms (e.g., business (process) models, manuals), this chapter focuses on ontologies, which are versatile knowledge artifacts created through externalization, with the power to fuel Nonaka's knowledge spiral. Nonaka's knowledge spiral visualizes how a body of unique corporate knowledge, and hence a competitive advantage, is developed through a collaborative and iterative knowledge creation process that involves iterative cycles of externalization, combination,[4] and internalization.[5] When corporate knowledge is documented with ontology, a knowledge spiral leads to ontology evolution.[6]

The next section of this chapter defines ontologies, discussing their level of context dependency and maturing process. The third section of this chapter discusses the state of the art in a business context, while the fourth section introduces directions for future research and development. A summary is presented in the last section.

WHAT IS ONTOLOGY?

The term *ontology* can refer to a philosophical discipline that deals with the nature and the organization of reality.[7] Ontology, as a philosophical discipline, is usually contrasted with epistemology, which is a branch of philosophy that deals with the nature and sources of our knowledge. However, an *ontology* is an artifact—more precisely, an intentional semantic structure that encodes the set of objects and terms that are presumed to exist in some area of interest (i.e., the universe of discourse or semantic domain), the relationships that hold among them, and the implicit rules constraining the structure of this (piece of) reality.[8,9] In this definition, *intentional*

The Complete Business Process Handbook. http://dx.doi.org/10.1016/B978-0-12-799959-3.00006-9

2.2　本体论的价值

Mark von Rosing, Wim Laurier, Simon M. Polovina

2.2.1　介绍

人们普遍认为,创造附加价值需要组织内部和组织之间的协作[1]。协作需要在贸易伙伴和同事之间共享知识(如共享对业务流程的理解)。组织建立竞争优势的基础是同事之间共享和创造(独特)知识[2]。为了充分利用这一知识,应在组织内尽可能广泛地传播。I. Nonaka(野中郁次郎,知识创造理论之父)区分了隐性知识与显性知识,他认为隐性知识是一种个人的、特定于背景的、不太容易交流的知识(如直觉、非人工智能模型、具体的技术技能),显性知识是用清晰的语言表达的有意义的信息,包括数字和图表[3]。

隐性知识可以通过社会化(如面对面交流、经验分享会议等方法)进行传播,这意味着其传播速度较慢,或者它可以被外化,即隐性知识转化为显性知识。虽然显性知识可以有多种形式[如业务(流程)模型、手册],但本节着重于介绍通过外化创建的多功能知识产物——本体,本体具有推动"野中知识螺旋"(Nonaka's knowledge spiral)的能力。野中知识螺旋图显示了一个独特的企业知识体系,以及由此产生的竞争优势,是如何通过一个包括外化、组合[4]和内化的迭代周期的协作和迭代的知识创造过程发展出来的[5]。当企业知识被记录在本体中时,知识螺旋将导致本体的演化[6]。

本节的第二部分(2.2.2～2.2.4)将定义本体论,讨论其对语境的依赖程度、成熟度及成熟过程。本节的第三部分(2.2.5)讨论商业背景下的技术现状,而第四部分(2.2.6)介绍未来研究和发展的方向并在2.2.6节中进行总结。

2.2.2　什么是本体论?

本体论这个术语可以指涉及现实实体的本质和组织[7]的哲学学科。本体论作为一门哲学学科,通常与认识论形成对比。认识论是哲学的一个分支,它涉及我们知识的本质和来源;而本体论是一种产物,更准确地说,是一种有意向性的语义结构,它对假定存在于某个所关注的领域(即论域或语义域)中的一组对象和术语进行编码,对它们之间的关系进行构建,以及对约束这个现实结构的隐含规则进行描述[8,9]。在这个定义中,意向性是指描述各种可能的事务状态的结构,

refers to a structure describing various possible states of affairs, as opposed to extensional, which would refer to a structure describing a particular state of affairs. The word *semantic* indicates that the structure has meaning, which is defined as the relationship between (a structure of) symbols and a mental model of the intentional structure in the mind of the observer. This mental model is often called a *conceptualization*.[10] Semantics are an aspect of semiotics, like syntax, which distinguishes valid from invalid symbol structures, and like pragmatics, which relates symbols to their meaning within a context (e.g., the community in which they are shared).[11]

ONTOLOGY CLASSIFICATION BASED ON CONTEXT DEPENDENCY

Ontologies can be classified according to their level of context dependency.[12] *Top-level* or *foundational ontologies* are context independent because they describe very general concepts, such as space, time, and matter, which are ought to be found in any context. *Task* and *domain ontologies* all relate to the context of a specific domain (e.g., banking, industry) or task (e.g., accounting, sales). Domain and task ontology terms are specializations of top-level ontology terms or terms used in a domain or task ontology with a wider scope (e.g., business-to-business (B2B) sales is a subcontext of sales), which means that they are directly or indirectly founded on top-level ontology terms. Finally, *application ontologies* relate to a very specific context (e.g., accounting in the banking industry, B2B sales in a single sales department). Their terms can be defined as specializations of domain and task ontology terms.

ONTOLOGY MATURITY AND THE MATURING PROCESS

Ontologies can also be classified according to their level of maturity. At the lowest level of maturity, we find *emerging ontologies*, which are rather ad-hoc, not well-defined, individually used, and informally communicated natural-language artifacts.[13] Within the ontology spectrum, which ranges from highly informal to formal ontologies, controlled vocabularies, glossaries, and thesauri, are suitable ontology formats for such informal ontologies. A *controlled vocabulary*, which is a finite list of terms with a unique identifier, is the most rudimentary ontology. A *glossary*, which is a controlled vocabulary in which each term's meaning is given using natural language statements, is a slightly richer ontology.[14] Both controlled vocabularies and glossaries provide a list of unrelated or implicitly related terms. Some of these emerging ontologies mature to become *folksonomies* or common vocabularies, which are shared within and collaboratively improved by a community. Like most emerging ontologies, folksonomies use an extensional notion of conceptualization, which means that the terms are defined through examples rather than through descriptions.[15]

Folksonomies can mature to become *formal ontologies* through the organization of their terms using relationships. These relationships can be ad hoc, as in thesauri,

这与外延性相反。外延性是指描述特定的事务状态的结构。语义是指这个结构具有意义，即在观察者的头脑中，符号（结构）与意向结构的心理模型之间存在关系，这种心理模型通常被称为概念化[10]。语义是符号学的一个方面，如句法，它可以区分有效符号和无效符号结构，以及语用学，它将符号与语境（如共享符号的团体）中的意义联系起来[11]。

2.2.3　基于语境依赖的本体分类

本体可以根据其语境依赖程度进行分类[12]。顶级或基础本体是独立于其语境的，因为它们描述了如空间、时间和物质等非常通用的概念，这些概念可以在任何语境中找到。任务和领域本体与特定的领域（如银行业、产业）或任务（如会计、销售）的语境相关，它们的术语是顶级本体术语或范围更广的领域或任务术语的特化（如B2B销售是销售的子语境），这意味着它们直接或间接建立在顶级本体术语之上。最后，应用本体与非常具体的语境相关（如银行业中的会计、单个销售部门中的B2B销售），它们的术语可以定义为领域和任务本体术语的特化。

2.2.4　本体成熟度与成熟过程

本体也可以根据其成熟度进行分类。在最低成熟度上的是涌现本体，它们是有特定语境的、定义不明确的、单独使用的和非形式化交流的自然语言产物[13]。在从高度非形式的本体到形式本体的本体谱系内，受控词表、术语表和叙词表，都是适用于此类非形式本体的本体格式。受控词表是最基本的本体，它是一个具有唯一标识符的有限词汇表。术语表是一种受控词表，其中每个术语的含义是使用自然语言陈述的，它是一个稍微丰富的本体[14]。受控词表和术语表都提供了一个不相关或隐式相关的术语列表。这些涌现本体中的一些已经成熟，成为大众分类或通用词汇，被某个团体共享并协作优化。与大多数涌现本体一样，大众分类使用概念的外延意义，这意味着术语是通过实例而不是通过描述来定义的[15]。

通过使用关系来组织术语，大众分类可以成熟为形式本体。这些关系可以是特定的，如叙词表；也可以是分层的，如分类表[16]。叙词表通过在受控词表中的术

or hierarchical, as in classification schemes.[16] A *thesaurus* increases ontology expressiveness by adding relations (e.g., synonyms) between terms in a controlled vocabulary. However, thesauri do not necessarily provide an explicit term hierarchy (e.g., specialization-generalization), which is a feature of classification schemes. A *classification scheme* contains informatory "is-a" relations, a *class hierarchy* strict specialization–generalization relations. Strict specialization–generalization relations create a treelike hierarchy, with a generic term as the root and more specific terms, which inherit meaning from the more generic concepts (e.g., the root) they are related to, as branches and leafs. The formality level of a class hierarchy can be increased by adding instantiation as a relation. Instantiation distinguishes between a meaningful term, which is often called a class (e.g., a car), and the terms that are examples of this class, which are often called instances (e.g., my car, your car). Inference rules can be derived from classification schemes and class hierarchies. For example, if a car is a kind of vehicle, then my car is also a vehicle. This implies that everything that can be said about vehicles can be said about my car. At a higher level of thesaurus expressiveness, *frames* include information about potential properties and relationships of classes and their instances (e.g., a car might have a price).[17]

The final phase in the maturing process is called the axiomatization.[18] An *axiom* is a statement for which there is no counter-example or exception.[19] *Value restrictions*, which increase the expressiveness of a frame by discriminating valid from invalid relationships between properties of classes and their instances,[20] are examples of axioms. Other examples of axioms include mathematical equations that relate properties, or logical restrictions on classes and their instances (e.g., disjointness constraint). Some ontologies also provide heuristic value restrictions (e.g., most cars consume fuel, most cars have one owner).

When an ontology was not formalized earlier, the axiomatization phase is often combined with the articulation in a formal language. This formality is a critical aspect of a well-known ontology definition, which dictates that an ontology needs to be a "formal specification of a shared conceptualization."[21] In this definition, the word *specification* requires that an ontology is an appropriate representation of its universe of discourse, which is typically referred to as but not limited to a (semantic) domain. The word *shared* refers to the need for social agreement about and shared understanding of the terms in the ontology. *Formal* refers to the fact that ontologies are frequently written in a formal (and often also machine-readable) language, which is a set of finite symbol structures taken from a finite alphabet of symbols[22] and defined by syntax.

STATE OF THE ART

Building and maturing an ontology is a collaborative and iterative process that requires thought and effort. The process also produces several valuable byproducts, including a better understanding of the organization.[23] Through documentation, structuring, and analysis of business process information, ontology development has been found to support business process detection,[24] continuous

语间添加关系(如同义词)来加强本体的表达。然而,叙词表不一定提供一个清晰的术语层次(如特化–泛化),而这是分类表所擅长的。分类表包含:信息性的"is–a"关系、类层次结构严格的特化–泛化关系。严格的特化–泛化关系创建了一个树状层次结构:以一个通用术语为根,而作为分支和叶的更具体的术语则继承了比它们通用的概念(如根)的含义。可以通过添加实例来提高一个类层次结构的完成度。实例化能有效区分一个有意义的术语(通常称为类,如汽车)和这个类的例子(通常称为实例,如我的车、您的车)。推理规则可以从分类表和类层次结构中得出,例如,如果一辆车是一种交通工具,那么我的车也是一种交通工具。这意味着关于车辆的一切属性也适用于我的车。在更高程度的叙词法表达中,框架包括了类及其实例的潜在属性和关系的信息(例如,一辆汽车可能有价格)[17]。

成熟过程的最后阶段称为公理化[18]。公理是一种没有反例或例外的陈述[19]。取值约束是公理的一个例子[20],它通过区分类及其实例属性之间的有效与无效关系来提高框架的表达能力。公理的其他例子包括定义属性间关系的数学方程,或对类及其实例的逻辑限制(如不相交约束)。一些本体也提供启发式的取值约束(例如,大多数汽车消耗燃料,大多数汽车只有一个所有者)。

当一个本体在前期的阶段没有被形式化时,公理化阶段通常与形式语言的表达相结合。这种形式化是本体的一个广为人知的定义中的一个重要方面,即本体应当是一种"对于共享概念模型的形式化规范说明"[21]。在这个定义中,规范说明一词要求本体是其论域的一个合适的表达,通常被视为但又不限于一个(语义)领域;共享一词是指需要对本体中的术语达成社会共识和共同理解;形式化是指本体经常用形式化的(通常也是机器可读的)语言编写,这是一组取自有限的符号字母表的、由句法定义的符号结构集[22]。

2.2.5   当前发展状况

构建一个本体并提升其成熟度是一个需要思考和努力的协作与迭代过程。这个过程还产生一些有价值的副产品,包括对组织的更好理解[23]。通过文档化、结构化和分析业务流程信息,本体可以帮助我们发现业务流程[24]、持续优化流程[25]和定义流程绩效指标(process performance indicators, PPI)[26]。借助对本体工程的

process refinement,[25] and defining process performance indicators.[26] Ontology engineering also requires discussion, which may yield valuable feedback, to reach consensus and obtain a conceptualization that is shared by all stakeholders. LEGO refers to this shared conceptualization as "One truth for all."[27] Several domain ontologies for business have been developed. Their main purposes are knowledge exchange[28] and knowledge management covering and bridging[29] several subdomains of business (e.g., business plans and other strategies,[30-36] operations,[37,38] finance,[39] accounting[40,41]), and auxiliary disciplines (e.g., information management,[42,43] requirements engineering,[44,45] information systems design,[46] and the development of the semantic web[47]).

Corporate knowledge is often visualized using conceptual modeling grammars (e.g., business process modeling notation (BPMN)). It has been demonstrated that an ontological assessment of such a modeling grammar (through semantic mapping) increases the perceived usefulness and ease of use.[48] An ontological assessment uses the knowledge embedded in ontologies to assess the expressiveness of modeling grammars by mapping grammar constructs to concepts of a relevant ontology.[49] The resulting mapping is called a semantic mapping and is proof of a modeling grammar's ontological commitment. The grammar constructs can be textual, iconic, or diagrammatic and are often referred to as symbols.[50] A semantic mapping can reveal grammar incompleteness, construct redundancy, excess, and overload.[51] *Construct deficit* occurs when one or more ontology concepts lack an equivalent grammar construct, which signals that the grammar is incomplete. *Construct redundancy* occurs when an ontology concept corresponds with two or more grammar constructs. *Construct excess* can be observed when one or more grammar constructs lack an equivalent ontology concept. *Construct overload* occurs when a grammar construct matches with two or more ontology concepts.

Next to symbols that can be mapped to ontological concepts, *modeling grammars* provide rules that prescribe how symbols, which refer to ontology concepts, can be combined to model real-world phenomena.[52] In formal languages, these rules are embedded in a proof theory, which consists of a set of inference rules.[53] These inference rules prescribe how new combinations of symbols can be derived from existing combinations of symbols. Consequently, a proof theory, together with the syntax and semantic mapping, permits a mathematical evaluation of a grammar's correspondence with the semantic domain.[54,55] In the ideal scenario, the set of all valid models generated from a modeling grammar's symbols and its (inference) rules covers the entire domain and nothing but the domain (i.e., every real-world phenomenon from the semantic domain can be modeled, and it is impossible to create a model that does not belong to the set of intended models).

Semantic mappings can also be applied to validate and integrate ontologies. When two or more ontologies that cover the same semantic domain share the same concept, the concept is more likely to belong to the semantic domain. When a concept occurs in only one of several ontologies that share the same domain, the concept is more likely to be redundant. The LEADing Practice community has applied semantic mappings to validate its ontology and integrate it with the ontologies of other frameworks and methods (e.g., The Open Group Architecture Framework (TOGAF), Control

讨论可能会产生有价值的反馈,以达成共识,并获得被所有利益攸关方共享的概念。乐高集团将这种共享的概念称为"所有人的真理"[27]。当前已有一些商用的领域本体被开发出来,它们的主要目的是知识交流[28]和知识管理,涵盖并连接[29]业务的几个子领域(如商业计划和其他战略[30-36]、运营[37,38]、财务[39]、会计[40,41])和辅助学科(如信息管理[42,43]、需求工程[44,45]、信息系统设计[46]以及语义网[47]的发展)。

通常可使用概念建模语法(如业务流程建模标记法BPMN)对企业知识进行可视化。这种建模语法的本体评估(通过语义映射)经证明可以提高感知的有用性和易用性[48]。本体评估使用嵌入在本体中的知识,通过将语法构造映射到一个相关本体的概念中来评估建模语法的表达能力[49]。生成的映射称为语义映射,是建模语法本体承诺的证明。语法构造可以是文本形式的、图符形式的或图解形式的,通常被称为符号[50]。语义映射可以显示语法的不完整、构造的冗余、过度和重载[51]。当一个或多个本体概念缺少对应的语法构造时,就会出现构造缺陷,这表明语法不完整;当一个本体概念与两个或多个语法构造对应时,就会出现构造冗余;当一个或多个语法构造缺少对应的本体概念时,可以观察到构造过度;当一个语法构造与两个或多个本体概念匹配时,就会发生构造重载[52]。

除可以映射到本体概念的符号之外,建模语法还提供了一些规则,规定了如何将符号(指本体概念)组合成模拟现实世界现象的模型[52]。在形式语言中,这些规则嵌入到由一组推理规则组成的证明理论中[53]。这些推理规则规定了如何从现有的符号组合中派生新的符号组合。因此,证明理论,连同句法和语义映射,允许对语法与语义域的对应关系进行精确评估[54,55]。在理想情形中,从建模语法的符号及其(推理)规则生成的所有有效模型的集合涵盖整个域,且仅涵盖这个域(即可以对来自语义域的每个现实世界现象进行建模,并且不可能创建不属于预期模型集的模型)。

语义映射也可以应用于验证和集成本体。当覆盖同一语义域的两个或多个本体共享同一概念时,该概念更可能属于语义域。当一个概念只出现在共享同一领域的几个本体中的一个时,这个概念更可能是冗余的。领导实践(The LEADing Practice)社区已经应用语义映射来验证其本体,并将其与其他框架和方法的本体[如开放组架构框架(TOGAF)、信息和相关技术控制目标(COBIT)、信息技术基础

Objectives for Information and Related Technology (COBIT), Information Technology Infrastructure Library (ITIL), Layered Enterprise Architecture Development (LEAD)).[56] However, most semantic mappings are applied in data management for the purpose of enterprise application integration[57] or database integration,[58] or to build a semantic[59,60] or pragmatic web.[61] Additionally, the semantic mappings allow for an automated translation of a concept from one ontology (e.g., applied in database A) to another equivalent ontology (e.g., applied in database B).

A lot of corporate knowledge is documented using diagrammatic languages (e.g., BPMN). An ontological evaluation of such languages through a semantic mapping of their symbols has been observed to improve their expressiveness and clarity. Consequently, semantic mappings between domain ontologies and domain-specific modeling languages[62,63] would allow organizations to improve these languages for the purpose of interorganizational communication. Semantic mapping might also allow organizations to develop unique intraorganizational languages based on an organization's application ontology for strategic information. Such an intraorganizational language might be defined as an extension of BPMN (e.g., extended business process modeling notation (X-BPMN)[64]) or as a completely independent language, which might need to respect the "physics of notation."[65]

Markup languages such as the Ontology Web Language (OWL), Resource Description Framework (RDF) and Knowledge Interchange Format (KIF) allow ontologies to be processed and distributed by computers, which allows for an automated combination and evaluation of (inter)organizational ontological knowledge.[66-68] Although some efforts have been made to formalize enterprise ontologies (e.g., REA (economic Resources, economic Events, and economic Agents)[69]) or best practices,[70] most applications of ontology in an organizational context are currently limited to building less formal ontologies (e.g., folksonomies). Therefore, organizations should invest in formalizing shared knowledge (e.g., big data).

CONCLUSIONS AND DIRECTIONS FOR FUTURE RESEARCH

This chapter defined ontology engineering as a discipline that can support corporate knowledge creation through the definition of fundamental concepts, as well as semantic relations and correlations between these concepts. Such definitions are not incorporated in contemporary BPM practice in an integrated and standardized way. Therefore, it would be advisable to develop an ontology for BPM.

A BPM ontology could include, among others, the following:

- It should state the primary concepts,[71] such as the entities/objects involved in BPM.
- It should define each of these primary involved concepts.
- It should define the relationships between these concepts.
  - It should preferably describe these relationships using class hierarchies.
  - These class hierarchies should preferably be based on existing classifications.
- It should describe the properties of the concepts and relationships above.

架构库(ITIL)、分层企业架构开发(LEAD)]进行整合[56]。然而,大多数语义映射都被应用于数据管理,以实现企业应用程序集成[57]或数据库集成[58],或者构建语义[59,60]或语用网[61]。此外,语义映射允许将概念从一个本体(如应用于数据库A)自动转换为另一个等效本体(如应用于数据库B)。

许多企业知识都是使用图表语言(如BPMN)记录的。通过对这些语言符号的语义映射,对它们进行本体评估,可以提高它们的表达能力和清晰程度。因此,领域本体和该领域专有的建模语言之间的语义映射[62,63]允许组织为了跨组织交流改进这些语言。语义映射还允许组织基于组织的应用本体开发独特的组织内语言,以获取战略信息。这种组织内语言可以定义为BPMN的扩展[如扩展的业务流程建模标记法(X-BPMN)[64]]或完全独立的语言,后者可能需要遵守"符号物理学"[65]。

诸如Web本体语言(ontology Web language,OWL)、资源描述框架(resource description framework,RDF)和知识交换格式(knowledge interchange format,KIF)等标记语言允许计算机处理和分发本体,从而实现(跨)组织本体知识的自动组合和评估[66-68]。尽管在使企业本体[如经济资源、经济事件和经济代理(economic resources, economic events, and economic agents,REA)[69]]或最佳实践形式化[70]方面已经做出了一些努力,但在组织环境中,本体的大多数应用目前仅限于构建不那么形式化的本体(如大众分类)。因此,组织应该投资于对共享知识(如大数据)的形式化。

2.2.6   结论和未来研究的方向

本章将本体工程定义为一门学科,它通过对基本概念的定义以及这些概念之间的语义关系和相互关系来支持企业知识的创造。这些定义并没有以一种集成和标准化的方式纳入到当代的BPM实践中。因此,最好为BPM开发一个本体。

BPM本体可以包括以下内容。

- 它应声明主要概念[71],如BPM中涉及的实体/对象;
- 它应定义这些主要涉及的概念;
- 它应该定义这些概念之间的关系:
  - 最好使用类层次结构描述这些关系;
  - 这些类层次结构最好基于现有分类。
- 它应描述上述概念和关系的属性;

- It should define a set of value restrictions, such as how and where can the process objects be related (and where not).
- It should be supported by as large a user community as possible.
- It should be vendor neutral and agnostic, therefore allowing it to be used with most existing frameworks, methods, and/or approaches that have some of its mentioned meta-objects.
- It should be practical.
- It should have fully integrated and standardized relationship attributes.

Within the context of a BPM ontology, what are the properties of process (meta) objects and how do they relate to other (meta) objects?

- It should define how to organize and structure viewpoints and concept associations.
- It should structure process knowledge.
- It should establish guiding principles for creating, interpreting, analyzing, and using process knowledge within a particular (sub) domain of business and/or layers of an enterprise or an organization.

End Notes

1. Brandenburger A. and Nalebuff B., *Co-opetition* 1st ed., xiv (New York: Doubleday, 1996), 290.
2. Nonaka I., Umemoto K., and Senoo D., "From Information Processing to Knowledge Creation: A Paradigm Shift in Business Management," *Technology in Society* 18, no. 2 (1996): 203–218.
3. Practice L., *Hands-on Modelling Templates* (2014); Available from: http://www.leadingprac tice.com/tools/hands-on-modelling/.
4. Combination aggregates explicit knowledge to create new explicit knowledge (e.g., analysis, reporting).
5. Internalization transforms shared explicit knowledge into personal tacit knowledge (e.g., learning, studying).
6. Liu J. and Gruen D. M., "Between Ontology and Folksonomy: A Study of Collaborative and Implicit Ontology Evolution," in *Proceedings of the 13th International Conference on Intelligent User Interfaces* (Gran Canaria, Spain: ACM, 2008), 361–364.
7. Guarino N. and Giaretta P., "Ontologies and Knowledge bases: Towards a terminological clarification," in *Towards Very Large Knowledge Bases: Knowledge Building and Knowledge Sharing*, ed. N. Mars (IOS Press, 1995), 314.
8. Ibid.
9. Genesereth M. and Nilsson N., *Logical Foundations of Artificial Intelligence* (Los Altos, CA: Morgan Kaufmann, 1987).
10. Gruber T. R., "A Translation Approach to Portable Ontology Specifications," *Knowledge Acquisition* 5, no. 2 (1993): 199–220.
11. Cordeiro J. and Filipe J., "The Semiotic Pentagram Framework – A Perspective on the Use of Semiotics within Organisational Semiotics." in *7th International Workshop on Organisational Semiotics* (Setúbal, Portugal, 2007).
12. Guarino N., "Semantic Matching: Formal Ontological Distinctions for Information Organization, Extraction, and Integration," in *SCIE* (1997), 139–170.

- 它应该定义一组价值约束，如流程对象的关联方式和位置（以及不关联的位置）；
- 它应该得到尽可能大的用户群体的支持；
- 它应该是不限供应商并开放的，因此它能与大多数现有的框架、方式、方法一起使用，这些框架、方式、方法都有它提到的一些元对象；
- 它应切实可行；
- 它应具有完全集成和标准化的关系属性。

在BPM本体的语境中，流程（元）对象的属性是什么？它们如何与其他（元）对象相关？

- 它应该定义如何组织和构造视图和概念关联；
- 它应该构建流程知识；
- 它应确立在企业或组织的特定（子）业务领域或层次内创建、解释、分析和使用流程知识的指导原则。

参考文献

[1] Brandenburger A. and Nalebuff B., Co-opetition 1st ed., xiv (New York: Doubleday, 1996), 290.

[2] Nonaka I., Umemoto K., and Senoo D., "From Information Processing to Knowledge Creation: A Paradigm Shift in Business Management," Technology in Society 18, no. 2 (1996): 203−218.

[3] Practice L., Hands-on Modelling Templates (2014); Available from: http://www.leadingpractice.com/tools/hands-on-modelling/.

[4] Combination aggregates explicit knowledge to create new explicit knowledge (e.g., analysis, reporting).

[5] Internalization transforms shared explicit knowledge into personal tacit knowledge (e.g., learning, studying).

[6] Liu J. and Gruen D. M., "Between Ontology and Folksonomy: A Study of Collaborative and Implicit Ontology Evolution," in Proceedings of the 13th International Conference on Intelligent User Interfaces (Gran Canaria, Spain: ACM, 2008), 361−364.

[7] Guarino N. and Giaretta P., "Ontologies and Knowledge bases: Towards a terminological clarification," in Towards Very Large Knowledge Bases: Knowledge Building and Knowledge Sharing, ed. N. Mars (IOS Press, 1995), 314.

[8] Ibid.

[9] Genesereth M. and Nilsson N., Logical Foundations of Artificial Intelligence (Los Altos, CA: Morgan Kaufmann, 1987).

[10] Gruber T. R., "A Translation Approach to Portable Ontology Specifications," Knowledge Acquisition 5, no. 2 (1993): 199−220.

[11] Cordeiro J. and Filipe J., "The Semiotic Pentagram Framework — A Perspective on the Use of Semiotics within Organisational Semiotics." in 7th International Workshop on Organisational Semiotics (Setúbal, Portugal, 2007).

[12] Guarino N., "Semantic Matching: Formal Ontological Distinctions for Information Organization, Extraction, and Integration," in SCIE (1997), 139−170.

13. Braun S., et al., "The Ontology Maturing Approach for Collaborative and Work Integrated Ontology Development: Evaluation Results and Future Directions," in *International Workshop on Emergent Semantics and Ontology Evolution ESOE, ISWC 2007* (Busan, Korea, 2007).

14. Lassila O. and McGuinness D. L., "The Role of Frame-based Representation on the Semantic Web," *Nokia Research Center* (2001).

15. See note 13 above.

16. See note 14 above.

17. See note 14 above.

18. See note 13 above.

19. Bahrami A., *Object Oriented Systems Development* (Boston, Mass, London: Irwin/McGraw-Hill, 1999), 411.

20. See note 14 above.

21. Borst W. N., "Construction of Engineering Ontologies for Knowledge Sharing an Reuse," in *Center for Telematics and Information Technology* (Enschede: Universiteit Twente, 1997), 227.

22. Gold E. M., "Language Identification in the Limit," *Information and Control* 10 (1967): 447–474.

23. Lamsweerde A. V., "Formal Specification: A Roadmap," in *Proceedings of the Conference on the Future of Software Engineering* (Limerick, Ireland: ACM, 2000), 147–159.

24. Damme C., Coenen T., and Vandijck E., "Turning a Corporate Folksonomy into a Lightweight Corporate Ontology," in *Business Information Systems*, ed. W. Abramowicz and D. Fensel (Springer Berlin Heidelberg, 2008), 36–47.

25. Prater J., Mueller R., and Beaugard B., *An Ontological Approach to Oracle BPM* (Oracle, 2011).

26. del-Río-Ortega, Resinas A. M., and Ruiz-Cortés A., "Defining Process Performance Indicators: An Ontological Approach," in *On the Move to Meaningful Internet Systems: OTM 2010*, ed. R. Meersman, T. Dillon, and P. Herrero (Springer Berlin Heidelberg, 2010), 555–572.

27. von Rosing M.,von Scheel H., and Falk Bøgebjerg A., *LEADing BPM Practice – Case Story* (2013).

28. ISO/IEC, "Information Technology – Business Operational View Part 4: Business Sransaction Scenario – Accounting and Economic Ontology," in *ISO/IEC FDIS 15944-4: 2007(E)* (2007).

29. Antunes G., et al., "Using Ontologies to Integrate Multiple Enterprise Architecture Domains," in *Business Information Systems Workshops*, ed. W. Abramowicz (Springer Berlin Heidelberg, 2013), 61–72.

30. Yu E.S-K., *Modelling Strategic Relationships for Process Reengineering* (University of Toronto, 1995), 181.

31. Osterwalder A. and Pigneur Y., *Business Model Generation: A Handbook for Visionaries, Game Changers, and Challengers* (Hoboken, NJ: Wiley, 2010).

32. Osterwalder A., Pigneur Y., and Tucci C. L., "Clarifying Business Models: Origins, Present and Future of the Concept," *Communications of AIS* 2005, no. 16 (2005): 1–25.

33. Hulstijn J. and Gordijn J. "Risk Analysis for Inter-organizational Controls," in *12th International Conference on Enterprise Information Systems (ICEIS 2010)* (Funchal, Madeira, Portugal, 2010).

34. Kort C. and Gordijn J., "Modeling Strategic Partnerships Using the E3value Ontology – A Field Study in the Banking Industry," in *Handbook of Ontologies for Business Interaction*, ed. P. Rittgen (Hershey, PA: Information Science Reference, 2007).

［13］Braun S., et al., "The Ontology Maturing Approach for Collaborative and Work Integrated Ontology Development: Evaluation Results and Future Directions," in International Workshop on Emergent Semantics and Ontology Evolution ESOE, ISWC 2007 (Busan, Korea, 2007).

［14］Lassila O. and McGuinness D. L., "The Role of Frame-based Representation on the Semantic Web," Nokia Research Center (2001).

［15］See note 13 above.

［16］See note 14 above.

［17］See note 14 above.

［18］See note 13 above.

［19］Bahrami A., Object Oriented Systems Development (Boston, Mass, London: Irwin/McGraw-Hill, 1999), 411.

［20］See note 14 above.

［21］Borst W. N., "Construction of Engineering Ontologies for Knowledge Sharing and Reuse," in Center for Telematics and Information Technology (Enschede: Universiteit Twente, 1997), 227.

［22］Gold E. M., "Language Identification in the Limit," Information and Control 10 (1967): 447–474.

［23］Lamsweerde A. V., "Formal Specification: A Roadmap," in Proceedings of the Conference on the Future of Software Engineering (Limerick, Ireland: ACM, 2000), 147–159.

［24］Damme C., Coenen T., and Vandijck E., "Turning a Corporate Folksonomy into a Lightweight Corporate Ontology," in Business Information Systems, ed. W. Abramowicz and D. Fensel (Springer Berlin Heidelberg, 2008), 36–47.

［25］Prater J., Mueller R., and Beaugard B., An Ontological Approach to Oracle BPM (Oracle, 2011).

［26］del-Río-Ortega, Resinas A. M., and Ruiz-Cortés A., "Defining Process Performance Indicators: An Ontological Approach," in On the Move to Meaningful Internet Systems: OTM 2010, ed. R. Meersman, T. Dillon, and P. Herrero (Springer Berlin Heidelberg, 2010), 555–572.

［27］von Rosing M., von Scheel H., and Falk Bøgebjerg A., LEADing BPM Practice - Case Story (2013).

［28］ISO/IEC, "Information Technology - Business Operational View Part 4: Business Transaction Scenario - Accounting and Economic Ontology," in ISO/IEC FDIS 15944-4: 2007(E) (2007).

［29］Antunes G., et al., "Using Ontologies to Integrate Multiple Enterprise Architecture Domains," in Business Information Systems Workshops, ed. W. Abramowicz (Springer Berlin Heidelberg, 2013), 61–72.

［30］Yu E.S.-K., Modelling Strategic Relationships for Process Reengineering (University of Toronto, 1995), 181.

［31］Osterwalder A. and Pigneur Y., Business Model Generation: A Handbook for Visionaries, Game Changers, and Challengers (Hoboken, NJ: Wiley, 2010).

［32］Osterwalder A., Pigneur Y., and Tucci C. L., "Clarifying Business Models: Origins, Present and Future of the Concept," Communications of AIS 2005, no. 16 (2005): 1–25.

［33］Hulstijn J. and Gordijn J. "Risk Analysis for Inter-organizational Controls," in 12th International Conference on Enterprise Information Systems (ICEIS 2010) (Funchal, Madeira, Portugal, 2010).

［34］Kort C. and Gordijn J., "Modeling Strategic Partnerships Using the E3value Ontology — A Field Study in the Banking Industry," in Handbook of Ontologies for Business Interaction, ed. P. Rittgen (Hershey, PA: Information Science Reference, 2007).

35. Gordijn J., Yu E., and van der Raadt B., "E-service Design Using i* and e/sup 3/value Modeling," *Software, IEEE* 23, no. 3 (2006): 26–33.

36. Gordijn J. and Akkermans H., "Designing and Evaluating E-Business Models," *IEEE Intelligent Systems* 16, no. 4 (2001): 11–17.

37. Geerts G. L. and O'Leary D., "RFID, Highly Visible Supply Chains, and the EAGLET Ontology," in *Working Paper* (Newark, DE: University of Delaware, 2008).

38. Haugen R. and McCarthy W. E., "REA, a Semantic Model for Internet Supply Chain Collaboration," in *Business Object Component Workshop VI: Enterprise Application Integration (OOPSLA 2000)* (2000).

39. Council E, *Financial Industry Business Ontology* (2014); Available from: http://www.edmcouncil.org/financialbusiness.

40. Geerts G. L. and McCarthy W. E., "An ontological Analysis of the Economic Primitives of the Extended-REA Enterprise Information Architecture," *International Journal of Accounting Information Systems* 3, no. 1 (2002): 1–16.

41. Geerts G. L. and McCarthy W. E., "Augmented Intensional Reasoning in Knowledge-Based Accounting Systems," *Journal of Information Systems* 14, no. 2 (2000): 127.

42. Fensel D., "Ontology-based Knowledge Management," *Computer* 35, no. 11 (2002): 56–59.

43. Benjamins V. R., Fensel D., and Perez A. G., "Knowledge Management Through Ontologies," *PAKM 98. Practical Aspects of Knowledge Management. Proceedings of the Second International Conference* (1998), 5/1–12.

44. Gordijn J., "Value-based requirements Engineering: Exploring Innovative E-commerce Ideas," in *Exact Sciences* (Amsterdam: Free University of Amsterdam, 2002), 292.

45. Gordijn J., Akkermans H., and Van Vliet H., "Value Based Requirements Creation for Electronic Commerce Applications," in *Proceedings of the 33rd Hawaii International Conference on System Sciences-Volume 6-Volume 6* (IEEE Computer Society, 2000).

46. Hruby P., *Model-driven Design Using Business Patterns* xvi (Berlin: Springer, 2006), 368.

47. Obrst L., et al., "The Evaluation of Ontologies, Towards Improved Semantic Interoperability," in *Semantic Web: Revolutionizing Knowledge Discovery in the Life Sciences*, ed. C. Baker and K-H. Cheung (New York: Springer Verlag, 2006), 139–158.

48. Rosemann M., et al., "Do Ontological Deficiencies in Modeling Grammars Matter?" *MIS Quarterly* 35, no. 1 (2011): 57–A9.

49. Shanks G., Tansley E., and Weber R., "Using Ontology to Validate Conceptual Models," *Communications of the ACM* 46, no. 10 (2003): 5–89.

50. Harel D. and Rumpe B., "Meaningful Modeling: What's the Semantics of 'Semantics'?" *Computer* 37, no. 10 (2004): 64–72.

51. Wand Y. and Weber R., "Research Commentary: Information Systems and Conceptual Modeling–A Research Agenda," *Information Systems Research* 13, no. 14 (2002): 363–376.

52. Ibid.

53. See note 23 above.

54. See note 12 above.

55. See note 23 above.

56. Practice L., *Interconnects with Existing Frameworks* (2014); Available from: http://www.leadingpractice.com/about-us/interconnects-with-main-existing-frameworks/.

57. Uschold M. and Gruninger M., "Ontologies and Semantics for Seamless Connectivity," *Sigmod Record* 33, no. 4 (2004): 58–64.

58. Liu Q., et al., "An Ontology-Based Approach for Semantic Conflict Resolution in Database Integration," *Journal of Computer Science and Technology* 22, no. 2 (2007): 218–227.

［35］ Gordijn J., Yu E., and van der Raadt B., "E-service Design Using i* and e/sup 3/value Modeling," Software, IEEE 23, no. 3 (2006): 26–33.

［36］ Gordijn J. and Akkermans H., "Designing and Evaluating E-Business Models," IEEE Intelligent Systems 16, no. 4 (2001): 11–17.

［37］ Geerts G. L. and O'Leary D., "RFID, Highly Visible Supply Chains, and the EAGLET Ontology," in Working Paper (Newark, DE: University of Delaware, 2008).

［38］ Haugen R. and McCarthy W. E., "REA, a Semantic Model for Internet Supply Chain Collaboration," in Business Object Component Workshop Ⅵ: Enterprise Application Integration (OOPSLA 2000) (2000).

［39］ Council E, Financial Industry Business Ontology (2014); Available from: http://www.edmcouncil.org/financialbusiness.

［40］ Geerts G. L. and McCarthy W. E., "An ontological Analysis of the Economic Primitives of the Extended-REA Enterprise Information Architecture," International Journal of Accounting Information Systems 3, no. 1 (2002): 1–16.

［41］ Geerts G. L. and McCarthy W. E., "Augmented Intensional Reasoning in Knowledge-Based Accounting Systems," Journal of Information Systems 14, no. 2 (2000): 127.

［42］ Fensel D., "Ontology-based Knowledge Management," Computer 35, no. 11 (2002): 56–59.

［43］ Benjamins V. R., Fensel D., and Perez A. G., "Knowledge Management Through Ontologies," PAKM 98. Practical Aspects of Knowledge Management. Proceedings of the Second International Conference (1998), 5/1–12.

［44］ Gordijn J., "Value-based requirements engineering: Exploring Innovative E-commerce Ideas," in Exact Sciences (Amsterdam: Free University of Amsterdam, 2002), 292.

［45］ Gordijn J., Akkermans H., and Van Vliet H., "Value Based Requirements Creation for Electronic Commerce Applications," in Proceedings of the 33rd Hawaii International Conference on System Sciences (IEEE Computer Society, 2000).

［46］ Hruby P., Model-driven Design Using Business Patterns xvi (Berlin: Springer, 2006), 368.

［47］ Obrst L., et al., "The Evaluation of Ontologies, Towards Improved Semantic Interoperability," in Semantic Web: Revolutionizing Knowledge Discovery in the Life Sciences, ed. C. Baker and K-H. Cheung (New York: Springer Verlag, 2006), 139–158.

［48］ Rosemann M., et al., "Do Ontological Deficiencies in Modeling Grammars Matter?" MIS Quarterly 35, no. 1 (2011): 57–A9.

［49］ Shanks G., Tansley E., and Weber R., "Using Ontology to Validate Conceptual Models," Communications of the ACM 46, no. 10 (2003): 5–89.

［50］ Harel D. and Rumpe B., "Meaningful Modeling: What's the Semantics of 'Semantics'?" Computer 37, no. 10 (2004): 64–72.

［51］ Wand Y. and Weber R., "Research Commentary: Information Systems and Conceptual Modeling—A Research Agenda," Information Systems Research 13, no. 14 (2002): 363–376.

［52］ Ibid.

［53］ See note 23 above.

［54］ See note 12 above.

［55］ See note 23 above.

［56］ Practice L., Interconnects with Existing Frameworks (2014); Available from: http://www.leadingpractice.com/about-us/interconnects-with-main-existing-frameworks/.

［57］ Uschold M. and Gruninger M., "Ontologies and Semantics for Seamless Connectivity," Sigmod Record 33, no. 4 (2004): 58–64.

［58］ Liu Q., et al., "An Ontology-Based Approach for Semantic Conflict Resolution in Database Integration," Journal of Computer Science and Technology 22, no. 2 (2007): 218–227.

59. Berners-Lee T., Hendler J., and Lassila O., "The Semantic Web – A New Form of Web Content That is Meaningful to Computers will Unleash a Revolution of New Possibilities," *Scientific American* 284, no. 5 (2001): 34.

60. Shadbolt N., Hall W., and Berners-Lee T., "The Semantic Web Revisited," *IEEE Intelligent Systems* 21, no. 3 (2006): 96–101.

61. Schoop M., De Moor A., and Dietz J. L. G, "The Pragmatic Web: A Manifesto," *Communications of the ACM* 49, no. 5 (2006): 75–76.

62. Walter T., Parreiras F. S., and Staab S., "OntoDSL: An Ontology-Based Framework for Domain-Specific Languages," in *Proceedings of the 12th International Conference on Model Driven Engineering Languages and Systems* (Denver, CO: Springer-Verlag, 2009).

63. Guizzardi G. and Wagner G., *Towards Ontological Foundations for Agent Modelling Concepts Using the Unified Fundational Ontology (UFO)*, in *Agent-Oriented Information Systems II* (Springer, 2005), 110–124.

64. Practice L., *The LEADing Practice – EXtended BPMN Standard* (2013).

65. Moody D., "The Physics of Notations: Toward a Scientific Basis for Constructing Visual Notations in Software Engineering," *Software Engineering, IEEE Transactions on* 35, no. 6 (2009): 756–779.

66. Kalfoglou Y. and Schorlemmer M., "Ontology Mapping: The state of the Art," *Knowledge Engineering Review* 18, no. 1 (2003): 1–31.

67. See note 47 above.

68. Rijgersberg H., Wigham M., and Top J. L., "How Semantics can Improve Engineering Processes: A Case of Units of Measure and Quantities," *Advanced Engineering Informatics* 25, no. 2 (2011): 276–287.

69. Gailly F., Laurier W., and Poels G., "Positioning and Formalizing the REA Enterprise Ontology," *Journal of Information Systems* 22, no. 2 (2008): 219–248.

70. Polovina S., Von Rosing M., and Laurier W., "Conceptual Structures in LEADing and Best Enterprise Practices," in *21st International Conference on Conceptual Structures (ICCS 2014)* (Iasi, Romania, 2014).

71. Within the context of BPM, these concepts are also called entities or objects.

［59］ Berners-Lee T., Hendler J., and Lassila O., "The Semantic Web —— A New Form of Web Content That is Meaningful to Computers will Unleash a Revolution of New Possibilities," Scientific American 284, no. 5 (2001): 34.

［60］ Shadbolt N., Hall W., and Berners-Lee T., "The Semantic Web Revisited," IEEE Intelligent Systems 21, no. 3 (2006): 96−101.

［61］ Schoop M., De Moor A., and Dietz J. L. G, "The Pragmatic Web: A Manifesto," Communications of the ACM 49, no. 5 (2006): 75−76.

［62］ Walter T., Parreiras F. S., and Staab S., "OntoDSL: An Ontology-Based Framework for Domain-Specific Languages," in Proceedings of the 12th International Conference on Model Driven Engineering Languages and Systems (Denver, CO: Springer-Verlag, 2009).

［63］ Guizzardi G. and Wagner G., Towards Ontological Foundations for Agent Modelling Concepts Using the Unified Fundational Ontology (UFO), in Agent-Oriented Information Systems Ⅱ (Springer, 2005), 110−124.

［64］ Practice L., The LEADing Practice — EXtended BPMN Standard (2013).

［65］ Moody D., "The Physics of Notations: Toward a Scientific Basis for Constructing Visual Notations in Software Engineering," Software Engineering, IEEE Transactions on 35, no. 6 (2009): 756−779.

［66］ Kalfoglou Y. and Schorlemmer M., "Ontology Mapping: The state of the Art," Knowledge Engineering Review 18, no. 1 (2003): 1−31.

［67］ See note 47 above.

［68］ Rijgersberg H., Wigham M., and Top J. L., "How Semantics can Improve Engineering Processes: A Case of Units of Measure and Quantities," Advanced Engineering Informatics 25, no. 2 (2011): 276−287.

［69］ Gailly F., Laurier W., and Poels G., "Positioning and Formalizing the REA Enterprise Ontology," Journal of Information Systems 22, no. 2 (2008): 219−248.

［70］ Polovina S., Von Rosing M., and Laurier W., "Conceptual Structures in LEADing and Best Enterprise Practices," in 21st International Conference on Conceptual Structures (ICCS 2014) (Iasi, Romania, 2014).

［71］ Within the context of BPM, these concepts are also called entities or objects.

The BPM Ontology

Mark von Rosing, Wim Laurier, Simon M. Polovina

INTRODUCTION

Many business process management (BPM) and/or process frameworks, methods, or approaches (e.g., Lean, Six Sigma, Business Process Reengineering (BPR), Total Quality Management (TQM), Zero Defect, Business Process Modeling Notation (BPMN), Business Process Execution Language (BPEL) have their own vocabulary. Each of these vocabularies has its own definition of terms, such as business process, process step, process activity, events, process role, process owner, process measure, and process rule. This chapter introduces a BPM ontology that can be applied within the area of process modeling, process engineering, and process architecture. It provides fundamental process concepts that can be used to document corporate knowledge and structure process knowledge by defining relation process concepts (e.g., the order of process steps). The BPM ontology is presented as a shared vocabulary (i.e., folksonomy) that structures knowledge in two ways. First, it allows practitioners to structure their business knowledge by adding meaningful relationships between the vocabulary terms. Second, it organizes concepts in hierarchic "is-a" relationships that allow a polymorphic inheritance of properties.

The BPM ontology presented in this chapter should help to remedy the inconsistent use of these terms by providing benchmark terms and definitions and mapping those terms and definitions to the terms in the vocabularies of other existing frameworks. As these mappings demonstrate the shared use of terms in the BPM ontology and several business standards and reference frameworks, we could argue that the BPM ontology documents (i.e., externalizes) a tacit business folksonomy that was mainly shared through socialization before.[1] Part of the BPM ontology presented here is an explicit business folksonomy that is supported by a wide community of practitioners and academics.[2]

This explicit business folksonomy is presented in the next section of this chapter. The BPM Ontology as a Thesaurus: Structuring Process Knowledge by Defining Relations then presents the BPM ontology as a thesaurus, focusing on the meaningful relationships that exist between the concepts of this business folksonomy. The BPM Ontology as a Frame: The Ontological Structure of the LEADing Practice Process Meta Model demonstrates how this thesaurus can be formalized as a frame, using conceptual graphs (CGs). The BPM ontology is discussed in Discussion of the BPM Ontology, and its advantages are summarized in the final section.

THE BPM ONTOLOGY AS A FOLKSONOMY: SHARING FUNDAMENTAL PROCESS CONCEPTS

All ontologies have a controlled vocabulary as a foundation. Because the BPM ontology is an extensive ontology that has the ambition to cover all aspects of

The Complete Business Process Handbook. http://dx.doi.org/10.1016/B978-0-12-799959-3.00007-0

2.3　BPM本体论

Mark von Rosing, Wim Laurier, Simon M. Polovina

2.3.1　介绍

许多BPM或流程框架、方式方法（如精益、六西格玛、BPR、TQM、零缺陷、BPMN、BPEL）都有自己的词汇表。每个词汇表都有自己的术语定义，如业务流程、流程步骤、流程活动、事件、流程角色、流程所有人、流程度量和流程规则。本节介绍一个可以应用于流程建模、流程工程和流程架构领域的BPM本体，它提供基本的流程概念，可用于通过定义相关的流程概念（如流程步骤的顺序）来记录公司知识和构造流程知识。BPM本体是作为一个共享词汇表（即大众分类）呈现的，它以两种方式构造知识。首先，它允许从业者通过在词汇表术语之间添加有意义的关系来构造他们的业务知识；其次，它将概念组织成有层次结构的"is-a"关系，并且允许属性的多态继承。

本节介绍的BPM本体通过提供参照术语和定义，并将这些术语和定义映射到其他现有框架词汇表中，帮助纠正对这些术语使用不一致的情况。由于这些映射展示了BPM本体以及一些业务标准和参考框架中术语的共享使用情况，我们可以认为BPM本体记录了（即外化了）一种以前主要通过社会化来共享的隐性的业务大众分类[1]。这里介绍的BPM本体部分是一种明确的业务大众分类，它得到了从业者和学者团体的广泛支持[2]。

这一明确的业务大众分类将在2.3.2小节中介绍。"作为叙词表的BPM本体：通过定义关系构建流程知识"将BPM本体呈现为叙词表，重点关注此业务大众分类中的概念之间存在的有意义的关系。"作为框架的BPM本体：领导实践流程元模型（LEADing practice process meta model）的本体结构"演示如何使用概念图（conceptual graphs, CGs）将该叙词表形式化为框架。在"关于BPM本体的讨论"中讨论了BPM本体，并在2.3.6小节总结了其优点。

2.3.2　作为大众分类的BPM本体：基本流程概念的共享

所有的本体都有一个受控词表作为基础。因为BPM本体论是一个广泛的本

business (as opposed to academic ontologies), its terms are organized in a top-level domain and multiple intersecting subdomains. The top-level ontology is kept relatively simple, consisting of four main terms: object, meta-object, object group, and object meta-model. *Objects* refer to something that is within the grasp of the senses and that which a subject relates to. They represent a piece of reality in a model or a document. *Meta-objects* create, describe, or equip objects. A meta-object defines an object's type, relation attributes, functions, control structures, etc. *Object groups* serve to group objects with a common purpose, goal, aim, target, objective, and sets. In the BPM ontology, object groups collect meta-objects related to a subdomain. *Object meta-models* are precise definitions of meta-objects, the semantics[3] of the relationships they are involved in, and the rules that apply to them.[4]

BPM ontology terms are assembled into two groups: composition and decomposition (meta-objects). The decomposition meta-objects are presented in Table 7.1 and allow modelers to structure processes. Categorizations assemble

Table 7.1 *Decomposed Process Meta-Objects*

Process Object	Description
Process area (categorization)	The highest level of an abstract categorization of processes.
Process group (categorization)	A categorization and collection of processes into common groups.
Business process	A set of structured activities or tasks with logical behaviors that produce a specific service or product.
Process step	A conceptual set of behaviors bound by the scope of a process that, each time it is executed, leads to a single change of inputs (form or state) into a single specified output. Each process step is a unit of work normally performed within the constraints of a set of rules by one or more actors in a role, which are engaged in changing the state of one or more resources or enterprise objects to create a single desired output.
Process activity	A part of the actual physical work system that specifies how to complete the change in the form or state of an input, oversee, or even achieve the completion of an interaction with others actors and which results in the making of a complex decision based on knowledge, judgment, experience, and instinct.
Event	A state change that recognizes the triggering or termination of processing.
Gateway	Determines the forking and merging of paths, depending on the conditions expressed.
Process rule	A statement that defines or constrains some aspect of work and always resolves to either true or false.
Process measurement (process performance indicator)	The basis by which the enterprise evaluates or estimates the nature, quality, ability, and extent as to whether a process or activity is performing as desired.

体论,它的目标是涵盖业务的所有方面(这一点与学术本体相反),它的术语组织在一个顶级领域和多个相交的子域中。顶层本体保持相对简单,它包含四个主要术语:对象、元对象、对象组和对象元模型。对象指的是感官所能感知的并与一个主体相关的事物,它们代表模型或文档中的一部分现实。元对象创建、描述或装备对象。元对象定义对象的类型、关系属性、函数、控制结构等。对象组用于对具有共同目的、目标、指向、对象和集合的对象进行分组。在BPM本体中,对象组收集与子域相关的元对象。对象元模型是对元对象、元对象所涉及的关系语义[3]以及适用于它们的规则[4]的精确定义。

BPM本体术语分为两组:组合和分解(元对象)。分解元对象如表1所示,它允许建模人员构造流程。分类组织由不同对象形成组,而分级组织由对象形成次序(如通过使用严格的部分−整体或顺序语义)。例如,一个流程域可以包含相互独立的流程,流程步骤需要上下相接。

表1　分解流程元对象

流程对象	描　　述
流程域(分类)	抽象的流程分类的最高级别
流程组(分类)	将流程分类并收集到公共组中
业务流程	一组结构化的活动或任务,具有产生特定服务或产品的逻辑行为
流程步骤	受流程范围约束的一组概念性行为,每次执行该行为时,都会将输入(形式或状态)更改为单个指定的输出。每个流程步骤都是一个工作单元,通常由一个角色中的一个或多个施动者在一组规则的约束下执行,这些参与者参与更改一个或多个资源或企业对象的状态以创建单个所需的输出
流程活动	实际物理工作系统的一部分,它规定了如何完成输入的形式或状态的变化,监督甚至完成与其他施动者的交互,从而根据知识、判断、经验和直觉做出复杂的决定
事　件	一种状态变化,识别流程启动或终止
网　关	根据所表示的条件确定流程分支的创建和合并
流程规则	一种定义或约束工作某些方面的陈述,并且总是解析为真或假
流程度量(PPI)	企业评估流程或活动的性质、质量、能力和程度是否按预期执行的基础

Table 7.1 *Decomposed Process Meta-Objects—cont'd*

Process Object	Description
Process owner	A role performed by an actor with the fitting rights, competencies, and capabilities to take decisions to ensure work is performed.
Process flow (including input/output)	A stream, sequence, course, succession, series, or progression, all based on the process input/output states, where each process input/output defines the process flow that together executes a behavior.
Process role	A specific and prescribed set of expected behavior and rights (authority to act) that is meant to enable its holder to successfully carry out his or her responsibilities in the performance of work. Each role represents a set of allowable actions within the organization in terms of the rights that are required for the enterprise to operate.

heterogeneous groups, whereas classifications assemble objects into order (e.g., through the use of strict part-whole or sequencing semantics). For example, a process area can cluster otherwise independent processes; process steps need to follow each other.

The decomposed process meta-objects listed in Table 7.1 can be used in process architecture and process engineering, as they allow for process decomposition. These fundamental concepts can be combined with auxiliary concepts to produce the semantic richness needed by practitioners. These auxiliary concepts are called process composition meta-objects and represent various process aspects such as strategy, goals, critical success factors, performance indicators, reporting, services, applications, and/or data. Together, process composition and decomposition meta-objects provide a structuring mechanism that facilitates the developments of corporate ontologies (e.g., combining the decomposition meta-object process step with the decomposition meta-object risk invites practitioners to think about the risks of each process step they identify). The composition meta-objects, which are shown in Table 7.2, intersect with several subdomains of business (e.g., process, strategy). Consequently, they can be reused for the elicitation of risks, costs, and other aspects of business in several subdomains of business next to processes.

In addition to the decomposed process meta-objects, other meta-objects relate to the concept of process modeling. The related meta-objects are called composed process meta-objects and are considered an essential part for any practitioner working with and around innovation and transformation across various relevant subjects (vs siloed process modeling, engineering and architecture view). The additional related meta-objects fundamental to the various process concepts shown in Table 7.2.

流程对象	描　　述
流程所有人	由具有适当的权利、能力和潜在能力的施动者承担的角色，作出决定以确保执行工作
流程流（包括输入/输出）	流、序列、过程、演替、系列或进程，所有这些都基于流程活动的输入/输出状态，其中每个流程活动的输入/输出都定义了流程，并最终一起执行某个行为
流程角色	一组特定的、规定的预期行为和权利（行动授权），旨在使其持有人能够成功履行其在工作中的责任。每个角色代表一组在组织内被允许的行动，表征为企业运营所需的权利

　　表1中列出的分解流程元对象可以用于流程体系结构和流程工程，因为它们可以将流程分解。这些基本概念可以与辅助概念相结合，进而产生从业者所需要的丰富语义。这些辅助概念称为流程组合元对象，代表策略、目标、关键成功因素、绩效指标、报告、服务、应用程序或数据等各流程方面。流程组合和分解元对象一起提供了一种结构化机制，有助于企业本体的发展（例如，将分解元对象流程步骤与分解元对象风险结合起来，让从业者思考他们识别的每个流程步骤的风险）。组合元对象（表2）与业务的几个子域（如流程、策略）相交，因此，可以在流程旁边的几个业务子领域中重新使用它们，以引出业务的风险、成本和其他方面。

　　除了分解流程元对象之外，其他的元对象还涉及流程建模的概念。相关的元对象被称为"复合流程元对象"，被认为是从业人员使用的创新和转型相关主题（与流程建模、工程和架构视图相比）的基本组成部分。表2显示了各种流程概念的基本附加相关元对象。

Table 7.2 Process Composition Meta-Objects

	Composed Process Meta-Object Descriptions
Goal (e.g., business, application, technology)	A desired result considered a part of the organizational direction, aims, targets, and aspirations.
Objective (critical success factor)	Time-bounded milestones to measure and gauge the progress towards a strategy or goal.
Value indicator (critical success factor)	Any of a series of metrics used by an enterprise, to indicate its overall ability to achieve its mission.
Performance indicator	Any of a series of metrics used by an enterprise, to indicate its overall success or the success of a particular area in which it is engaged.
Performance expectation	The manner in which, or the efficiency with which, something reacts or fulfills its intended purpose as anticipated by a specific stakeholder.
Performance driver	Those variables that are critical to develop the means and overall performance of an enterprise.
Quality	A state of excellence or worth, specifying the essential and distinguishing individual nature and the attributes based on the intended use.
Risk	The combined impact of any condition or events, including those cause by uncertainty, change, hazards, or other factors that can affect the potential for achieving these objectives.
Security	The objects or tools that secure, make safe, and protect through measures that prevent exposure to danger or risk.
Business measure	Any type of measurement used to gauge some quantifiable component of an enterprise's performance.
Report	The exposure, description, and portrayal of information, about the status, direction, or execution of work within the functions, services, processes, and resources of the enterprise.
Timing	A plan, schedule, or arrangement when something should happen or be done or to take place.
Business area	The highest level meaningful grouping of the activities of the enterprise.
Business group	An aggregation within an enterprise, which is within an enterprise area.
Business competency	An integrated and holistic set of related knowledge, skills, and abilities, related to a specific set of resources (including persons and organizations) that combined enable the enterprise to act in a particular situation.
Business resource/actor	A specific person, system, or organization that initiates or interacts with the defined functions and activities. Actors may be internal or external to an organization.
Business role	A part that someone or something has in a particular defined function, activity, or situation. A resource/actor may have a number of roles.
Business function	A cluster of tasks creating a specific class of jobs.

表2　流程组合元对象

流程组合元对象	描　　述
目标（如业务、应用、技术）	期望的结果被认为是组织方向、目的、目标和愿望的一部分
目的（关键成功因素）	有时间限制的里程碑，用来衡量战略或目标的进展
价值指标（关键成功因素）	企业使用的一系列指标中的任何一个，用以表明其完成任务的总体能力
效益指标	企业使用的一系列指标中的任何一个，用以表明企业的整体成功或其所从事的特定领域的成功
绩效期望	特定利益相关者预期的某种事物反应或实现其预期目的的方式或效率
绩效动因	那些对于开发企业的手段和整体性能至关重要的变量
质　量	一种卓越的或有价值的状态，规定了本质和区别于个人的本质和基于预期用途的属性
风　险	任何条件或事件的综合影响，包括不确定性、变化、危害或其他可能影响实现这些目标的潜在因素
安全性	通过防止暴露于危险或风险中的措施来防护、确保安全和保护的物体或工具
业务度量	用于衡量企业绩效的可量化部分的任何类型的度量
报　告	关于企业的功能、服务、流程和资源中工作的状态、方向或执行信息的公开、描写和叙述
时间计划	当某件事出现、完成或发生时的计划、时间表或安排
业务域	企业活动的最高级别、有意义的分组
业务组	企业内部某个业务域中的业务聚合
业务能力	与一组特定资源（包括人员和组织）相关的一组综合的、整体的相关知识、技能和能力，这些资源的组合使企业能够在特定情况下采取行动
业务资源/施动者	启动或与已定义的功能和活动交互的特定人员、系统或组织。施动者可以是组织内部的或外部的
业务角色	某人或某物在特定的功能、活动或情况下所具有的部分。资源/施动者可以具有多个角色
业务职能	创建特定工作类的任务集群

Table 7.2 *Process Composition Meta-Objects—cont'd*

	Composed Process Meta-Object Descriptions
Business owner	A role performed by an actor with the rights, rules, competencies, and capabilities to take decisions for the part of enterprise for which stewardship responsibilities have been assigned.
Cost	An amount that has to be paid or given up to obtain the use or access to something.
Revenue	The realized income of an enterprise or part thereof.
Object (business and information)	A real-world thing of use by or which exists within the enterprise and information objects reveal only their interface, which consists of a set of clearly defined relations. In the context of the business competency, the relevant objects are only those which relate to the enterprise's means to act.
Product	A result and output generated by the enterprise. It has a combination of tangible and intangible attributes (features, functions, usage).
Contract	An agreement between two or more parties that establishes conditions for interaction.
Business rule	A statement that defines or constrains some aspect of behavior within the enterprise and always resolves to either true or false.
Business compliance	The process or tools for verifying adherence to rules and decisions.
Location	A facility, place, or geographic position.
Business channel	A means of access or otherwise interacting within an enterprise or between an enterprise and its external partners (customers, vendors, suppliers, etc.).
Business workflow	A stream, sequence, course, succession, series, and progression as well as order for the movement of information or material from one enterprise function, enterprise service, or enterprise activity (worksite) to another.
Business service	The externally visible ("logical") deed or effort performed to satisfy a need or to fulfill a demand, meaningful to the environment.
Service flow (including output/input)	A set of one or more service input output states, where each service state defines a step in the service flow that, when entered, executes a behavior.
Service measurement (Service Performance Indicator (SPI) and Service Level Agreement (SLA))	The basis by which the enterprise evaluates or estimates the nature, quality, ability, or extent of the services. The commitments of a business service are assessed.
Logical application component	An encapsulation of application functionality that is independent of a particular implementation.
Physical application component	A deployable part of a software product, providing identifiable functions and existing within a specific version of the product.
Application function	The specification of a significant aspect of the internal behavior of the application, which acts as a broader description of a set of application features.
Application task	The automated behavior of a process activity performed by an application.

Continued

（续表）

流程组合元对象	描　　述
业务所有人	一种角色,由具有权力、规则、素质和能力的参与者执行,该参与者可以为分配了管理职责的企业部分做出决策
成　本	为了获得或使用某种事物而必须支付或花费的金额
收　益	企业或其部分的已实现收入
对象(业务和信息)	企业中使用的或存在于企业中的真实事物以及只显示它们接口的信息对象,接口由一组明确定义的关系组成。在业务能力的语境中,相关的对象仅仅是那些与企业的行为方式相关的对象
产　品	企业产生的成果和输出。它同时具有有形和无形属性(特性、功能、用途)
合　同	双方或多方之间建立交互条件的协议
业务规则	定义或约束企业内某些行为方面的语句,并始终解析为真或假
业务合规性	验证遵守规则和决策的流程或工具
位　置	设施、地点或地理位置
业务渠道	企业内部或企业与其外部合作伙伴(客户、卖方、供应商等)之间的合作或交互方式
业务工作流	从一个企业功能、企业服务或企业活动(工作场所)到另一个企业功能、企业服务或企业活动(工作场所)的信息或材料移动的流、序列、过程、演替、系列和进展以及顺序
业务服务	为满足需要或满足对环境有意义的需求而进行的外部可见(合乎逻辑)的行为或努力
服务流(包括输出/输入)	一组一个或多个服务的输入输出状态,其中每个服务状态定义为服务活动流中的一个步骤,当输入该步骤时,该步骤执行一种行为
服务度量[服务绩效指标(SPI)和服务水平协议(service level agreement, SLA)]	企业评估或估计服务的性质、质量、能力或程度的基础。评估业务服务的承诺
逻辑应用组件	独立于具体实现的应用功能的封装
物理应用组件	软件产品的可部署部分,提供可识别的功能并存在于产品的特定版本中
应用功能	应用程序内部行为的一个重要方面的规范,它用作对一组应用程序特性的更广泛描述
应用任务	由应用程序执行的流程活动的自动化的行为

Table 7.2 *Process Composition Meta-Objects—cont'd*

	Composed Process Meta-Object Descriptions
Application service	An externally visible unit of functionality, provided by one or more components, exposed through well-defined interfaces, and meaningful to the environment.
Application/system flow	The specification of the sequence in which two application tasks processes (or an application task and an application event or gateway) are executed, one of which provides an output, which is an input to the other.
System measurement	Measures that are defined and implementable within an application.
Application/system report	Reports that are defined and implementable or implemented within or by an application.
Application roles	A role performed by an actor with the rights, competencies, and capabilities to take decisions about an application, its behavior, and properties.
Application rule	A business rule implemented within and able to be executed by an application.
Data object	A logical cluster of all sets of related data representing an object view of a business object.
Data table	A physical specification of the means of arranging data in rows and columns while being stored in physically persistence structures.
Data flow	The specification of the sequence in which data moves from one state to another.
Data owner	A role performed by an actor with the rights, competencies, and capabilities to take decisions about the aspects of data for which stewardship responsibilities have been assigned.
Data rule	Criteria used in the process of determining or verifying values of data or generalizing certain features of data.
Platform device	A set of platform components configured to act as a modular part of a platform.

THE BPM ONTOLOGY AS A THESAURUS: STRUCTURING PROCESS KNOWLEDGE BY DEFINING RELATIONS

The process objects that have been defined through a search for process composition and decomposition meta-object instances in an organization require additional structure.

Structuring the process knowledge includes identifying the existing classes and groups of process objects and the relations between them and the characteristics that unite or differentiate them. The following criteria facilitate grouping:

- *Identity*: allows users to distinguish an object from any other object and distinguishes objects from meta-objects, which have no identity.
- *State*: is the aggregate of an object's properties, including its relations with other object, meta-objects, classes, etc.
- *Behavior*: distinguishes between legal and illegal state changes.

（续表）

流程组合元对象	描　　述
应用服务	由一个或多个组件提供的一种外部可见的功能单元,通过定义良好的公共接口满足环境需要
应用/系统流	执行两个应用任务进程(或应用任务和应用事件或网关)的序列的规范,其中一个的输出是另一个的输入
系统度量	在应用程序中定义并可实现的评估
应用/系统报告	在应用程序内或由应用程序实现或定义的报告
应用角色	由施动者执行的角色,具有对应用程序及其行为和属性进行决策的权限、能力和素质
应用规则	在应用程序内实现并能够由应用程序执行的业务规则
数据对象	表示业务对象的对象视图的所有相关数据集的逻辑集群
数据表	在物理持久性结构(如数据库)中存储数据并将数据排列成行和列的方法的物理规范
数据流	数据从一种状态移动到另一种状态的序列的说明
数据所有人	由一个施动者执行的角色,该施动者有权、有能力和有素质对分配了管理职责的数据方面做出决策
数据规则	在确定或验证数据值或概括数据某些特征的过程中使用的标准
平台设备	配置为作为平台的模块化部分的一组平台组件

2.3.3　作为叙词表的BPM本体论：通过定义关系构建流程知识

在一个组织中,通过搜索流程组合和分解元对象实例来定义的流程对象需要额外的结构。

构建流程知识包括识别现有的流程对象的类别和组合,以及它们之间的关系及统一或区分它们的特征。以下标准有助于分组。

- 标识：允许用户将对象从其他对象中区分出来,并将其从没有标识的元对象中区分出来。
- 状态：是对象属性的集合,包括与其他对象、元对象、类等的关系。
- 行为：区分符合规则和违背规则的状态的变化。

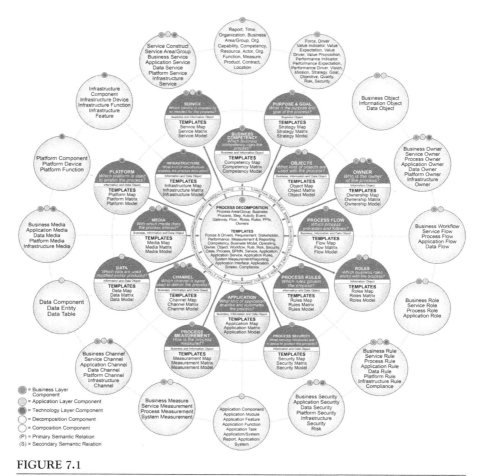

FIGURE 7.1

The 16 basic process classes and groups.[5]

Although relations are mainly defined that the level of meta-objects (e.g., in corporate ontologies), the BPM ontology contains a set of archetypal relations that have been observed to apply to almost any process. These relations have been defined at the level of meta-object groups, which means that they apply to object groups in corporate ontologies, elicited using these meta-objects. Sixteen meta-object groups can be identified. Although these groups contain meta-objects, they are not meta-objects. Their relations with the process meta-object group are summarized in Figure 7.1, which is an overview of these 16 classes and how they relate to the process objects. These 16 groups assemble composition meta-objects, which can be observed several areas of business other than processes. Consequently, this template can be reused to represent the relations between these 16 groups and other aspects of business.[6]

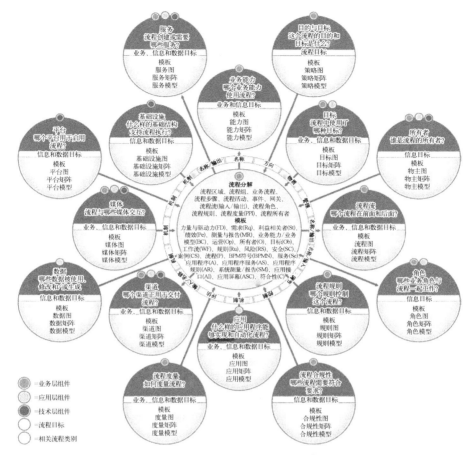

图1 16个基本流程类别和组别[5]

　　尽管关系主要是在元对象级别(如在公司本体中)定义的,但是BPM本体包含一组原型关系,这些关系可以应用于几乎所有流程。这些关系是在元对象组的层次上定义的,也就是说,它们应用于企业本体中的对象组,由这些元对象引出。BPM共有16个可识别的元对象组。尽管这些组包含元对象,但它们不是元对象。它们与流程元对象组的关系如图1所示,图1概述了这16个类以及它们与流程对象的关系。这16个组组成了元对象,可以在流程以外的几个业务域中观察到这些元对象。因此,可以重用此模板来表示这16个组与业务的其他方面之间的关系[6]。

Next to the meta-object relations visualized in Figure 7.1, process composition meta-objects do not only have relationships to the central concept of a business process, but also with multiple other areas. These relations provide an important tool to assess the details of a corporate business ontology, as each object that belongs to one of these 16 meta-object groups is expected to be related to any business process object in order to obtain a complete business process specification. Consequently, a process specification that is missing one or more of these essential objects in its relationships will be considered to be malformed and incomplete. This approach is expected to provide a powerful tool to assist in the identification and capture of all relevant process aspects.

The following process meta-objects and relations are expected to exist within most organizations:

1. The *business competency meta-object group* relates to the following meta-objects: organizational construct, business capability, business resource/actor, business function, product, location, report, timing, revenue, and cost. They intersect with the process meta-object groups as a business calls upon its *business competencies*, or organizational skills and knowledge, which are part of its business model and thereby the organizational structure, to create value within the organization and for its customers via its processes, events, and decisions, or gateways, which are decomposed process meta-objects.

   The relations between business competency and process meta-objects include descriptions of relations between cost and the process objects, of which some examples are given below:
   a. Cost occurs when executing a task within a business process, cost can therefore be related to a business process.
   b. Cost accrued at an event can be associated and tracked.
   c. Cost measures can be specialized within a process measurement (process performance indicator).
   d. Cost control is, among others, the responsibility of a process owner.
   e. Cost compliance can be ensured through process rules.
   f. Cost flow can be found in various process flows.

2. The *purpose and goal meta-object group* contains the following meta-objects: driver (value/performance), strategy, goal, objective, value indicator, value expectation, value proposition, performance indicator, performance expectation, quality, risk, and security. They intersect with the meta-objects of the purpose and goal meta-object groups as business strategies will dictate the *purpose and goals (value)* that provide directions for the process objects. This includes business process objectives, performance expectations, and performance indicators, which can be measured and linked back to the strategy through process performance indicators (PPIs).

   Below is an example of the semantic relations between performance drivers, which belong to the purpose and goal meta-object group and the process objects:
   a. Performance driver influences choices of process owner.
   b. The categorization of process areas and groups can be influenced by performance drivers.
   c. Performance drivers influence the design of business processes.

在如图1所示的元对象关系旁边,流程组合元对象不仅与业务流程的中心概念有关系,而且与多个其他领域也有关系。这些关系为评估企业业务本体的细节提供了重要的工具,因为属于这16个元对象组之一的每个对象都应该与任何业务流程对象相关,以便获得完整的业务流程规范。因此,在流程规范的关系中缺少一个或多个这些基本对象的流程规范将被视为格式错误且不完整。该方法有望为识别和捕获所有相关流程方面提供强大的工具。

以下流程元对象和关系预计将存在于大多数组织中。

(1)业务能力元对象组与以下元对象相关:组织结构、业务能力、业务资源/施动者、业务功能、产品、位置、报告、时间计划、收入和成本。它们与流程元对象组相交,因为业务需要其业务能力或组织技能和知识,这是其业务模型的一部分,所以以上元对象是组织结构的一部分,并通过流程、事件和决策或接口为客户提供服务,这些都是分解的流程元对象。

业务能力与流程元对象之间的关系包括成本与流程对象之间关系的描述,下面给出一些示例:

a. 成本发生在业务流程中执行任务时,因此成本可能与业务流程相关;

b. 可以关联和跟踪事件的应计成本;

c. 成本度量可以专门用于流程度量(PPI);

d. 成本控制是流程所有人的责任;

e. 可通过流程规则确保成本合规性;

f. 成本流可以在各种流程中找到。

(2)目的和目标元对象组包含以下元对象:驱动因素(价值/绩效)、策略、目标、目的、价值指标、价值期望、价值主张、绩效指标、绩效期望、质量、风险和安全。它们与目的元对象和目标元对象组的元对象相交,因为业务策略将表示为流程对象目的和目标(价值)。这包括业务流程目标、绩效期望和绩效指标,这些指标可以通过PPIs衡量并链接到战略。

下面是属于目的和目标元对象组的性能驱动程序与流程对象之间语义关系的示例:

a. 绩效驱动因素影响流程所有人的选择;

b. 流程域和流程组的分类会受到绩效驱动因素的影响;

c. 绩效驱动因素影响业务流程的设计;

  d. Events realize the various performance drivers.

  e. Performance driver sets criteria for the direction of the gateways.

  f. Performance driver set criteria for the execution of the process flow (including input/output).

  g. Performance drivers set presentation criteria for the process role.

  h. Process rules are set based on various performance drivers.

  i. Process measurements (PPI) can be tracked and reported against the performance drivers.

3. The *object meta-object group* has the following members: business objects, information objects, and data objects. They need to be considered because (parts of) business, information, and data objects give *substance* to business process tasks and services. A business process uses, modifies, and/or produces business information and data objects on several hierarchical levels; data objects with data components, business processes with information objects, and business process tasks with data services.

  Below is an example of the semantic relations between the information objects and the process objects:

  a. Business process areas and groups consume and develop information objects relevant for decision making.

  b. Business processes use, produce, and store information objects.

  c. Information objects change the state of an event.

  d. Gateways produce and consume information objects.

  e. Information objects are produced and consumed by process roles.

  f. Process rules regulate the compliance of specific information objects.

  g. Information objects are a part of any process measurement (PPI).

  h. Process owners have the responsibility for the information objects involved in the process.

4. The *owner meta-object group* contains the following: Business owner, process owner, service owner, application owner, data owner, platform owner, and infrastructure owner. They are important because multiple owners can have the authority to steward or *manage* business processes. All owners have specific responsibilities that result in different desires, demands, and various performance and value expectations. In the context of business processes, the business process owners have the responsibilities connected to business tasks, process flow, service, creating value, achieving performance goals set by the strategy adhere to security, and maintaining compliance standards within the "work system."

  Below is an example of the semantic relations between the business owner and the process objects:

  a. Business owners define through the business goals the direction of the business process.

  b. Business owners set performance criteria for the business process.

  c. Business owners create and specify the performance indicators within the process measurements (PPI).

d. 事件实现各种绩效驱动因素；

e. 绩效驱动因素为网关的方向设置标准；

f. 绩效驱动因素为流程的执行设定标准（包括输入/输出）；

g. 绩效驱动因素为流程角色设置展示标准；

h. 根据各种绩效驱动因素设置流程规则；

i. 流程量度（PPI）可根据绩效驱动因素进行跟踪和报告。

（3）目标元对象组的构成要素为：业务对象、信息对象和数据对象。由于部分业务、信息和数据对象为业务流程任务和服务提供了实质性的内容，需要对它们进行考虑。我们还要认识到：业务流程在多个层次上使用、修改、生成业务信息和数据对象，同时，数据对象关联着数据组件，业务流程关联着信息对象，业务流程任务关联数据服务。

下面是信息对象和流程对象之间语义关系的示例：

a. 业务流程域和流程组使用和开发与决策相关的信息对象；

b. 业务流程使用、生成和存储信息对象；

c. 信息对象改变事件的状态；

d. 接口产生和使用信息对象；

e. 信息对象由流程角色生成和使用；

f. 流程规则、规范符合特定信息对象的特性；

g. 信息对象是任何流程量度（PPI）的一部分；

h. 流程所有人对流程中涉及的信息对象负责。

（4）所有人元对象组包含以下内容：业务所有人、流程所有人、服务所有人、应用所有人、数据所有人、平台所有人和基础设施所有者。因为多个所有人都有权管理或经营业务流程，每一个所有人都有特定的责任，导致不同的愿望、需求以及不同的绩效和价值期望，所以，所有人元对象组包含的内容都很重要。在业务流程的背景下，业务流程所有人承担与业务任务、流程、服务、创造价值、实现战略设定的绩效目标相关的责任，并在工作系统内维护合规标准。

下面是业务所有人和流程对象之间语义关系的示例：

a. 业务所有人通过业务目标定义业务流程的方向；

b. 业务所有人为业务流程设置绩效标准；

c. 业务所有人在流程量度（PPI）中创建并指定绩效指标；

　　d. Process owners work with business owners.

　　e. Business owners govern the process flow.

　　f. Business owners are involved in the verification and conformance of the process rules.

5. The *flow meta-object group* consist of the following: business workflow, process flow, service flow, information flow, data flow, and application/system flow. They should be considered because business processes *call and provide output* to the business process flow, which interacts with several different flows within the business. These flows include the business workflow, information flow, data flow, etc., all interacting with the process flow.

　　Below is an example of the semantic relations between the information objects and the process objects:

　　a. Business processes are found within the information flow.

　　b. Events sequence the information flow.

　　c. Information flows have gateways.

　　d. information flow crosses the process flow (including input/output).

　　e. Process measurements (PPI) are a part of the information flow.

　　f. Process owners are involved with the creation of certain information flows.

　　g. Information flows and their rules can be derived from process rules.

6. The *roles meta-object group* has the following members: business role, process role, service role, and application role. It is important to consider them because the enacted business process roles *input and call* the processes through the process steps and activities so as to be supported by the roles of the respective business functions and tasks.

　　Below is an example of the semantic relations between the business roles and the process objects:

　　a. The process group categorizes business roles into its groups.

　　b. Business roles execute the tasks in the business process and activities.

　　c. The process role is a form of the business role.

　　d. Process owners interact with various business roles.

　　e. Business roles participate within the process flow.

　　f. Business roles abide by the process rules.

7. The *rules meta-object group* contains the following: business rule, process rule, service rule, application rule, data rule, platform rule, and infrastructure rule. Business process rules *regulate* the processes, which are then instantiated in services and implemented within applications that enable these processes, data that they consume or produce, and security behavior. This must also both be adhered to and embedded within the different parts of the planning, creation, realization, and governance processes of the business processes.

　　Below is an example of the semantic relations between the business rules and the process objects:

　　a. Business rules regulate business process tasks.

　　b. Business rules ensure the correctness of process flow (including input/output).

d. 流程所有人与业务负责人合作；

e. 业务所有人治理流程；

f. 业务所有人参与流程规则的验证和一致性检查。

（5）流元对象组由以下部分组成：业务工作流、流程流、服务流、信息流、数据流和应用/系统流。我们考虑它们是因为业务流程调用并向业务流程流提供输出，而业务流程流与业务中的几个不同流（工作流、信息流、数据流等）交互。

下面是信息对象和流程对象之间语义关系的示例：

a. 业务流程位于信息流中；

b. 依据事件对信息流进行排序；

c. 信息流中存在接口；

d. 信息流与工作流交叉（包括输入/输出）；

e. 流程度量（PPI）是信息流的一部分；

f. 流程所有人参与创建某些信息流；

g. 信息流及其规则可以从流程规则中派生出来。

（6）角色元对象组包括：业务角色、流程角色、服务角色和应用角色。考虑它们很重要，因为已制定的业务流程角色通过流程步骤和活动输入和调用流程，从而得到各自业务职能和任务的角色的支持。

下面是业务角色和流程对象之间语义关系的示例：

a. 流程组将业务角色分类至其组中；

b. 业务角色执行业务流程和活动中的任务；

c. 流程角色是业务角色的一种形式；

d. 流程所有人与各种业务角色交互；

e. 业务角色参与流程；

f. 业务角色遵守流程规则。

（7）规则元对象组包含以下内容：业务规则、流程规则、服务规则、应用规则、数据规则、平台规则和基础设施规则。业务流程规则控制流程，然后在服务中实例化这些流程，并在启用这些流程的应用、它们使用或生成的数据以及安全行为中实现这些流程。这也必须同时被坚持并嵌入到业务流程的规划、创建、实现和治理的不同部分中。

下面是业务规则和流程对象之间语义关系的示例：

a. 通过业务规则规范业务流程的任务；

b. 业务规则确保流程（包括输入/输出）的正确性；

    c. Business rules apply to gateways.

    d. Business rules relate to process roles.

    e. Business rules are contained within process rules.

    f. Business rules are measured by process performance indicators (PPIs).

    g. Business rules are also a part of the responsibility of process owners.

8. The *compliance meta-object group* contains the following: business compliance, application compliance, data compliance, platform compliance, and infrastructure compliance. When designing, building, implementing, updating, working with, or terminating business process tasks, events, and services, it is essential to demonstrate the level of *control* necessary to demonstrate process compliance with respect to applicable policies, guidelines, standards, and regulations through the use of governance controls, risk management, audits, evaluation, security, and monitoring.

    Below is an example of the semantic relations between business compliance and the process objects:

    a. Business compliance verifies execution of business processes.

    b. Business compliance verifies execution of the gateway.

    c. Process flow (including input/output) conforms to business compliance.

    d. Business compliance assesses the performance process role.

    e. Business compliance verifies conformance to the design of the process rule.

    f. Business compliance evaluates process measurements (PPIs).

    g. Business compliance assesses the performance of process owners.

9. The *application meta-object group* contains the following: logical application component, physical application component, application module, application feature, application function, application task, application/system report, and application/system. An *application* is a *mechanism* used to automate a business process, and/or its steps, activities, events, and flows. Applications are also used to automate process reporting through the use of system measurements and system reporting.

    Below is an example of the semantic relations between the application tasks and the process objects:

    a. The application task partially or fully automates the business process and process activities.

    b. Gateways are automated by application tasks.

    c. The application task partially or fully automates process flow (including input/output).

    d. Process rules are partially or fully automated by application tasks.

    e. Process owners desire application task automation.

10. The *measurement meta-object group* contains the following: business measure, service measure, process measure, and system measure. The measurement indicators are the basis by which we *evaluate* the business processes; their outputs and results can all be measured. Process measurements or their automated equivalent, the system measurements, are linked to business reporting (at the strategic, tactical, and operational levels) through scorecards, dashboards, and cockpits, which aid in this assessment.

c. 业务规则适用于接口；

d. 业务规则与流程角色相关；

e. 业务规则包含在流程规则中；

f. 业务规则由PPIs衡量；

g. 业务规则也是流程所有人责任的一部分。

（8）合规性元对象组包含以下内容：业务合规性、应用合规性、数据合规性、平台合规性和基础设施合规性。在设计、构建、实施、更新、启动或终止业务流程任务、事件和服务时，必须通过使用治理控制、风险管理、审计、评估、安全和监控来证明流程符合适用政策、指导方针、标准和法规的必要控制水平。

下面是业务合规性和流程对象之间语义关系的示例：

a. 通过业务合规性验证业务流程的执行；

b. 业务合规性验证接口的执行；

c. 流程流（包括输入/输出）符合业务合规性；

d. 业务合规性评估绩效流程角色；

e. 业务合规性验证设计是否符合流程规则；

f. 业务合规性评估流程度量（PPI）；

g. 业务合规性评估流程所有人的绩效。

（9）应用元对象组包含以下内容：逻辑应用组件、物理应用组件、应用模块、应用特征、应用功能、应用任务、应用/系统报告和应用/系统。应用是用于自动化业务流程及其步骤、活动、事件和流的一种机制。应用还用于通过使用系统度量和系统评估来实现流程报告的自动化。

下面是应用任务和流程对象之间语义关系的示例：

a. 应用任务可以将业务流程和流程活动部分或完全自动化；

b. 接口由应用任务自动化实现；

c. 应用任务部分或全部自动化流程流（包括输入/输出）；

d. 流程规则由应用任务部分或完全自动化；

e. 流程所有人希望应用任务自动化。

（10）度量元对象组包含以下内容：业务度量、服务度量、流程度量和系统度量。度量指标是我们评估业务流程的基础，它们的输出和结果都可以被度量。通过记分卡、仪表盘和驾驶舱与业务报告（在战略、战术和运营级别）的关联，可对流程进行评价。

Below is an example of the semantic relations between the business measurements and the process objects:

a. Business process performance is tracked by business measures.
b. Events can be tracked against business measures.
c. Business measures are found within the process flow (including input/output).
d. Process roles are evaluated against business measures.
e. Process rules are tracked and reported by business measures.
f. Process owners report part of the business measures.

11. The *channel meta-object group* contains the following: business channel, service channel, application channel, data channel, platform channel, and infrastructure channel. The value delivery to those that benefit from the output of a process occurs through business and technology *channels*. The business channel stages can range from marketing, sales, distribution, business service, and so on.

Below is an example of the semantic relations between the business channel and the process objects:

a. Business Channels require execution of business processes and process activities.
b. Business channels require execution of gateways.
c. Business channels involved within the process are the responsibility of process owners.
d. Business process flow participates in the business channel.
e. Process rules regulate the business channel.

12. The *data meta-object group* contains the following: data component, data entity, data objects, and data table. Process execution is the mechanism by which *data* are created, used, or consumed.

Below is an example of the semantic relations between the data objects and the process objects:

a. Data objects are related to business processes and activities.
b. Data objects change state at an event.
c. Data objects abide by process rules.
d. Data objects are within process measurements (PPIs).

13. The *media meta-object group* contains the following: business media, application media, data media, platform media, and infrastructure media. *Media* is the mechanism that is part of any process by which inputs or outputs of a process are held. There are many kinds of media involved within a process, such as paper, visual, or auditory for manual processes; screens, memory, or disks may act as media for automated processes.

Below is an example of the semantic relations between the business media and the process objects:

a. Business media are supplied or consumed by business processes and process activities.
b. Gateways use business media.
c. Business media are produced at events.
d. Process owners have the responsibility for the business media.

下面是业务度量和流程对象之间语义关系的示例：

a. 业务流程绩效可通过业务度量进行跟踪；

b. 可以根据业务度量跟踪事件；

c. 在流程流（包括输入/输出）中可找到业务度量；

d. 根据业务度量评估流程角色；

e. 通过业务度量跟踪和报告流程规则；

f. 部分业务度量通过流程所有人报告提供。

（11）渠道元对象组包含以下内容：业务渠道、服务渠道、应用渠道、数据渠道、平台渠道和基础设施渠道。通过业务和技术渠道向那些从流程输出中受益的人提供价值。业务渠道阶段可以包括营销、销售、分销、业务服务等。

下面是业务渠道和流程对象之间语义关系的示例：

a. 业务渠道需要执行业务流程和流程活动；

b. 业务渠道需要执行接口；

c. 流程中涉及的业务渠道由流程所有人负责；

d. 流程活动参与在业务渠道中；

e. 流程规则规范业务渠道。

（12）数据元对象组包含以下内容：数据组件、数据实体、数据对象和数据表。流程执行是创建、使用或消费数据的机制。

下面是数据对象和流程对象之间语义关系的示例：

a. 数据对象与业务流程和业务活动相关；

b. 数据对象在事件中更改状态；

c. 数据对象遵守流程规则；

d. 数据对象在流程度量（PPIs）中。

（13）媒介元对象组包含以下内容：业务媒介、应用媒介、数据媒介、平台媒介和基础设施媒介。媒介是任何流程的一部分，通过它来保持流程的输入或输出。一个流程涉及多种媒介，例如，手动流程中的纸张、视觉或听觉等，自动化流程中作为媒介的屏幕、内存或磁盘等。

下面是业务媒介和流程对象之间语义关系的示例：

a. 业务媒介由业务流程和流程活动提供或使用；

b. 接口使用业务媒介；

c. 业务媒介产生于事件中；

d. 流程所有人对业务媒介负有责任。

14. The *platform meta-object group* consists of the following: physical platform component, and platform device. A *platform* is a mechanism used to enable process automation; for example, a platform component enables an application component, and a platform service enables an application service and thereby a business service. Platforms such as laptops, smart phones, or tablets are used to access processes.

Below is an example of the semantic relations between the platform devices and the process objects:

a. Platform devices generate and participate in a business process.
b. Platform devices are used by process roles.
c. Platform devices participate within the process flow.
d. Platform devices change the state of events.

15. The *infrastructure meta-object group* contains the following: physical infrastructure component and infrastructure device. From a process architecture perspective, processes are automated with dedicated technology, which use a mechanism to draw on *infrastructure* for their ability to execute. For example, a process rule engine resides on infrastructure components, and infrastructure services support the platform services.

Below is an example of the semantic relations between the infrastructure and the process objects:

a. Automated business processes reside on physical infrastructure components.
b. Physical infrastructure components host business process engines (rules, measures, etc.).

16. The *service meta-object group* contains the following: business service, application service, data service, platform services, and infrastructure services. Business services are what actually deliver value within the organization and to its customers. They do this when they call upon and provide output to the processes necessary to instantiate them. This is because value creation is subject to the relationships between business processes and their resources, tasks, events, and the services they deliver. Although there is a distinction between manual and automated services, the division is captured within the process notations, which relate the automated service to the relevant web services, application services, data services, platform services, and infrastructure services and the business services to their manual counterparts.

Below is an example of the semantic relations between the business compliance and the process objects:

a. Business services are realized by business processes.
b. Business services resolve events.
c. Business services are provided to process roles.
d. Business services are regulated by process rules.
e. Business services are measured by process measures.
f. Business services are governed by process owners.

（14）平台元对象组由以下部分组成：物理平台组件和平台设备。平台是用于实现流程自动化的载体，例如，平台组件启用应用组件，平台服务启用应用服务，进而启动业务服务。此外，诸如笔记本电脑、智能手机或平板电脑等平台设备可用于流程访问。

下面是平台设备和流程对象之间语义关系的示例：

a. 平台设备生成并参与业务流程；

b. 平台设备由流程角色使用；

c. 平台设备参与工作流；

d. 平台设备改变事件状态。

（15）基础设施元对象组包含以下内容：物理基础设施组件和基础设施设备。从流程架构的角度来看，流程是通过专用技术实现自动化的，该技术使用一种机制来利用基础设施实现其执行能力。例如，流程规则引擎驻留在基础设施组件上，而基础设施服务支持平台服务。

下面是基础设施和流程对象之间语义关系的示例：

a. 业务流程自动化以物理基础设施组件为基础；

b. 物理基础设施组件承载业务流程引擎（规则、度量等）。

（16）服务元对象组包含以下内容：业务服务、应用服务、数据服务、平台服务和基础设施服务。业务服务实际上是为组织内部及其客户提供有价值的东西。当它们调用并向实例化它们所需的流程提供输出时，它们就会这样做。这是因为价值创造取决于业务流程与其资源、任务、事件及其提供的服务之间的关系。虽然手动服务和自动服务之间存在区别，但这一划分是在流程标记中得到的，流程标记将自动服务与相关的网络服务、应用服务、数据服务、平台服务和基础设施服务，以及业务服务与它们对应的手动服务相关联。

下面是业务合规和流程对象之间语义关系的示例：

a. 业务服务通过业务流程实现；

b. 业务服务解析事件；

c. 业务服务为流程角色服务；

d. 业务服务由流程规则管理；

e. 业务服务由流程度量评估；

f. 业务服务由流程所有人治理。

THE BPM ONTOLOGY AS A FRAME: THE ONTOLOGICAL STRUCTURE OF THE LEADING PRACTICE PROCESS META MODEL

The business process ontology that is embedded in the LEADing Practice narratives, models, tables, and diagrams can be explicated and interrelated by bringing them together in a universal conceptual structure, such as conceptual graphs (CGs).[7–9] CGs provide a graphical interface for first-order logic that enables the visualized objects and relations in the ontology to be articulated as a (class) hierarchy and, by linking (meta)objects to each other through their object relations, the direct and indirect interrelationships in business processes can be discovered. It is the vehicle by which LEADing Practice's ontology and semantics foundation can be applied to enterprises wishing to understand and improve their own processes.[10]

Figures 7.2 and 7.3 show an extract from the meta-object ontology of the business layer and application layer, respectively, taken from LEADing Practice's *Process Architecture Reference Content*.[11] The objects are shown as a CG type hierarchy, linking subtypes (subobjects) to their supertypes (superobjects). Thus, in Figure 7.2

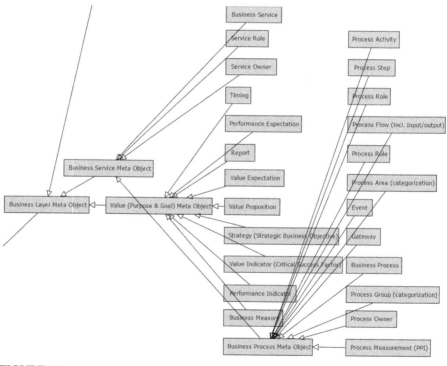

FIGURE 7.2

Extract from the business layer meta-object ontology.

2.3.4　作为框架的BPM本体论：LEADing Practice（领导实践）流程元模型的本体结构

嵌入在 LEADing Practice（领导实践）叙述、模型、表格和图表中的业务流程本体可以通过将它们组合在一个通用的概念结构（如CGs）中来解释和相互关联[7~9]。CGs 为一阶逻辑提供了一个图形化接口，使本体中的可视化对象和关系能够被表达为（类）层次结构，并且通过对象关系将（元）对象相互链接，可以发现业务流程中的直接和间接的相互关系。LEADing Practice 本体论和语义基础可以应用于希望了解和改进自身流程的企业[10]。

图2和图3分别显示了业务层和应用层的元对象本体的摘录，摘自 LEADing Practice流程架构参考内容[11]。对象显示为CG类型层次结构，将子类型（子对象）链接到其父类型（父对象）。因此，例如，在图2中，父类型路径的子类

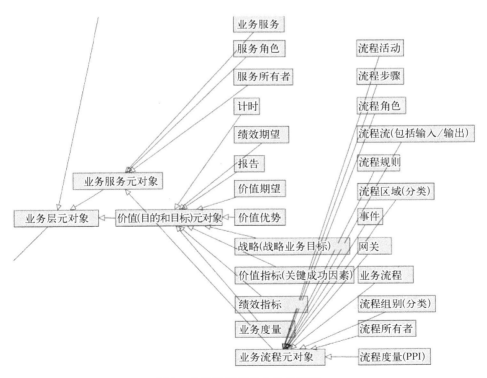

图2　业务层元对象本体的摘录

for example, the subtype to supertype path is Process Owner < Business Process Meta-Object < Business Service Meta-Object < Business Layer Meta-Object < Top (not shown). Another is Product < Business Competency Meta-Object < Business Layer Meta-Object < Top. In Figure 7.3, an example is Data Service < Data Meta-Object < Application Layer Meta-Object < Top. Another is Application Module < Application Meta-Object < Application Layer Meta-Object < Top. The "<" symbol can be read as "is a"; for example, product is a business competency meta-object. It is also transitive; thus, for example, process owner < (is a) business layer meta-object. Furthermore, it is polymorphic; properties affecting a superobject will cascade to *all* of its subobjects. Thus, if we make an assertion about the business layer meta-object, for example, then that assertion will also apply to all its subobjects; in this case, all the objects shown in Figure 7.2. Note that it does not apply the other way; thus, for example, an assentation made about the product will only affect that object. Otherwise, it would wrongly affect everything that comes under Business

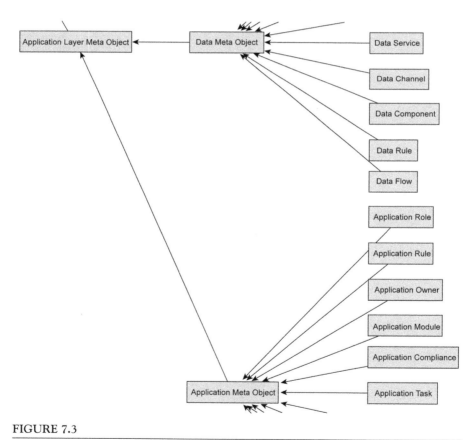

FIGURE 7.3

Extract from the application layer meta-object ontology.

型是流程所有者 ＜ 业务流程元对象 ＜ 业务服务元对象 ＜ 业务层元对象 ＜
Top(未示出)，另一个未示出的路径是Product ＜ 能力元对象 ＜ 业务层元对象 ＜
Top。在图7.3中，一个例子为数据服务 ＜ 数据元对象 ＜ 应用层元对象 ＜
Top，另一个是应用程序模块 ＜ 应用元对象 ＜ 应用层元对象 ＜ Top。＜符号可
以理解为"is a"（"是…的实例"）规则，例如，产品是业务能力元对象的实例。它
也是可传递的，因此，例如，流程所有者 ＜（is a）业务层元对象。此外，它是多
态的，影响父对象的属性将级联到它的所有子对象。因此，例如，如果我们对
业务层元对象进行声明，那么该声明也将应用于其所有子对象。在本例中，所
有对象如图2所示。请注意，它不采用另一种方式，因此，例如，关于产品的说
明将只影响该对象。否则，它将错误地影响业务能力元对象 ＜ 业务层元对

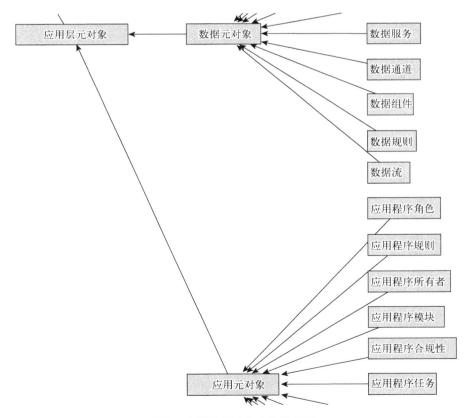

图3　应用层元对象本体的摘录

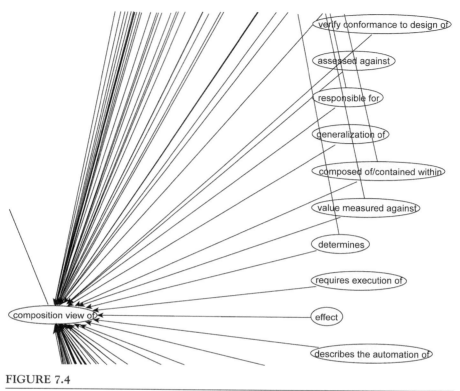

FIGURE 7.4

Extract from the composition relation ontology.

Competency Meta-Object < Business Layer Meta-Object < Top. Consequently, we have the ability to apply reasoning at multiple levels of the ontology.

Figures 7.4 and 7.5 similarly describe extracts of the object relations of the ontology as a CG relation hierarchy. The relations are structured to capture the composition-decomposition views in LEADing Practice's process architecture reference content.[12] The "<" (is a) rules also apply to object relations, such as requires execution of < decomposition view < link (not shown), in Figure 7.4. An example from Figure 7.5 would be participates in < decomposition view < link. Although not shown in these figures, some of the relations are subrelations of both the composition view and decomposition view. These are indicated by relations in the figures that have two lines going from them, one of which goes off the figure to the other view as its superrelation. One such example in Figure 7.4 is "participates in." That relation also has composition view as its superrelation. The "based on" relation in Figure 7.4 only has the decomposition view as its superrelation. Examples of both sorts of relations also appear in Figure 7.3.

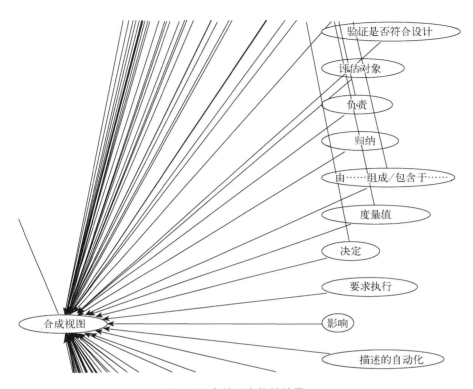

图4　组合关系本体的摘录

象 < Top下的所有内容。因此,我们有能力在本体的多个层次上应用推理。

　　图4和图5同样将本体的对象关系提取描述为CG关系层次,这些关系被构造以捕获LEADing Practice流程架构参考内容中的合成分解视图[12]。< (is a)规则也适用于对象关系,如图4中要求执行< 分解视图 < link(未示出)。图5中的一个示例将参与< 分解视图 < link。虽然这些在图4和图5中没有显示,但其中一些关系是合成视图和分解视图的子关系。图4和图5中有两条线,由图中关系所表示,其中一条线作为图的父关系从图中引出到另一个视图。图4中的一个例子是"参与",这个关系也有组合视图作为它的父关系。图4中的"基于"关系只有分解视图作为其父关系。这两种关系的示例也出现在图3中。

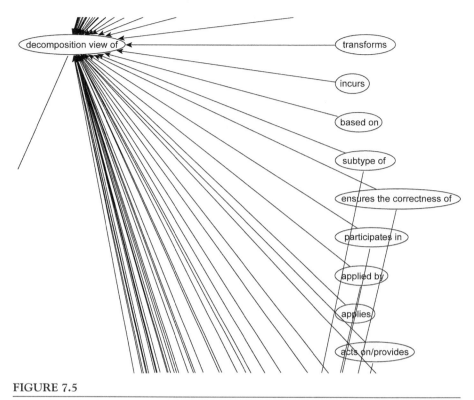

FIGURE 7.5

Extract from the decomposition relation ontology.

Figures 7.6 and 7.7 show extracts of the objects linked by their relations for the business process composition and decomposition attribute taxonomy, respectively, based on the object and relation ontology of the earlier figures. Semantics is thereby added to these taxonomies, as each object is described by its relation to other objects through the ontological structure defined in the CGs of Figures 7.2–7.5. Accordingly, for example, business process is delivered by business service. Business Service < Business Service Meta-Object; thus, properties (assertions) applied to this meta-object would cascade to business service (but not vice versa as explained earlier). Business process is not on this hierarchical path, so it would be unaffected. Of course, any properties applied to the business layer meta-object would affect them both. The same pattern applies to the object relations. Overall, we can see how this multilevel behavior affects the context of each object in relation to its others. Properties applied to the superobject and relations are thereby reused at their sublevels. Such relationship models also acts as the test that properties are not applied at too high a level, as that would highlight oversimplification through overgeneralization. Conversely, when common properties are discovered at a common sublevel, they can be generalized and reused over those objects. This generalization and specialization can be updated in

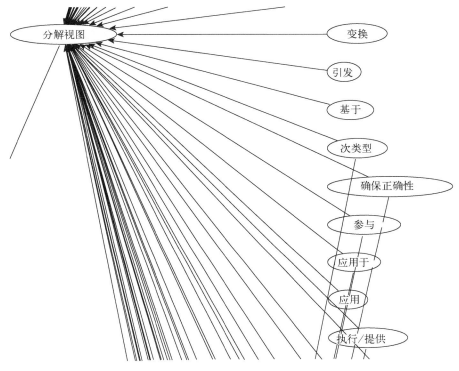

图5 分解关系本体的摘录

图6和图7分别基于前面图中的对象和关系本体,显示了由业务流程组合和分解属性分类关系链接的对象的提取。因此,语义被添加到这些分类中,每个对象通过图2～图5的CGs中定义的本体结构通过其与其他对象的关系来描述。因此,例如,业务流程由业务服务提供。因此,应用于此元对象的属性(声明)将传递给业务服务(但如前面所说,反之不然)。业务流程不在此分层路径上,因此不会受到影响。当然,任何应用于业务层元对象的属性都会同时影响它们。同样的模式也适用于对象关系。总的来说,我们可以看到这种多级行为如何影响每个对象相对于其他对象的前后关系。因此,应用于父对象和关系的属性在其子级中被重用,这种关系模型也可以作为一种测试,即属性在某个级别上不能应用得太高,因为这将通过过度泛化来强调过于简单化。相反,当在公共子级上发现公共属性时,可以对这些对象进行泛化和重用。这种泛化和特化可以根据新的最佳实

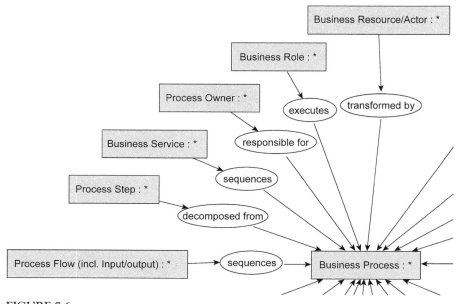

FIGURE 7.6

Extract from the business process composition attribute taxonomy.

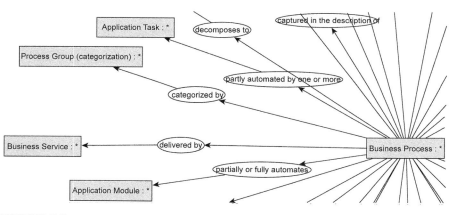

FIGURE 7.7

Extract from the business process decomposition attribute taxonomy.

the light of new best practices; notably, those best practices are being applied through CG logic rather than loosely on less formal foundations. Understanding is assisted by how objects are linked to other objects (directly and indirectly) through their relations, thus adding context to how the generalizations may be applied.

The CGs shown in each figure were drawn in the CoGui[13] software. As well as a CG editor, CoGui enables the first-order logic reasoning of CGs, as was outlined

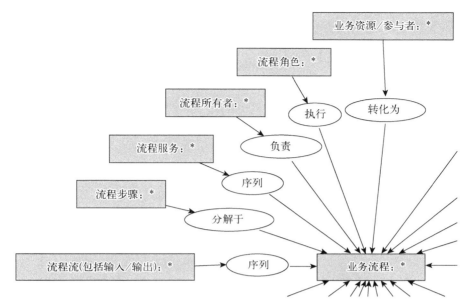

图6　业务流程组合属性分类的摘录

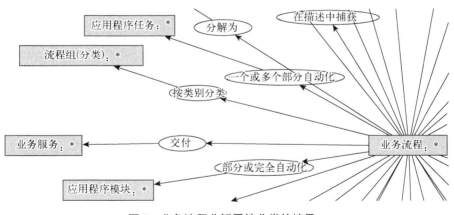

图7　业务流程分解属性分类的摘录

践进行更新,特别是,这些最佳实践通过CG逻辑应用,而不是松散地在不太正式的基础上应用。对象通过它们之间的关系(直接或间接)与其他对象联系起来,从而为如何应用泛化添加语境信息。

　　每个图中显示的CGs都是在CoGui[13]软件(一个免费的基于图形的可视化工具)中绘制的。如前所述,和CG编辑器一样,CoGui支持CGs的一阶逻辑推理。因

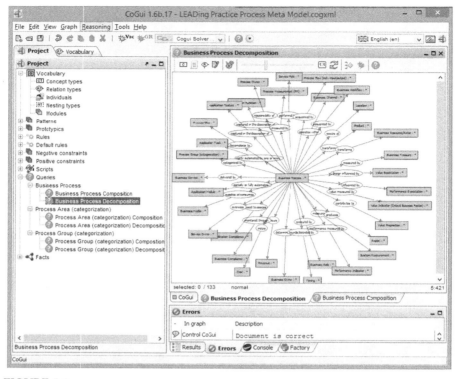

FIGURE 7.8

Example of querying the process meta-model.

earlier. Consequently, as Figure 7.8 indicates, the business process decomposition (and composition) meta-model can be used to query the models of a given enterprise. This enables the enterprise to test the conformity of their business models against the rich body of knowledge underpinned by LEADing Practice's ontology and semantics, identifying where the enterprise's own business processes might require further maturity.

DISCUSSION OF THE BPM ONTOLOGY

The BPM ontology is an empiric ontology, meaning that its roots lie in practice, as it was developed by practitioners documenting their practical knowledge of the field rather than having originated from theory and academics specialized in a restricted area of business. Consequently, it is one of the few ontologies that has the ambition to cover all aspects of business. To attain the desired level of completeness, the ontology is complemented with elicitation support, such as the guiding principles for creating, interpreting, analyzing, and using process objects within a particular domain and/or layers of an enterprise or an organization. The BPM ontology also

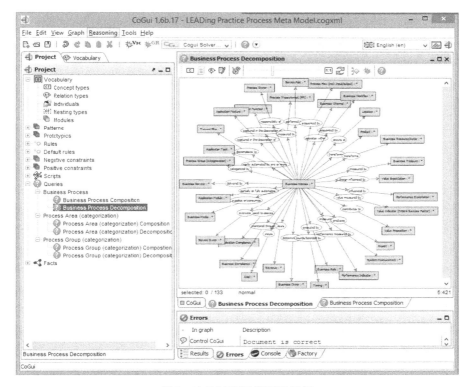

图8　查询流程元模型的示例

此，如图8所示，业务流程分解（和组合）元模型可用于查询给定企业的模型。这使企业能够根据LEADing Practice本体论和语义所支持的丰富知识体测试其业务模型的一致性，确定企业自己的业务流程可能需要进一步成熟的地方。

2.3.5　关于BPM本体论的讨论

BPM本体论是一种经验主义本体论，因为它是由实践者开发的，并且记录了他们对该领域的实际知识，而不是源自专门从事有限业务领域的理论和学术，因此它的根源在于实践。故此，BPM本体是少数涵盖业务所有方面的本体之一。为了达到所需的完整性水平，本体与启发式支持相补充，如在企业或组织的特定领域或层次中创建、解释、分析和使用流程对象的指导原则。BPM本体还提供了一组原

offers a set of principles, views, artefacts, and templates that have detailed meta-object relations and rules that apply to them, such as how and where can the process objects be related (and where not). Because the BPM ontology has the ambition to support a large community, it is open-source within the community and vendor neutral or agnostic, so it can be used with most existing frameworks, methods and or approaches that have any of the meta-objects mentioned in this document. The mapping can be found online.[14]

By sharing knowledge within the community, practitioners have found and documented repeatable patterns[15] for process-related objects, structures, and artefacts. This has led to the identification of 16 cross-domain meta-object groups that provide additional structure to the ontology, and it may lead to the development of 16 orthogonal task ontologies (e.g., describing costs or risks) that intersect with domain ontologies (e.g., business processes). However, further research is needed to determine whether or not such a decomposition is feasible and desirable.

The ontology is also complemented by a framework that helps practitioners transform their (ontological) knowledge of a process into process models and (new) working methods. To be able to cope with the complexity of the real world, the framework allows practitioners to (temporarily) simplify their (mental) models by taking partial views on their knowledge. These viewpoints are especially useful in the context of process engineering, process modeling, and process architecture.

SUMMARY

The BPM ontology's primary purpose is to provide a shared vocabulary for practitioners and academics in the business domain. This purpose was achieved by selecting terms from other business process ontologies embedded in existing frameworks, standards, and approaches and mapping them to their equivalent, which is often the exact same term, in the BPM ontology. Because practitioners need more than just a glossary to describe the aspects of business, this folksonomy is enriched with relationships between meta-objects to build a business thesaurus. This frame has been complemented with rules and a framework that should help practitioners to transform their process knowledge in competitive advantage. This will help practitioners to achieve the following:

- Identify the relevant process objects
- Decompose the process objects into the smallest parts that can, should, and need to be modeled, and then compose the process objects entities before building them (through mapping, simulation, and scenarios)
- Visualize and clarify process object relationships with the process artefacts by using maps, matrices, and models (alternative representation of information)
- Reduce and/or enhance the complexity of process modeling, process engineering, and process architecture principles by applying the process decomposition and composition standard (see decomposition and composition reference content) [16]
- Model the relevant process objects through the architectural layers
- Adding process requirements (see requirement reference content)

则、视图、原型和模板,这些原则、视图、原型和模板具有详细的元对象关系和应用于它们的规则,如流程对象可以如何和在何处关联(以及在何处不关联)。因为BPM本体有支持大型社区的雄心,所以它在社区内是开放源代码的,并且厂商是中立的或难以认知的,因此它可以与大多数现有的框架、方法或途径一起使用,这些框架、方法或途径具有本书中提到的任何元对象,并可以在线找到映射[14]。

通过在社区内共享知识,从业者发现并记录了与流程相关的对象、结构和原型的可重复模式[15],这导致了16个跨域元对象组的识别,这些元对象组为本体提供了额外的结构,并且可能导致16个正交的任务本体(如描述成本或风险)的开发,这些本体与领域本体(如业务流程)相交。然而,还需要进一步的研究来确定这种分解是否可行和可取。

本体论还得到了一个框架的补充,该框架帮助从业者将他们对流程的(本体)知识转化为流程模型和(新的)工作方法。为了能够应对现实世界的复杂性,该框架允许从业者(暂时)通过对自己知识的不完整视图来简化他们的(心理)模型。这些视图在流程工程、流程建模和流程架构的语境中特别有用。

2.3.6　总结

BPM本体的主要目的是为业务领域的从业者和学者提供一个共享的词汇表。这一目标是通过从嵌入在现有框架、标准和方法中的其他业务流程本体中选择术语,并将它们映射到它们的等价物(通常是BPM本体中完全相同的术语)来实现的。因为从业者需要的不仅仅是一个词汇表来描述业务的各个方面,这个大众分类还通过元对象之间的关系被丰富来构建业务词汇表。这一框架与规则和有助于从业者将其流程知识转换为竞争优势的框架相辅相成。这将有助于从业人员实现以下目标。

- 识别相关的流程对象;
- 将流程对象分解为可以、应该和需要建模的最小部分,然后在构建流程对象实体之前(通过映射、模拟和场景)组合流程对象实体;
- 通过使用地图、矩阵和模型(信息的替代表示)来可视化和澄清流程对象与流程制品的关系;
- 通过应用流程分解和组合标准(见分解和组合参考内容[16]),降低和/或增强流程建模、流程工程和流程架构原则的复杂性;
- 通过架构层为相关流程对象建模;
- 添加流程需求(见需求参考内容);

- Provide a structured process blueprinting and implementation (see blueprint and implementation reference content).

This chapter also demonstrated how parts of this thesaurus (i.e., the BPM ontology) have been determined and how the entire thesaurus will be formalized as a frame, which allows for polymorphic property inheritance.

In the next chapters, the BPM ontology's meta-objects, groups, categorizations, strict specialization–generalization relations, and rules will be elaborated in detail, with examples. For further information on semantic process relations, process decomposition and composition, layered modeling, process engineering, and process architecture or how the BPM Ontology content can be used, we refer the reader to to the *Business Process Reference Content.*[17]

End Notes

1. Practice L., (2014b). Interconnects with Existing Frameworks, from: http://www.leadingpractice.com/about-us/interconnects-with-main-existing-frameworks/.
2. Practice L., (2014a). Community Open Source, from: http://www.leadingpractice.com/about-us/community-open-source/.
3. A Process Ontology & Process Semantic Description, Views, Stakeholders and Concerns.
4. Rosing M. v. (Producer), *Objects and Object Relations around Business Modelling and Business Architecture* (2014). Retrieved from: http://www.leadingpractice.com/wp-content/uploads/presentations/LEADing%20Practice%20&%20OMG%20Business%20Architecture%20and%20Business%20Modelling.pdf.
5. Practice L., (2014c). The Leading Practice Process Reference Content #LEAD-ES20012BC.
6. LEADing Practice Business Process Reference Content #LEAD-ES20005BP.
7. Chein M. and Mugnier M-L., *Graph-based Knowledge Representation: Computational Foundations of Conceptual Graphs, Advanced Information and Knowledge Processing* (Springer, 2008).
8. Polovina S., *An Introduction to Conceptual Graphs Conceptual Structures: Knowledge Architectures for Smart Applications* (Springer, 1997), 1–15.
9. Sowa J., *Conceptual Structures: Information Processing in Mind and Machine* (Addison-Wesley, 1984).
10. PDF of Diagram on www.leadingpractice.com Homepage, (2014). LEADing Practice. Retrieved from: http://www.leadingpractice.com/wp-content/uploads/2014/02/LEAD-Frameworks-Enterprise-Engineering-Enterprise-Modelling-Enterprise-Architecture.pdf.
11. Process Architecture Reference Content, (2014). Retrieved from: http://www.leadingpractice.com/enterprise-standards/enterprise-architecture/process-architecture/.
12. See note 11 above.
13. CoGui: A Conceptual Graph Editor, (2014). LIRMM. Retrieved from: http://www.lirmm.fr/cogui/.
14. See note 1 above.
15. The definition of a pattern used here is the description of the repeatable and mostly used/generic specifications and relations of a topic, not all theoretically possible specifications or relations.
16. LEADing Practice decomposition and composition reference content, LEADing Practice, 2012 content.
17. See note 6 above.

- 提供结构化的流程蓝图和实施（参见蓝图和实施参考内容）。

本节还演示了如何确定该叙词表的某些部分（即BPM本体），以及如何将整个叙词表形式化为一个框架，从而允许属性的多态继承。

在下一节中，将通过示例详细阐述BPM本体的元对象、组别、分类、严格的特化–泛化关系和规则。有关语义流程关系、流程分解和组合、分层建模、流程工程和流程架构或如何使用BPM本体论内容的更多信息，我们将读者引向业务流程参考内容[17]。

参考文献

［ 1 ］ Practice L., (2014b). Interconnects with Existing Frameworks, from: http://www.leadingpractice.com/about-us/interconnects-with-main-existing-frameworks/.

［ 2 ］ Practice L., (2014a). Community Open Source, from: http://www.leadingpractice.com/about-us/community-open-source/.

［ 3 ］ A Process Ontology & Process Semantic Description, Views, Stakeholders and Concerns.

［ 4 ］ Rosing M. v. (Producer), Objects and Object Relations around Business Modelling and Business Architecture (2014). Retrieved from: http://www.leadingpractice.com/wp-content/uploads/presentations/LEADing%20Practice%20&%20OMG%20Business%20Architecture%20and%20Business%20Modelling.pdf.

［ 5 ］ Practice L., (2014c). The Leading Practice Process Reference Content #LEAD-ES20012BC.

［ 6 ］ LEADing Practice Business Process Reference Content #LEAD-ES20005BP.

［ 7 ］ Chein M. and Mugnier M-L., Graph-based Knowledge Representation: Computational Foundations of Conceptual Graphs, Advanced Information and Knowledge Processing (Springer, 2008).

［ 8 ］ Polovina S., An Introduction to Conceptual Graphs Conceptual Structures: Knowledge Architectures for Smart Applications (Springer, 1997), 1−15.

［ 9 ］ Sowa J., Conceptual Structures: Information Processing in Mind and Machine (Addison-Wesley, 1984).

［10］ PDF of Diagram on www.leadingpractice.com Homepage, (2014). LEADing Practice. Retrieved from: http://www.leadingpractice.com/wp-content/uploads/2014/02/LEAD-Frameworks-Enterprise-Engineering-Enterprise-Modelling-Enterprise-Architecture.pdf.

［11］ Process Architecture Reference Content, (2014). Retrieved from: http://www.leadingpractice.com/enterprise-standards/enterprise-architecture/process-architecture/.

［12］ See note 11 above.

［13］ CoGui: A Conceptual Graph Editor, (2014). LIRMM. Retrieved from: http://www.lirmm.fr/cogui/.

［14］ See note 1 above.

［15］ The definition of a pattern used here is the description of the repeatable and mostly used/generic specifications and relations of a topic, not all theoretically possible specifications or relations.

［16］ LEADing Practice decomposition and composition reference content, LEADing Practice, 2012 content.

［17］ See note 6 above.

Process Tagging—A Process Classification and Categorization Concept

Mark von Rosing, Neil Kemp, Maria Hove, Jeanne W. Ross

INTRODUCTION

Categorization and classification have been around for a long time. Classical categorization first appeared in the context of Western philosophy in the work of Plato[1] who, in his Statesman dialogue, introduced the approach of grouping objects based on their similar properties.[2] This approach was further explored and systematized by Aristotle in his Categories treatise, where he analyzed the differences between classes and objects.[3] Aristotle also intensively applied the classical categorization scheme in his approach to the classification of living beings (which used the technique of applying successive narrowing questions such as "Is it an animal or vegetable?" "How many feet does it have?" "Does it have fur or feathers?" and "Can it fly?"), establishing the basis for the formulation of natural taxonomy. Classification is an important tool in science. It reduces the complexity of a body of work easily as it exposes patterns and structures to provide a clearer picture of the area of interest, serving to assist in understanding the relationships and acting as a baseline for subsequent work.

LOGICAL CLUSTERING: LEARNING FROM OTHER AREAS

Within many parts of enterprise modeling, enterprise engineering, and enterprise architecture there is clarity in the classification and categorization of the objects involved. For example, data practitioners distinguish between:

- Data types, in terms of master data, metadata, tacit data and transactional data.
- Data levels, where the decomposed and composed relationship of data components, data objects, data entities, data services and data tables is established in terms of their relationship and how they assemble by order.
- Data nature, where data are grouped into either structured or unstructured data. For the most part, structured data refers to information with a high degree of organization, such that inclusion in a relational database is seamless and readily searchable by simple, straightforward search engine algorithms or other search operations; unstructured data are essentially the opposite.

2.4 流程标记——流程分集和分类概念

Mark von Rosing, Neil Kemp, Maria Hove, Jeanne W. Ross

2.4.1 介绍

分集和分类已经存在很长时间了。古典分类最早出现在西方哲学中,是柏拉图[1]在其《政治家篇》一书中提出的根据对象的相似性质进行分组的方法[2]。

亚里士多德在《范畴篇》一书的论述中对这个方法进行了进一步探索和系统化,在书里他分析了分类和对象之间的差异[3]。亚里士多德还在他对生物分类的方法中大量应用了经典的分类方法(该方法使用了一种不断缩小问题的技术,如"它是动物还是植物?""它有几只脚?""它有毛或羽毛吗?"以及"它能飞吗?"),为自然分类学的制定奠定了基础。分类是科学中的一个重要工具,它可以很容易地降低工作的复杂性。它解释了模式和结构,从而为所关注的领域提供了一个更清晰的图谱,进而有助于理解实体间的关系,并为后续工作奠定了基础。

2.4.2 逻辑聚类:从其他领域学习

在企业建模、企业工程和企业架构的许多部分中,所涉及的对象的分集和分类是清晰的。例如,从业者对数据的分类如下所述。

- 数据类型:分为主数据、元数据、隐性数据和交易数据等方面;
- 数据级别:数据组件、数据对象、数据实体、数据服务和数据表的分解和组合关系是根据它们的关系以及它们如何按顺序组合而建立的;
- 数据性质:数据被分为结构化或非结构化数据。在大多数情况下,结构化数据是指具有高度组织性的信息,这样,在关系型数据库中包含的内容是无缝的,并且可以通过简单、直接的搜索引擎算法或其他搜索操作轻松搜索;非结构化数据本质上是相反的。

Figure 1 illustrates the data classification and categorization forms.

This data example shows logical clustering in terms of data categorization and classification. In each case, these data tags provide the data expert with important information about the properties and features of the data, and tools that give insight into how to organize the data to maximize their utility, protect them, and so on.

Data Classification / Categorization	Data class / category
Level *(decomposition)	Data Component
	Data Object
	Data Entity
	Data Service
	Data Table
Nature**	Structured
	Unstructured
Type**	Master
	Meta
	Transaction
*Classify = to assemble by order **Categorize = to divide into groups	

FIGURE 1

Data classification and categorization.

The reason such a tagging is beneficial is that the properties and features of data should be understood in terms of their relationship to the larger, overarching system or structure. Logical data clustering in this case works to uncover the data nature, data type, and data levels, and thereby the aspects of their structure. These relations constitute a combined structure, and behind the properties and features in the surface phenomena there are laws of context that are a part of the complex structure. This is exactly what the logical clustering of classification and categorization tries to identify. The concept of identifying properties and features in terms of type, nature, tiers, and levels has been applied in a diverse range of fields, including anthropology, sociology, psychology, engineering, economics, positivism, functionalism, conflict theories, mathematics, to name but a few. Therefore, while concepts of logical clustering exist in nearly all the mentioned areas and also nearly all areas of information technology (IT), from application modeling, to measurements, reporting, business intelligence, etc. this maturity does not exist to this degree within the process/business process modeling (BPM) world. Although some concepts exist regarding how to tag a process according to management, or a main or supporting process, there are almost no existing concepts for process nature or process decomposition, or at least, none that put it all together for an integrated and standardized process-tagging concept.

We believe this adds to process modeling, process engineering, and process architecture difficulties and the high cost of process work. The purpose of this chapter is

图1说明了数据分集和分类形式。

这个数据示例从数据分集和分类的角度展示了逻辑聚类。在各种情况下,这些数据标记都为数据专家提供关于数据的属性和特性的重要信息,并提供一些工具,这些工具可以帮助他们了解如何组织数据,使其效用最大化并保护数据,等等。

数据分类/归类	数据级别/类别
等级*(分解)	数据组件
	数据对象
	数据实体
	数据服务
	数据表
性质**	结构化的
	非结构化的
类型**	主控者
	元
	交易

* 分类=按顺序组装
**归类=分组

图1　数据分集和分类

这样标记之所以有益是因为数据的属性和特性应该根据它们与更大、更全面的系统或结构的关系来理解。在这种情况下,逻辑数据聚类可以揭示数据的性质、数据类型和数据级别,从而揭示其结构的各个方面。这些关系构成了一个组合的结构,在表面现象的性质和特征背后,存在着构成复杂结构一部分的背景法则,这正是分集和分类的逻辑聚类试图识别的内容。从类型、性质、层次和级别上识别属性和特征的概念已应用于包括人类学、社会学、心理学、工程学、经济学、实证主义、功能主义、冲突理论、数学等在内的各种领域。因此,虽然逻辑聚类的概念几乎存在于上述以及IT的从应用程序建模到度量、报告、业务智能等所有领域,但在流程/BPM世界中,这种程度的成熟度并不存在。尽管存在一些关于如何根据管理或主要/次要支持流程对流程进行标记的概念,但是几乎没有关于流程性质或流程分解的现成概念,或者至少没有一个概念将它们组合在一起以形成一个集成的和标准化的流程标记概念。

我们认为这增加了流程建模、流程工程和流程体系结构的难度以及流程工作的成本。因此,本节旨在制定流程分类和流程分集的流程标记方案,进而从多个维

thus to set out a process-tagging scheme for process categorization, and process classification, which will provide insight to the nature of process from multiple dimensions.

CONCEPTUAL AND LOGICAL PROCESS CLASSIFICATION AND CATEGORIZATION

The classical Aristotelian worldview claims that categories are discrete entities characterized by a set of properties that are shared by their members. In analytic philosophy, these properties are assumed to establish the conditions that are both necessary and sufficient to capture meaning. According to the classical view, categories should be clearly defined, mutually exclusive, and collectively exhaustive. This way, any entity of the given classification universe belongs unequivocally to one and only one of the proposed categories. However, both within enterprise engineering and within enterprise architecture there is a more modern variation of the classical approach that derives from attempts to explain how knowledge is represented. In this approach, classes (clusters or object entities) are generated by first formulating their conceptual descriptions and then classifying the object entities according to the descriptions and possibly relations. Figure 2 illustrates an example from the view of the enterprise layers, e.g., Business, application, and technology, and the conceptual representation, of how the categories with their object entities relate to each other.

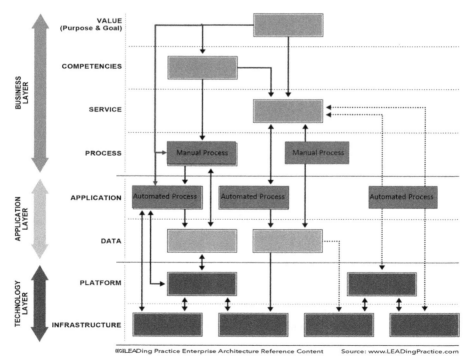

((O))LEADing Practice Enterprise Architecture Reference Content   Source: www.LEADingPractice.com

FIGURE 2

Example of layered architecture categories and their relations to each other.

度洞察流程的性质。

2.4.3　概念和逻辑流程分集与分类

经典的亚里士多德世界观认为类别是离散的实体,其特征是其成员共享一组属性。在哲学分析中,这些性质被假定为建立捕捉意义的必要和充分条件。按照古典主义的观点,分类应定义明确、彼此排斥,并共同穷尽。这样,给定分类领域的任何实体都明确地属于类别中的一个,而且只有一个。然而,在企业工程和企业架构中,经典方法有一个更现代的变体,它源自如何对知识进行解释。在这种方法中,类(集群或对象实体)的生成首先是通过构造它们的概念进行描述,其次根据描述和可能的关系对对象实体进行分类。图2从企业层(如业务层、应用程序层和技术层)以及概念表示的角度举例说明了类别及其对象实体如何相互关联。

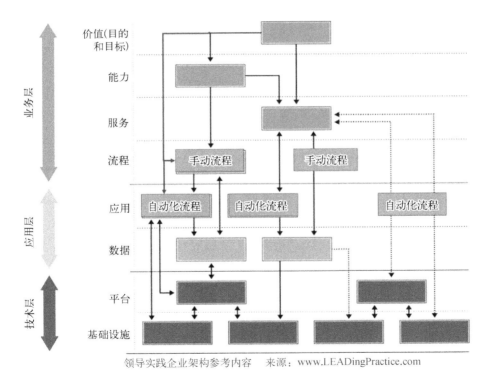

领导实践企业架构参考内容　来源:www.LEADingPractice.com

图2　分层体系结构类别及其相互关系示例

The layers within Figure 2 are in fact one classification of the parts of an enterprise that illuminates a relationship between the layered objects and the various subjects and provides the basis for the conceptual and logical clustering of the identified objects.

Conceptual and logical object clustering is the technique in which objects are identified, understood, and differentiated into their specific categories. Such concepts are also applied and presented in the BPM ontology categories (groups).

We realize that conceptual and logical object clustering is a paradigm shift toward traditional grouping, categorization, or even ordinary data clustering by generating a concept description for each generated categorization or classification potential, providing clustering of the labels for certain objects and enabling integrated and standardized classification and categorization within and across the layers allows accurate prediction of class labels as well as future-proofing areas that will be added, including the specification of object attribute relations, semantic rules, and sub-layer category labels. The task of clustering involves recognizing the inherent structure in a set of objects together by similarity into classes based on:

- Attribute relations
- Semantic rules
- Specific sub-layer categories and how the clustered objects can relate to other objects.

By using the classifications and categories such as presented in Figure 3 a logical cluster of the processes may be formed. On the other hand, clustering areas with no standardized classification and categorization labels are referred to as unsupervised classification, learning, or clustering.

Classification / Categorization	Class / Category
Level *(decomposition)	Process Area
	Process Group
	Process
	Steps
	Activities
Type*	Management
	Main
	Supporting
Tiers*	Strategic
	Tactical
	Operational
Nature*	Simple/Static
	Generic/Hybrid
	Complex/Dynamic
*Classify = to assemble by order	
**Categorize = to divide into groups	

FIGURE 3

Process classification and categorization.

　　图2中的分层实际上是对企业各个部分的分类,它阐明了分层对象和不同主题之间的关系,并为识别对象的概念和逻辑聚类提供了基础。

　　概念和逻辑对象的聚类是一种将对象识别、理解和区分为其特定类别的技术。这些概念也在BPM本体分类(组)中应用和呈现。

　　我们认识到,概念和逻辑对象的聚类是一种向传统分组、分类甚至普通数据聚类转变的范式,它通过为产生的分类或分集生成一个概念描述、为某些对象提供标签,并在层内和层间实现一体化和标准化的分集和分类,从而能够准确预测各类标签以及将要添加的未来区域,包括对象属性关系、语义规则和子层类别标签的规范。集群的任务包括通过相似性将一组对象中的固有结构识别为基于以下内容的类别:

- 属性关系;
- 语义规则;
- 特定的子层类别以及集群对象如何与其他对象相关联。

　　通过使用如图3所示的分集和分类可以形成流程的逻辑集群。另一方面,没有标准化分集和分类标签的聚类区域被称为无监督的分类、学习或聚类。

分类/归类	级别/类别
等级*(分解)	流程领域
	流程组别
	流程
	步骤
	活动
类型**	管理
	主要
	支撑
等级*	战略性
	战术性
	操作性
性质**	简单/静态
	通用/混合
	复杂/动态

*分类=按顺序组装
** 归类=分组

图3　流程分集分类

Below is an overview of the classification and categorization of the business processes:

- Process decomposition captures the manner by which the process objects are broken down into simpler forms of objects. The business process area can be decomposed into one or more business Process Group(s), while a Business Process Group is made up of multiple business processes. The business Process, in turn, can be broken down into process steps and then further into process activities.
- A process type categorizes the process based on the role of the process, differentiating between processes that focus on planning and control (management), those that are the main processes for the production of output, and those that are necessary for the main processes to execute (support).
- The process nature is categorized by the nature of the processes based on their complexity.
- Process tiers classify processes into one of three tiers, distinguishing between processes that are strategic, tactical, or operational.

In the BPM Ontology chapter, we learned about the importance of understanding the nature of a process object, its description, and its relations. The purpose of this chapter is to explore the nature of processes, the hierarchical semantic relations for identifying how the description of work can best be organized and used by process modelers, process engineers, and process architects. The intent is that resulting descriptions can be done independently of the particulars of implementation, but they can be used as the basis for relating both logical and physical descriptions the dynamic aspect of business operations, and that both the descriptions and their specification be described as a set of persistent classes and relationships.

This approach attempts to set out a repeatable pattern leading to a means to capture and represent work at levels that are typically the subject of business process analysis in BPMs and system models.

The expectation is that by making a method to describe business process behavior that is highly structured, the quality of models will improve, the result will be more accessible to stakeholders, the level of rework and process variances will reduce, and the results will be more predictable and testable. Process modeling will move from a condition where the result depends less on the ability of the specific worker and will become a more repeatable capability in the BPM team and or the BPM center of excellence (CoE).

The key difficulty that comes from traditional techniques for representing process work is that too many decisions are left in the hands of the business analyst. Specifically, when modeling work without a proper context, the start and stop points for any subject to be explored and represented can be arbitrary, resulting in depictions of work that cannot be compared with each other. Similarly, in these same circumstances, the level at which work is modeled can vary widely, and standards, if they exist, whether from Business Process Modelling Notation (BPMN), Unified Modelling Language (UML), or the Consortium for Advanced Management–International (CAM-I), are not particularly helpful. Although the modeler may strive to describe work at a specific level of abstraction, it is almost impossible to do and the resulting work will end up at different levels of detail. Fundamentally and finally, when one examines process at different levels of abstraction, one finds that processes at each of the levels have both different

以下是业务流程分集和分类的概述：

- 流程分解是将流程对象分解为更简单对象的方式。业务流程区域可以分解为一个或多个业务流程组，而业务流程组由多个业务流程组成。反过来，业务流程可以分解为流程步骤，然后进一步分解为流程活动。
- 流程类型是根据流程的角色对流程进行分类的，它不同于专注计划和控制（管理）的流程，这些流程既是生产输出的主要流程，也是主要流程运行（支撑）所必需的。
- 流程性质是根据流程的复杂性进行分类的。
- 流程层次将流程判定为战略、战术或操作流程这三个层次中的一个。

在BPM本体论一节中，我们讲到了关于理解流程对象的性质、描述及其关系的重要性。本节的目的是探讨流程的性质，以及流程建模者、流程工程师和流程架构师如何更好地组织和使用描述工作的层次语义关系。其目的是，生成的描述可以独立于现有的细节，它们可以作为逻辑和物理描述、业务操作的动态方面相关联的基础，并且这些描述和它们的规范都可以描述为一组持久的类和关系。

这种方法试图建立一个可重复的模式，从而获得一种捕获和表示级别的工作方法，而这些级别通常是BPM和系统模型中业务流程分析（BPA）的主题。

我们的期望是通过制定一种描述高度结构化的业务流程行为的方法，提高模型的质量使得结果更加易于被利益相关者接受、降低返工程度和流程差异、使结果更加可预测。流程建模将从结果较少依赖于特定工作者能力的情况开始，并将成为BPM团队和BPM CoE中更具可重复性的能力。

基于传统技术的流程工作的关键困难在于将太多的决策留给了业务分析人员。具体地说，当建模工作没有适当的工作环境背景时，会导致要研究和表示的任何主题的起始点和停止点是任意的，从而无法对工作描述进行相互比较。类似地，在这些相同的环境中，工作建模的级别可以有很大的不同，如果存在标准，无论是BPMN、统一建模语言（UML），还是国际高级管理联合会（Consortium for Advanced Management-International，CAM-I），都不是特别有用。尽管建模人员可能努力在特定的抽象级别上描述工作，但这几乎是不可能的，并且最终的工作将在不同的细节级别上完成。从根本上说，当我们在不同的抽象层次上检查流程时，我们发现

relationships and their own unique set of attributes, meaning nontrivial functional dependency of the item of interest fully depends on the name of each instance.

The result does not diminish existing work on structured methods of documenting and representing work, but it provides a framework that allows work on process design to be more accurately positioned in terms of the problems addressed and the value of the particular model used. It also provides a means both to keep the components of a specific process model at a common level of exposition and to permit models to be placed into a context to allow for a greater understanding of the context of the work.

CLASSIFICATION OF PROCESS BY METHOD OF EXECUTION

In the ground breaking book, "Enterprise Architecture as Strategy". Ross et al.[4] identify that the way processes can executed can be classified two ways. They observe that processes may by classified based on the extent to which parts of the organization perform the same process the same way and the extent to which these same processes share data.

- Process Integration – the extent to which processes within the same organization share data
- Process Standardization – the extent to which parts of the organization perform the same process the same way

Each category can be measured as being "high", or "low".

These classifications work together to create a set of extremely powerful second order categories that will inform the way the way the processes are implemented and how the technology to enable these processes can best be deployed. These secondary categories describe four possible operating strategies (Figure 4).

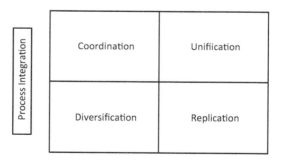

FIGURE 4

Application of process standardization and process integration to create an operating model.[4,5,6]

每个层次的流程都有不同的关系和它们自己独特的属性集,这意味着所关注的项目的重要功能依赖性完全取决于每个实例的名称。

尽管结果并没有减少现有的结构化工作方法的记录和表示工作,但它提供了一个框架,允许工作流程设计更准确地定位在解决的问题和所使用的特定模型的价值。它还提供了一种方法,既可以将特定流程模型的组件保持在公开的公共级别上,也可以将模型放在工作背景中,以便更好地理解工作环境。

2.4.4　通过执行方法对流程进行分类

在开创性的《企业架构策略》一书中,J. W. Ross 等[4]指出流程的执行方式可以分为两种。他们观察到,流程可以根据组织中哪些部分以相同的方式执行相同流程的程度以及这些相同的流程共享数据的程度进行分类。

- 流程一体化:同一组织内流程共享数据的程度;
- 流程标准化:组织中各部分以相同方式执行相同流程的程度。

每个类别都可以被评估为高或低。

这些分类共同创建了一组非常强大的二级分类,它们将预示流程的实现方式以及如何最好地部署以实现这些流程的技术。这些二级类别描述了四种可能的操作策略(图4)。

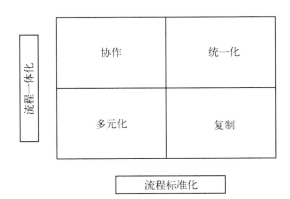

图4　应用流程标准化和流程一体化来创建操作模型[4-6]

What is ironic is that these classifications were developed to differentiate essential aspects of process but have rarely been applied in that context, but are frequently applied in many other contexts.

- Coordination – Low process standardization (interactions or processes are local, based on local requirements) but high process integration (data are shared across the enterprise).
- Unification – both high standardization and integration (data and processes are universally similar across the business). In this situation data and processes can be centrally managed to achieve economies of scale, and reduce cost and risk within the infrastructure.
- Diversification – businesses requiring low standardization (few data standards need exist) and low integration (operations are unique within the separate business units). Business units are largely autonomous.
- Replication – high standardization (transactions have limited variations and are designed centrally) but low integration (few customers are shared so data are locally owned). These conditions are seen in franchises, or retail chain stores.

These strategies are explicit, conscience design choices that lead to investment and operating decisions. The decision as to how company wants to operate with regard to standardizing and integrating processes across various organizational domains (e.g. business units, geographies, product lines, franchises). It doesn't speak to the implications of those decisions so much as it facilitates discussions as to how a company wants to pursue its business model. By examining how the data and process **should be used** the enterprise creates a "foundation for execution". By selecting the strategy, understanding the implications on the method of execution and therefore on the classification of process standardization and process integration enterprises can achieve higher profitability, faster time to market, and lower IT costs.[4]

THE NATURE OF PROCESS DECOMPOSITION

One of the fundamental difficulties with modern process practices for understanding and representing business with structured (graphical) process models revolves around the analysis and representation of work, the representation of actions involving mental or physical effort done to achieve a purpose or result. The core of the difficulty with describing work is that there is little clarity and less agreement about what is meant by work or with respect to how to organize its descriptions in ways that are meaningful, repeatable, and reusable. Methodologies, most of which are focused on software implementation of behaviour, talk about using a variety of methods to analyze and design workflows and processes within an organization, using such terms as "process," "sub-process," "activity," "task," "procedure," "transaction," or "step" to describe the way work is executed but without distinguishing the nature of one from another. Indeed, most attempts at clarification simply classify process into levels of detail. By failing to offer definitions whose perspective and context are clear, they unfortunately fail before they start.

具有讽刺意味的是,这些分类是为了区分流程的基本方面而开发出来的,但很少在这种背景下应用,而是经常在许多其他背景下应用。

- 协调:流程标准化程度低(交互或流程是本地的,基于本地需求)但是流程集成程度高(数据在整个企业中共享)。
- 统一:高度标准化和集成(数据和流程在整个业务中普遍相似)。在这种情况下,数据和流程可以集中管理,以实现规模经济,并降低基础设施内的成本和风险。
- 多样化:需要低标准化(几乎不存在数据标准)和低集成(在单独的业务单元中,操作是唯一的)的业务。业务单元在很大程度上是自发的。
- 复制:高度标准化(事务具有有限的变化,并且是集中设计的),但是低集成(很少有客户是共享的,因此数据是本地拥有的)。这些情况可以在特许经营或零售连锁店中看到。

这些策略是明确的良心设计选择,可用于投资和运营决策。但它没有涉及一家公司希望如何在不同组织领域(如业务部门、地理位置、产品线、特许经营)标准化和集成流程方面进行运营决策的含义,而仅仅在公司希望如何追求其商业模式方面进行了讨论。通过检查数据和流程应该如何使用,企业创造了“执行的基础”。通过选择策略、了解对执行方法的影响,从而对流程标准化和流程一体化分类,企业可以实现更高的盈利能力、更快的上市时间和更低的IT成本[4]。

2.4.5　流程分解的性质

用结构化(图形化)流程模型理解和表示业务的现代流程实践的一个基本困难在于工作的分析和表示,包括为达到目的或结果而进行的精神或身体努力的行为表示。描述工作的核心难点在于对于工作的含义或如何以有意义、可重复的方式组织其描述,几乎没有明确的内容,也没有达成一致。其中大部分的方法论侧重于行为的软件行为实现,讨论使用各种方法来分析和设计组织内的工作流和流程,使用诸如流程、子流程、活动、任务、程序、业务或步骤等术语来描述工作的执行方式,但不区分彼此的性质。事实上,大多数尝试只是简单地将流程划分为细节层次。因为未能提供视角和背景的清晰定义,所以它们在开始之前就失败了。

It may be that a factor with existing approaches to describing work is that they are based on the old assembly line thinking invented by automotive manufacturing, and that this approach does not directly scale to address their application in a different context.

When capturing the details of work on an assembly line, the place where the flow starts and stops is clear. Owing to the tangible nature of the work involved, whether modeling a workstation on the entire line, both the nature of the work and the scope are clear. Anyone can point out and relate to where the work starts and where it ends. Work starts at the factory door or at the start of the line and ends with a completed product. In addition, whether building subassemblies out of individual components or doing final assembly of a vehicle, the items being manipulated are clear. This clarity does not easily transfer from the physical world of transformational work to the more conceptual world that is aimed at processing information.

We have already covered this in the chapter entitled "What Is BPM," but we already see the first signs of this problem when we examine the literature, which is full of definitions for process. Although a small selection of the more commonly used definitions show a certain level of consistency, none of them provides the clarity needed to provide testable criteria, to ensure an analyst is able to be clear regarding where the work being documented is to start, where it is to end, or what level of representation is applicable to address the business problem at hand.

The core idea of a process is that any piece of definable work will always produce a specific product (or economic service); i.e., the reason for the existence of the process is the output of the product or service that it produces. In our context, an attribute or feature of a process is the output it is designed to create, each time the process is executed it will create a new instance of its product or service, and the thing that it is designed to produce remains consistent each time the process is performed.

Consider the following:

1. "A process is a group of related activities that together creates a result of value for clients."[7] This tells us that there is some sort of relationship between "process" and "activity" without giving any illumination as to what either is, or providing tests that we can use to any great effect.
2. "A business process is a series of steps designed to produce a product or service."[8] This definition tells us that there is a relationship between "process" and "steps" but leaves us no wiser.
3. "A business process is a series of logically connected business events and the logical connection is to a bigger scenario such as 'source to pay.'"[9] This view of the subject says that a process and an event are the same thing, while indicating that a series of processes is part of some larger idea.
4. "A process describes a sequence or flow of activities in an organization with the objective of carrying out work. Processes can be defined at any level from enterprise-wide processes to processes performed by a single person. Low-level processes can be grouped together to achieve a common business goal."[10] This

基于汽车制造业发明的老式流水线思想可能是现有的工作描述方法的一个因素,但这种方法不能直接扩展以解决其在不同背景下的应用。

在捕捉流水线上工作的细节时,流程开始和停止的位置是清晰的。由于所涉及工作的有形性质,不论是否在整个生产线上对工作站进行建模,工作的性质和范围都是明确的。任何人都可以指出并将工作开始的地点和结束的地方联系起来。工作从工厂开门或在生产线开动时开始,并以完成产品为结束。此外,无论是从单个部件构建子组件还是进行车辆的最终组装,被操纵的物品都是清楚的,这种清晰度不容易从工作的物理世界转移到旨在处理信息的更概念化的世界。

我们已经在题为“什么是BPM”的1.4节中讨论过这个问题,但是我们在研究文献时已经看到了这个问题的最初迹象,文献中充满了流程的定义。虽然一小部分更常用的定义显示出一定程度的一致性,但它们都不能提供可测试标准所需的清晰性,以确保分析员能够清楚地了解记录的工作从何处开始、在何处结束,或什么程度的表示法适用于解决手头业务问题。

一个流程的核心思想是:任何可定义的工作都将始终产生一个特定的产品(或经济服务),即流程存在的原因是它所产生的产品或服务的输出。在我们的工作场景中,流程的一个属性或特性就是它设计用来创建的输出,每次执行流程时,它都会创建一个新的产品或服务实例,并且每次执行流程时,它用来实现产生的东西都保持一致。

考虑以下事项:

(1)“流程是一组相关的活动,共同为客户创造价值结果。”[7]这告诉我们,流程和活动之间存在某种关系,但没对这种关系进行任何说明,也没有提供给我们可以使用并获得显著效果的测试方法。

(2)“业务流程是为生产产品或服务而设计的一系列步骤。”[8]这个定义告诉我们,流程和步骤之间存在关系,但没告诉我们更多内容。

(3)“业务流程是一系列逻辑上相互关联的业务事件,并能连接到一个如‘支付源’等的更大的场景。”[9]这个观点认为流程和事件是相同的,同时表明一系列流程是一些更大概念的一部分。

(4)“一个流程描述了一个组织中以开展工作为目标的一系列活动。流程可以在任何级别定义,从企业范围的流程到由单个人员执行的流程。低级别流程可以组合在一起以实现一个共同的业务目标。”[10]这只是表示流程可以是您想要的任

is just another way of saying that a process can be anything you want; it does not help if one is trying to create a testable, repeatable specification.

In the end, all of these definitions offer a view of process that is akin to a *matryoshka* or Russian nesting doll of decreasing size placed one inside the other with no way to distinguish where the dolls are in the hierarchy without having other dolls in the set to provide a basis for comparison.

Figure 5 shows a set of five nested *matryoshka* dolls and a set of five undifferentiated processes that are similarly nested. In the case of either the dolls or any of the processes, it is not possible to distinguish members of the set without additional information. This lack of information creates a challenge for making the documentation and representation of work a repeatable process, where the start and stop of the description is meaningful and results in a complete specification, and where the level of description is suited to the problem being addressed and the result desired.

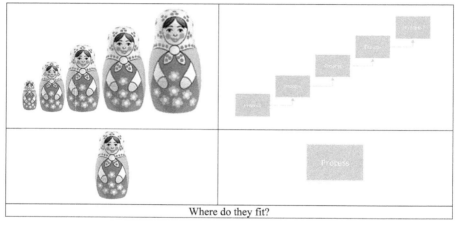

Where do they fit?

FIGURE 5

Nested dolls and nested processes.

At its simplest level, APQC's Process Classification Framework (PCF) is a list that organizations use to define work processes comprehensively and without redundancies. The goal of the PCF is to create an inventory of the processes practiced by most organizations, categorize them, and align them according to a standard system.

Processes are organized into levels. Level 1 processes are a simple categorization of service. Level 2 processes capture more detail within the same process, and so on.

In counterpoint to this, the SAP ASAP (Accelerated SAP) process hierarchy makes a partially useful attempt to distinguish among levels of complexity in processes. The SAP process hierarchy arguably offers the most advanced thinking on the classification of processes within a hierarchy, and therefore offers an excellent starting point. Key members of this hierarchy are as follows:

何东西的另一种方式,如果您试图创建一个可测试的、可重复的规范,那么它将毫无帮助。

最后,所有这些定义都提供了一个类似于嵌套娃娃或俄罗斯套娃流程的视图,一个放在另一个里面,如果在集合中没有其他娃娃,就无法区分娃娃在层次结构中的位置。

图5显示了一组5个嵌套的娃娃玩偶和一组5个类似嵌套的未分化流程。对于玩偶或任何流程,在没有附加信息的情况下,无法区分集合的成员。这种信息的缺乏使得文档和工作的表达成为一个可重复的过程,在这个过程中,描述的开始和结束是有意义的,并形成一个完整的规范,在这个过程中描述的水平适合正在处理的问题和期望的结果。

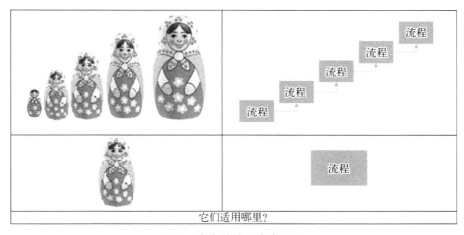

图5　嵌套娃娃和嵌套流程

在最简单的层次上,美国生产力和质量中心(APQC)的流程分类框架(process classification framework, PCF)是一个组织用来定义工作流程全面而没有冗余的列表。PCF的目标是创建大多数组织实践的流程的清单,并对其进行分类,以及根据标准系统对其进行调整。

流程被梳理成不同的级别。1级流程是对服务的简单分类,2级流程捕获同一流程中的更多细节,以此类推。

与此相反,SAP的ASAP(accelerated SAP,增强SAP)流程层次结构仅可以部分区分流程中的复杂程度。SAP流程层次结构可以提供关于层次结构内流程分类的最先进的思考,因此提供了一个很好的起点。此层次结构的主要内容如下。

1. Level 1 Process Areas: A high level aggregation of deliverable processes.
2. Level 2 Process Groups: "A bundle of processes that belong to the same area of responsibility dealing with similar tasks and activities for functional or other reasons".[11] This suggests that processes can be bundled based into arbitrary classification, i.e., "for functional or other reasons," a definition that might have been useful except that the bundling can be for any arbitrary reason.
3. Level 3 business processes: "The business process is the level that aggregates business-oriented functions or steps into a unit that is meaningful and comprehensive in the sense that the steps or functions incorporated are essential to fulfill a business mission-related task; i.e., a business process is defined by steps that transform an input into an enriched output."[12] Without knowing what a "business mission-related task" is and without clarity on how "steps or functions" are organized this definition does not directly advance our understanding.
4. Level 4 process steps: "An activity performed by a user or a piece of software together with other process steps forming a business process."
   The SAP process hierarchy specification offers specific guidance about the creations of process steps, specifying that:
   a. A process step is an activity related to exactly one object (e.g., a human, a sheet of paper, a purchase order (system), etc.).
   b. A process step is typically executed by one person and documented using an appropriate representation of the object (paper, data in an IT system, etc.).
   c. From a user interaction point of view, a process step is a single work task in a causal workflow without role change. A process step is typically identified by the fact that the task owner has all necessary responsibilities to execute the task. A process step can be performed by a single human being or by interaction between human/system and system/system."
   Although it should be evident that work can be performed by things other than "users" or "software," such as by machines, not only is the intent of this definition with its supporting tests exactly the type of guidance we are looking for, it appears to be extremely useful with regard to understanding the nature of process. We will also choose not to be confused by the use of the term "activity" in this definition and assume what is meant is "work."
5. Level 5 activities: "Activities are the lowest granularity for business process modeling and reflect the single actions a user or a system performs to fulfill the process step; i.e., filling in the fields of a special mask consists of activities as each field has to be filled to end the step." Again, other than being transactional work-centric and focused on "modeling," this definition provides clear guidance about the nature of work at this level of granularity.

The challenge, then, in part is that these definitions need to be expanded to embrace all forms of work—transformational, transactional, and tacit—and to provide greater clarity at all levels of this hierarchy so as to create an integrated set of

（1）一级流程域：可交付流程的高级集成。

（2）二级流程组："由于功能或其他原因，属于同一职责区域的处理类似任务和活动的流程分组"[11]，这表明流程可以基于任意分类进行捆绑，即"由于功能或其他原因"，这一定义只在捆绑原因非任意时有用。

（3）三级业务流程："业务流程是将面向业务的功能或步骤聚合到一个有意义和全面的单元中的级别，从这一意义上来说，所包含的步骤或功能对于完成与业务任务相关的任务是必不可少的，即业务流程是通过将输入转换为增值的输出的步骤来定义的。"[12]如果我们不知道"与业务任务相关的任务"是什么，也不清楚"步骤或功能"是如何组织的，这一定义就不能直接促进我们的理解。

（4）四级流程步骤："由用户或一个软件执行的活动以及形成业务流程的其他流程步骤。"

SAP流程层次结构规范提供了有关流程步骤创建的具体指导，其中规定了以下几条：

a. 流程步骤是指与一个对象［如一个人、一张纸、一张采购订单（系统）等］有关的活动。

b. 流程步骤通常由一个人执行，并使用对象的适当表示形式（纸张、IT系统中的数据等）进行记录。

c. 从用户交互的角度来看，流程步骤是没有角色更改的因果工作流中的单个工作任务。流程步骤通常由任务所有者具有执行任务的所有必要职责这一事实来识别。流程步骤可由单个人执行，也可由人/系统与系统/系统之间的交互执行。

虽然很明显，工作可以由用户或软件以外的事物（如机器）来完成，但这一定义的目的不仅在于支持测试，也就是我们正在寻找的指导类型，而且在理解流程的性质方面似乎非常有用。我们还将选择不要被这个定义中活动一词的使用所混淆，并假定所指的是工作。

（5）五级活动："活动是业务流程建模的最低粒度，反映了用户或系统执行流程步骤所执行的单个操作。也就是说，各个阶段都由活动组成，因为每个阶段都必须填充完整流程才能结束。"其次，除以事务工作为中心并专注于建模之外，该定义还提供了关于此粒度级别的工作性质的清晰指导。

因此，部分挑战在于这些定义需要扩展，以涵盖所有形式的工作、转换、交易和默许，并在该层次结构的所有级别上提供更清晰的定义，以便创建一组集成的可测试定义，这些定义带有支持标准，可用于生成一致的、可重复的定义。两个或两个

testable, definitions with supporting criteria that can be used to produce consistent, repeatable results that can be applied to obtain similar results by two or more analysts working to understand, document, and represent the decomposition of work. Table 1 summarizes the key classification schemes for classifying process detail/decomposition.

Table 1 *Overview of Different Views of Process Levels*						
		APQC PCF	**LEAD Process Levels**	**SAP Process Levels**	**SAP Solution Manager**	**SCOR**
Levels	1	Category	Process area	Business area		
	2	Process Group	Process Group	Process Group		
	3	Process	Business process	Business process	Scenario	Level 1
						Level 2
	4	Activity	Process step	Business process variant	Process	Level 3
	5		Process activity	Process step	Process step	Level 4
	6			Process activity		

DESCRIBING WORK

An important tool in setting out the components of any set of ideas is a diagram in the form of a structured model that shows the constituent parts of the idea and how they relate; i.e., in UML, a class model is organized within a larger framework that ensures clear boundaries are applied. Whereas class models are typically used in business analysis and software engineering to describe the structure of the information within the domain of interest, they can also be used to set out the structure of the ideas that make up the standard for describing how to express these ideas. This type of structural model is, or at least it should be, an important tool in specifying any notation-centric standard, or in any situation where the exchange of ideas or insight has a significant role.

> The challenge in attempting to bring clarification to subjects such as this is one of semantics. It is important to distinguish between the label put on ideas and the idea. The fact that "activity" as a label within BPMN or used within the LEAD process hierarchy or the SAP process hierarchy has a particular meaning in the case and the way in which "activity" is used in this document should not be taken to mean they are the same thing just because the labels are coincidently the same.

The class model for describing work that this chapter applies is set out in Figure 6 – Simplified Class Model for Describing Work. Note that "activity" is the "coal face" where work is first truly exposed; when information, for example, is created, updated, and consumed; where a transaction is finalized or where inputs complete their transformation into a final output; and where decisions are made. The remaining aspects, process area, Process Group, process, and process step, are logical and conceptual components that provide an organizing context to the level, manage the way we

以上的分析员通过工作来理解、记录和表示工作分解，从而获得类似的结果。表1总结了用于对流程细节/分解进行分类的关键分类方案。

表1　流程级别的不同视图概述

		APQC 模型的 PCF	LEAD 流程级别	SAP 流程级别	SAP 解决方案管理器	供应链运作参考模型（SCOR）
级别	1	分类	流程领域	业务域		
	2	流程组	流程组	流程组		
	3	流程	业务流程	业务流程	方案	1级
						2级
	4	活动	流程步骤	业务流程变量	流程	3级
	5		流程活动	流程步骤	流程步骤	4级
	6			流程活动		

2.4.6　描述工作

在设置任何一组思想的组成部分时，一个重要的工具是以结构化模型的形式呈现的图表，该图表显示了思想的组成部分以及它们之间的关系。也就是说，在UML中，一个类模型被组织在一个更大的框架内，以确保应用清晰的边界。虽然类模型通常用于在业务分析和软件工程中描述感兴趣领域内的信息结构，但它们也可以用来建立想法的结构，用来构成标准以描述如何表达想法。这种类型的结构模型是（或者至少应该是）指定任何以符号为中心的标准的重要工具，或者在任何情况下，在交换思想或洞察中发挥重要作用。

试图对诸如此类的主题进行明晰的挑战之一是语义。区分提出想法的标签和想法是很重要的。活动作为BPMN中的标签或在LEAD流程层次结构或SAP流程层次结构中使用的事实在该案例中具有特定含义，尽管标签是相同的，但是本节中活动的使用方式不应视为与它们一致，即便是相同也是一种巧合。

本节所应用的描述工作的类模型如图6所示——描述工作的简化类模型。请注意，活动是工作首先真正暴露的"工作面"，例如，在创建、更新和使用信息时；在交易完成或输入完成转换为最终输出的情况下；在哪里做决定。剩下的方面、

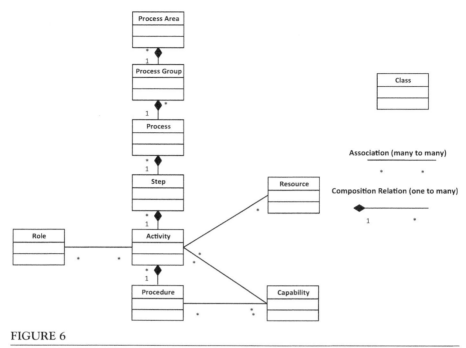

FIGURE 6

Simplified class model for describing work.

describe and understand how value or products are being achieved, and are part of the means by which we connect work at the level of process activity to strategy, which is where its value is actually exposed.

The most general classification of a process is a process area, the highest possible level of aggregation of processes. A process area may be broken down into Process Groups, logical sets of work necessary and sufficient to produce an output that is in a form that has value to the enterprise. Process Groups represent a categorization and collection of processes into a set that covers the full set of work needed to plan, provision, deliver, and decommission the resources over which they operate so as to create the desired valued output. Note that the reason for a Process Group is for its ability to create something of value. While this falls within the possible reasons set out within the SAP hierarchy, this is much more specific and has significant implications that will be discussed later.

Each instance of a process within a Process Group transforms a set of inputs into a specific instance of an output that is complete and useful in that it leads specifically to the value that shapes the scope of its Process Group. At its core, a process is a set of structured actions with logical behavior that produce a specific economic service or product. Processes have a specific typing that identifies the steps needed for its completion in a controlled manner. A process step represents a conceptual set of behaviors bound by the scope of a process which, each time it is executed (exceptions aside), leads to a single change of inputs (form or state) into a single specified output. Each

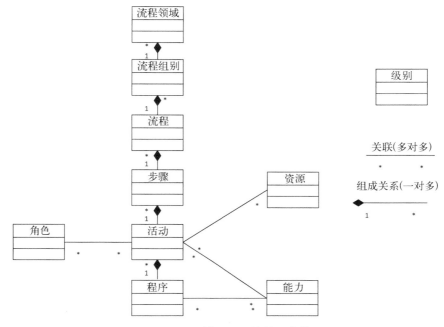

图6   用于描述工作的简化类模型

流程领域、流程组、流程和流程步骤是逻辑和概念上的组成部分,它们提供了一个层次的组织背景,管理我们描述和理解价值或产品如何实现的方式,并且是意味着我们将流程活动级别的工作与战略联系起来,而战略的价值实际上是暴露的。

　　流程的最一般分类是流程域,即流程聚合的最高级别。流程域可以分为流程组,即产生对企业有价值的输出所必需和足够的逻辑业务集。流程组将流程的分类和集合表示为一个集合,该集合涵盖了规划、提供、交付和解除资源使用所需的全部工作,从而创建所需的有价值的输出。请注意,流程组存在的原因在于它能够创建有价值的东西,虽然这属于SAP层次结构中列出的可能原因,但这更具体、具有重要的含义,稍后将对此进行讨论。

　　流程组中的每个流程实例都将一组输入转换为输出的特定实例,该实例能够刻画出流程组范围值,因此是完整且有用的。从本质上讲,流程是一组具有逻辑行为的结构化动作,能产生特定的经济服务或产品。流程有一个特定的类型,它以受控的方式标识完成过程所需的步骤。流程步骤表示一组概念性行为,这些行为受流程范围的约束,每次执行流程步骤时(抛开异常),都会导致将输入(形式或状

process step is a unit of work normally performed within the constraints of a set of rules by one or more actors in roles that are engaged in changing the state of one or more resources or business objects to create a single desired output.

A process step is in turn composed of activities. An activity is a part of the actual physical work system that specifies how to complete the change in the form or state of an input, oversee, or even achieve the completion of an interaction with others actors, and which results in the making of a complex decision based on knowledge, judgment, experience, and instinct. Each activity is a single repeated, complete, and discernible action required to complete the process based on the process type and on the policies that specify the requirements for its completion. The completion of an activity can be tested for conformance to some standard and will always lead to the completion of a specific output. Through the completion of an activity an actor in a role may change or record the state of a resource.

At an additional level of detail, the actions required to complete an activity may be captured in procedure as a set of contiguous actions, i.e., in a common "swim lane"[13] with no intervening activities performed by another role, which are part of the set of actions required to complete the process. Procedures are based on or shaped by individual or organizational capability, or by leverage-specific capabilities. That is, the features of the resources that act to enable the work will shape how that work is executed, leading to a different script.

A final layer exists; this layer is not the subject of "Architecture", but of analysis. This layer is not included in Figure 6 because this is the realm of "Procedure". This level of skilled work, where human and other capabilities connect to the work. Figure 7 shows how treating the components of a vehicle as being different classes of things

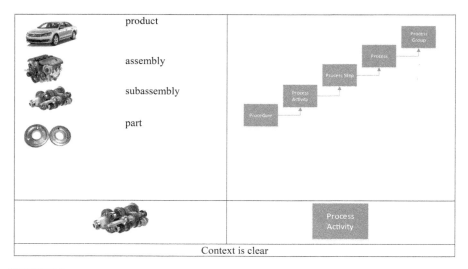

FIGURE 7

Components and processes existing in differing contexts.

态)更改为单个指定的输出。每个流程步骤都是一个工作单元,通常由角色中的一个或多个参与者在一组规则的约束下执行,这些参与者参与更改一个或多个资源或业务对象的状态以创建单个所需的输出。

流程步骤又由活动组成。活动是实际物理工作系统的一部分,它指定如何完成输入形式或状态的更改、监督甚至完成与其他参与者的交互,从而根据知识、判断、经验和直觉做出复杂的决策。每个活动都是基于流程类型和指定其完成要求的策略完成流程所需的单个重复、完整和可识别的操作。活动可以通过测试是否符合某些标准来判断是否完成,并且始终会导致特定输出的完成。通过完成活动,角色中的参与者可以更改或记录资源的状态。

在更详细的细节上,完成活动所需的动作可以在流程中被捕获为一组连续的动作,即在一个共同的"通道"[13]中,没有由另一个角色执行的干预性活动,这是完成该过程所需的一组动作的一部分。程序以个人或组织能力为基础或由个人或组织能力形成,或通过利用特定能力形成。也就是说,用于启用工作的资源的特性将影响工作的执行方式,从而导致不同的脚本。

最后的一层,这一层不是体系结构的主题,而是分析的主题。图6中没有包含这一层,因为这是程序的领域。这是一种人的能力和其他能力互相协调完成的熟练工作。图7显示了如何将车辆部件视为不同种类的事物,并以类似的方式查看流程的类型,我们不能将其视为相同的事物,而是将其视为离散和独立的事物,从

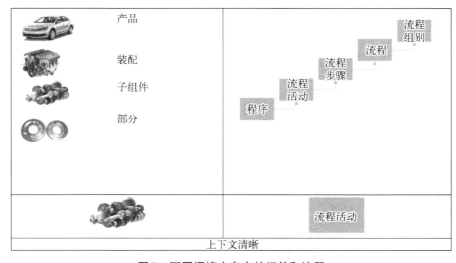

图7　不同语境中存在的组件和流程

and similarly viewing the types of processes, not as the same things, but as discrete and separate things makes the context of each clear.

A significant result of this framework is that no process will appear in more than one progress group (but will appear in the Process Group in which it adds to the ability to create value). Similarly, a step will appear in exactly one process, an activity will be within exactly one step, and a procedure will provide detail for exactly one activity. Economies of scale and reuse of resources are possible, but how these opportunities are found or expressed is not shown in this model.

It would seem clear from the above that each layer of process is actually in a different conceptual space in which each view of what constitutes process is different from the others, with different relationships, properties, and indeed purpose. In other words, breaking down a process is not about more detail, but about identifying the sub-assemblies from which the parent is constructed. In this view, the parent component is an assembly of components with features that are different and distinct from them; they are not a *matryoshka*, but are in a container–piece relationship.

Not focusing on apparently shared labels used to name these things, but paying attention to the essential features, attributes, and relationships of each level, we see that process is not a single thing but a set of things that must be viewed as different objects. The result of this thinking is that we move away from the idea that processes exist in a hierarchy where differentiation is about levels of granularity, but that they are actually different things with unique features, relationships, and behavior.

By recognizing the separate context of the various elements, it is possible to exploit the similarities within each layer of decomposition to identify better, more repeatable methods of capturing and representing the different aspects in a standard form.

PROCESS AREA
What Is a Process Area?

A process area is a high-level, abstract aggregation of a set of Process Groups that frames or positions the context of the Process Group as to its nature. A business competency may be described by the process areas needed for that competency to perform as expected.

How to Identify Process Areas

A process area is categorized according to either:

- Enterprise business areas, business units, or divisions
- End-to-end flow of process areas

How Is a Process Area Documented?

The documentation of a process area is produced in a Process Map, illustrated in Figure 8; (more on this is in the following chapter on process templates). There are various ways to graphically illustrate process areas; however, there are no standards of how to present

而使每种事物的语境变得更加清晰。

这个框架的一个重要结果是,没有一个流程会出现在多个流程组中(但会出现在能够增加创造价值能力的流程组中)。类似地,一个步骤将出现在某一个流程中,一个活动将恰好出现在某一个步骤中,并且一个程序将提供恰好某一个活动的详细信息。规模经济和资源再利用是可能的,但在这个模型中并没有显示如何发现或表达这些机会。

从上面看来,流程的每一层实际上都处于一个不同的概念空间中,在这个空间中,构成流程的每一种观点都不同于其他观点,具有不同的关系、属性和真正的目的。换句话说,分解一个流程不是为了更详细,而是为了识别从中构造父级的子程序集。在此视图中,父组件是具有不同于子集的特性的组件的组合,子集不是俄罗斯套娃,而是容器-糖块的关系。

我们不关注用于命名这些事物的明显的共享标签,而是关注每个层次的基本特征、属性和关系,我们看到流程不是一件单一的事情,而是一组必须被视为不同对象的事物。这种思想的结果是,我们不再认为流程存在于一个层次结构中,在这个层次结构中,区分是关于粒度级别的,但是这些流程实际上是不同的东西,具有独特的特性、关系和行为。

通过识别不同元素的独立语境,可以利用每个分解层中的相似性来识别以标准形式捕获和表示流程不同方面的更好、更可重复的方法。

2.4.7 流程域

1. 什么是流程域?

流程域是一组流程组的高级抽象聚合,它们根据流程组的性质构建或定位流程组的上下文关系。业务能力可以由该能力按照预期执行所需的流程域来描述。

2. 如何识别流程域

流程域按照以下两种方式进行分类:

- 企业业务域、业务单元或部门;
- 流程域的端到端流程。

3. 如何记录流程域?

流程域在流程地图中记录表述,如图8所示(有关更多信息请参阅后续流程模板一节)。有各种各样的方法来图形化地说明流程域,但是,没有关于如何呈现它们的标准。在这种情况下,重要的是要知道BPM符号不会在其符号中显示流程

Process Area
Defense Navy Administration
Naval Force Planning
Force Generation
Navy Force Employment

FIGURE 8

Illustration of defense Navy process area example of a Process Map (list).

them. In this context, it is vital to know that BPM notations do not show Process Areas in their notations. The first example of this documentation presented will be value chain diagrams from ARIS (Architecture of Integrated Information Systems) (see section on Prof. Scheer); then other variations of that we see organizations use to illustrate their process areas or the flow of process areas and groups.

In the following, we present various ways that we see organizations represent their process areas:

- Representation of process areas in a Process Map (list), with a defense Navy process area example
- Value-added chain diagram: Figure 9 – Illustration of a Value added Chain Diagram-example ITIL V3 Process Areas, shows the IT service centric process areas and the supporting Process Groups.

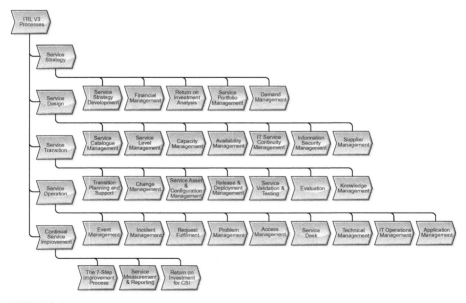

FIGURE 9

Illustration of a value-added chain diagram: example of ITIL V3 process areas.

流程域
国防海军管理局
海军规划
军兵力生成
海军部队就业

图8　以国防海军流程域为例的流程地图(清单)图示

域。本文的第一个例子是来自ARIS的价值链图(见1.3.2小节);然后在其他变体中,我们看到组织如何说明其流程域或流程域和组的流程。

在下面的文章中,我们将介绍组织表示其流程域的各种方式。

- 以国防海军流程域为例,在流程地图(清单)中的流程域表示;
- 增值链图——图9增值链图示意图:ITIL V3流程域示例显示了以IT服务为中心的流程域和支持的流程组;
- 流程域可以通过组织实体及其相互关系加以说明。这可以在国防组织实

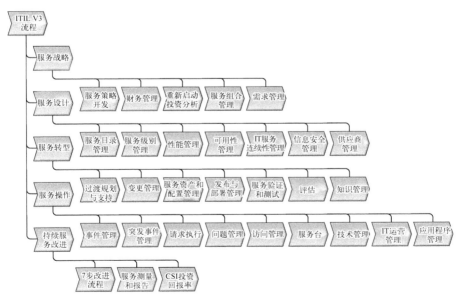

图9　增值链图示意图:ITIL V3流程域示例

- A Process Area could be illustrated by the organizational entities and their relationship to each other. This can be seen in the example of defense organizational entities and the command, reporting, and information relationship (Figure 10). While we also do not say in the example that this is a good way of illustrating process areas, nevertheless we also see such illustrations graphically showing the process areas and how they in their command, reporting, and information relationship work together

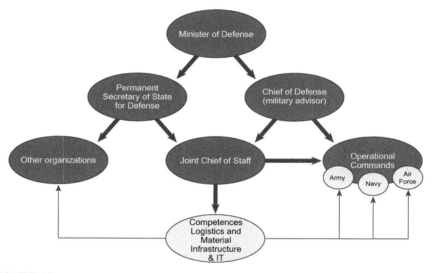

FIGURE 10

Process area based on enterprise/business high-level relations.

PROCESS GROUP

What Is a Process Group?

A Process Group is a bundle of processes that acts as a container to combine a coherent and complete set of processes, which together produce a final output that provides a specific benefit or value to a specific set of stakeholders. These stakeholders may be either internal or external to the enterprise. A well-formed Process Group will consist of a complete set of processes that describes the activity required to carry out the full set of work required for the business to deliver both the valued output that the Process Group is intended to produce and all the business objects within its operational cycle. The requirement to establish a Process Group is determined by exploring options to achieve economies of scale within the work of the enterprise and by exploring the chain of value necessary for the enterprise to achieve its purpose and address the needs and requirements of its marketplace.

体和指挥、报告和信息关系的示例中看到(图10)。虽然我们在示例中也没有说这是说明流程域的一种好方法,但是我们也看到了这些图示以图形方式显示流程域,以及它们在指挥、报告和信息关系中如何协同工作。

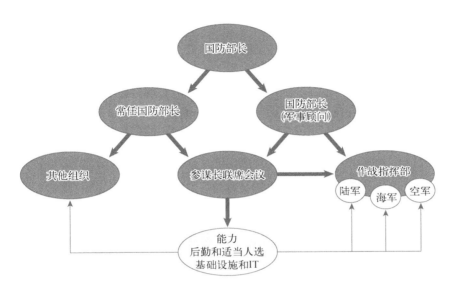

图10   基于企业/业务高层关系的流程域

2.4.8   流程组

1. 什么是流程组?

流程组是充当容器的一组流程,能够将一组连贯完整的流程组合在一起,共同产生最终输出,为特定的利益相关者提供特定的利益或价值。这些利益相关者可以是企业内部的,也可以是企业外部的。良好的流程组将由一套完整的流程组成,这些流程描述了执行业务所需的全套工作所需的活动,以便交付流程组打算生成的有价值的输出以及其运营周期内的所有业务对象。建立流程组的要求是通过探索在企业工作中实现规模经济的各种选择,以及探索企业实现其目标和满足市场需求所必需的价值链来确定的。

Within a Process Group, there is exactly one process (and, as we will see, one process step and one process activity) that delivers the output for which the Process Group is valued. All other processes that are within the group and as part of the larger package are there to ensure that the main process has everything it needs to ensure the value desired from this work is delivered.

Generally, a Process Group provides a specific capability to the enterprise to operate a delegated center of operation within an enterprise, or to operate the infrastructure of the enterprise itself. Examples of what is achieved by the processes within a Process Group include any well-bounded set of processes that provide the enterprise with the ability to produce a cohesive output, for internal or external consumption, that solve a specific business problem.

A key feature of a Process Group is that it contains the set of work supported by relevant capabilities necessary to deliver the entire set of work required to deliver the desired value over all phases of value creation for the output it exists to produce. In formulating the way work can be performed in an enterprise, the opportunity to achieve an economy of scale through the sharing of resource that encapsulates this value is critical. Although it is not obvious, this relation is in fact significant. By scoping a Process Group not based on functional or other reasons, but solely to the type of value produced, two results are achieved: First, a repeatable basis for identifying the purpose of and processes within a Process Group is established, and secondly, a Process Group becomes the means by which a Business Service is provided the resources it needs to fulfill its intended purpose.

When it is determined that a product produced by a process within a Process Group can be organized so the enterprise can benefit from the result in a sharable fashion, two things happen: First, the product ceases to be viewed for what it is, but is judged for its value, thus being separated from the other members of its current Process Group to become the final valued output of a new Process Group. Second, this simultaneously necessitates its being supported by its own unique set of capabilities, including the full set of processes necessary to ensure it is able to deliver the value expected of it within its own value chain.

Why Separate Process Group and Business Service as Unique Concepts?

The idea of a service in this sense (which should not be confused with the idea of an economic service, which is a class of intangible commodities and has no meaningful part in these concepts, except tangentially) is that the complexity of resources used to produce an output can be hidden from the consumer and can then be reused for different purposes, together with the policies that should control usage. Once the complexity is hidden, the consumer of the service, whether data, application, infrastructure, platform, or business, is able to access the output of a service with no understanding or knowledge of anything that happens behind the interface. The whole point of a service is that the consumer of the service does not need to know any of the magic needed to provide the value they are looking for. As a consumer

　　在一个流程组中,恰好有一个流程(如我们将看到的,还有一个流程步骤和一个流程活动)提供了对流程组进行估值的输出。集团内的所有其他流程以及作为大部门的一部分的所有其他流程都在那里,以确保提供此工作所需的价值。

　　通常,流程组通过为企业提供特定的能力以在企业内部实现对授权业务和基础设施的运行,流程组内实现的示例包括任何边界良好的流程集,这些流程集为企业提供了为内部或外部消费提供输出的能力,从而解决特定的业务问题。

　　流程组的一个关键特性是它包含由相关功能支持的一组工作,这些功能是交付整个工作集所必需的,这些工作可以在为其生产的产出创造价值的所有阶段中,交付所需的价值。在制定企业中工作的执行方式时,通过共享封装此价值的资源来实现规模经济的机会至关重要。虽然这并不明显,但这种关系实际上意义重大。通过不基于功能或其他原因确定流程组的范围,而仅仅根据所产生的价值类型,可以获得两个结果:第一,确定流程组内目标和流程的可重复基础;第二,流程组成为业务服务提供达到其预期目的所需资源的手段。

　　当确定由流程组内的流程产生的产品可以被组织,使企业能够以可共享的方式从结果中受益时,有两件事将会发生:第一,产品不再仅仅被视为是一个产品,而是根据它的价值进行判断,从而与当前流程组的其他成员分离,成为新流程组的最终价值输出;第二,这同时需要其自身独特的能力支持,包括全套必需的流程,以确保其能够在其自身价值链中实现其预期价值。

　　2. 为什么将流程组和业务服务分开视作独特的概念?

　　从这个意义上讲,服务的概念(不应与经济服务的概念混淆,经济服务是一类无形商品,在这些概念中没有任何有意义的部分,只是相关的)是指可以对消费者隐藏并用于产生输出的资源的复杂性,以及应该进行使用控制的策略。一旦隐藏了复杂性,无论是数据、应用程序、基础设施、平台还是业务,服务的使用者都能够访问服务的输出,而不需要了解接口后面发生的任何事情。服务的全部意义在于:消费者不需要知道用于提供他们正在寻找的价值所需的任何隐藏的东西。

of a Business Service, I am simply looking for something to be achieved; the details of who, what, where, when, how, etc. are not of interest to me, and actually none of my business. To put it another way, I am interested in the value accrued from the relationship and leave the details to the service provider

A Business Service in this context is "The externally visible ("logical") deed, or effort performed to satisfy a need or to fulfill a demand, meaningful to the environment." To be instantiated, a Business Service requires a full set of capabilities, processes, information, major systems, people, organizational structures, etc. Thus, the concept of service allows for both simplification of the relationship between provided and consumer and separation of value and effort. It does not actually do away with the need to account for, describe and ensure the enterprise has access to the capabilities while exploring how to design the enterprise so as to deliver value.

To be clear, as the customer of a process I am in the process and therefore am aware of what is happening, whereas when I am a customer of the service, I simply make requests of the service and receive that which I seek and therefore acquire the value I am seeking without gaining significant insight into who is doing the work, how the work is done, how exceptions are handled by the service provider, where the work is performed, and so on (or at least, should not be made aware of these if the service is designed and operating properly).

How to Identify Process Groups

A Process Group must meet all the following conditions:

1. Every Process Group will have a single output identified that it is accountable to produce.
2. The output of the Process Group must be in a final valued form (recognized through the fact that the output is recognized as value imparted from within the enterprise though a Business Service, which held as accountable for value through a contract, service level agreement, or similar vehicle).
3. The Process Group output must address the recognized needs of an identified group and be received by at least two categories of recipients.
4. A single, unique owner is accountable for the output.
5. The Process Group must be represented by a Business Service created to encapsulate the set of resources and business processes needed to produce the output.
6. The output Business Service must be either consumed by two or more other Business Services, two or more recipients of the Business Service, or some combination.
7. A Process Group must be independent of all other Process Groups such that if any other Process Group or groups for some reason cease to exist the Process Group in question continues to exist and remains unchanged in terms of its output. (Though it will be required to perform additional processes to address the shortfall.)

Further, a Process Group will consist of multiple processes that are each of one of five types; all five types must be represented in the processes of a Process Group

作为一个商业服务的消费者,企业只是在寻找可以实现的东西,关于谁、什么、在哪里、何时、如何等的细节对客户来说不感兴趣,实际上与企业的业务无关。换一种说法,消费者对关系产生的价值感兴趣,并将细节留给服务提供商。

在这种情况下,业务服务是"外部可见的(逻辑的)行为,或为满足需求所做的努力或满足对环境有意义的需求"。如果要实现实例化,那么业务服务需要一整套功能、流程、信息、核心系统、人员、组织结构等。因此服务的概念既可以简化提供者和消费者之间的关系,也可以分离价值和努力。它实际上并没有消除在探索如何设计企业以实现价值的同时解释、描述和确保企业能够使用这些功能的需要。

明确地说,流程的客户参与流程因而知道正在发生的事情,然而对于服务的客户来说,其只需提出服务请求并接收其所寻求的服务,从而获得其所寻求的价值,而不必对谁在做工作、如何完成工作、服务提供者如何处理异常、在何处执行工作等有深入的了解(或者至少不应了解到这些,如果服务设计和运行正常的话)。

3. 如何识别流程组

流程组必须满足以下所有条件:

(1)每个流程组都将有一个标识为它负责生成的输出;

(2)流程组的输出必须是最终的有价值的形式(通过以下事实来确认:输出被确认为通过业务服务从企业内部传递的价值,业务服务通过合同、服务协议或类似的工具对价值负责);

(3)流程组输出必须满足已标识组的已识别需求,并且至少由两类接收人接收;

(4)一个唯一的所有人对输出负责;

(5)流程组必须由一个业务服务来表示,该服务是为封装生成输出所需的一组资源和业务流程而创建的;

(6)输出业务服务必须由两个或多个其他业务服务、两个或多个业务服务接受者或其组合使用;

(7)流程组必须独立于所有其他流程组,这样,如果任何其他流程组因某种原因停止,相关流程组则继续存在,并且在其输出方面保持不变。不过,它将需要执行额外的流程来弥补这一缺口。

此外,流程组将由五种类型中每一种的多个流程组成,所有五种类型都必须在流程组的流程中表示。

1. Planning processes that describe the work of determining how the Business Service will respond to demands. Planning processes operate on a planning cycle, or in response to a contingency.

2. Provisioning processes that describe the work of preparing the Business Service to respond to demands in accordance with plans. Provisioning processes operate in response to a routine drawdown of resources, or in response to a contingency or required protective or stewardship activity.

3. Delivery processes that operate repeatedly when each request for a Business Service output is received.

4. Deregister/decommission processes that recognize the lifecycles of resources, suppliers, Business Service outputs, or Business Service recipients and operate according to the lifecycle stages of these elements

5. Oversight processes monitor, provide feedback, and thus control the performance of the other processes within the Process Group.

In the simplest organization there will only be three Process Groups: those needed to manage the organization itself; those needed to operate the organization; and those needed to plan, provision, deliver, and terminate the creation of the single thing that represents the value proposition central to its purpose by meeting the needs of the target community that it services. That being said, even the smallest organization will draw on services beyond its doors to give itself the wherewithal to achieve its purpose and carry out the production of its valued output, no matter what it is. As the organization grows in size, it will seek to differentiate itself within its environment and to optimize cost and value. This leads to the need and ability to the need and ability to hollow out aspects of the initial process set so as to specialize, standardize, achieve economies of scale, or use specialized resources capable of addressing high-complexity/low-value Business Services, in the process perhaps changing the mix of services it provides for itself and those it draws from the market. Every time this occurs, the affected process will disappear from the set of processes that existed before to be replaced by a Business Service that delivers not the product, but the value of that product, while simultaneously adding a new and complete Process Group to instantiate for the Business Service the necessary planning, provision, delivery, and deregistration processes, while addressing the need to ensure oversight of this new feature.

How Are Process Groups Documented?

Documentation of Process Group is done in a Process Map, illustrated in Figures 11 and 13 also (more on this is in the following chapter on process templates). A Process Group is named based on the valued output it is accountable for producing, prefixed with a verb that imparts the finality of what is accomplished. Verbs such as "provide" (provide funding, provide food), "furnish" (furnish car, furnish payment), "address" (address question, address complaint), and "steward" (steward funds, steward buildings) are appropriate, as are others in Figure 11.

（1）描述确定业务服务将如何响应需求工作的计划流程。计划流程在计划周期内运行，或响应突发事件；

（2）描述根据计划准备业务服务以响应需求的供应流程，供应流程的运作是为了响应日常的资源减少，或响应应急或必要的保护或管理活动；

（3）接收到业务服务输出的每个请求时重复运行的传递流程；

（4）取消注册/取消授权流程识别资源、供应商、业务服务输出或业务服务接受者的生命周期，并根据这些元素的生命周期阶段进行操作；

（5）监督流程监控、提供反馈，从而控制流程组内其他流程的绩效。

在最简单的组织中，只有三个流程组：管理组织本身所需的流程组；运营组织所需的流程组；以及计划、提供、交付和终止创建单一事物所需的流程组，这些事物通过满足其目标群体的需要来代表其目标的核心价值主张。这就是说，即使是最小的组织也会利用其额外的服务来为自己提供实现其目的的必要资金，并实现其有价值的产出，无论它是什么。随着组织规模的扩大，它将寻求在其环境中实现差异化并优化成本和价值。这意味着需要对初始流程集的各个方面进行挖掘，或使用能够解决高复杂性/低价值业务服务的专业资源，以便实现专业化、标准化、规模经济。在此过程中，它为自己提供的服务组合以及从市场中提取的服务组合可能会发生变化，当发生这种情况时，受影响的流程将从之前存在的流程中消失，而由业务服务替换，需要说明的是，这种业务服务不是产品，而是产品的价值，同时添加新的完整流程组以实例化业务服务、规划、供应、交付和注销流程，同时要对此新功能进行监督。

4. 如何记录流程组？

流程组的文档是在流程地图中完成的，如图11和图13所示（关于这方面的更多信息，请参阅后续2.5.3小节"什么是流程模板？"）。流程组是根据其负责生成的有价值的输出来命名的，前面加上一个动词，表示所完成工作的最终结果。诸如提供（提供资金、提供食物）、供应（供应汽车、供应付款）、解决（解决问题、解决投诉）和管理（管理资金、管理大楼）等动词是适当的，也如图11中的其他动词。

Defense Navy Process Map	
Process Area	**Process Group**
Defense Navy Administration	Exercise Controls and Analysis
	Execute Navy Budget
	Grants Management - Grantee
	Manage Employee Resources
	Manage Position Plan
	Formulate Budget
	Locality Management
	Manage Vehicle Fleet
	Provide Medical and Health Service
	Conduct Analytics
	Manage Financials
	Manage Human Capital
	Deliver Corporate Services
	Provide Operations Support
Naval Force Planning	Develop Strategic Plan
	Develop Naval Force Requirement
	Develop Naval Force Goal
	Develop Naval Force Plan
Force Generation	Create Navy Master Data
	Create Basic Navy Organization
	Manage Navy Personnel
	Manage Navy Equipment
	Develop Organization
	Conduct Individual Training
Navy Force Employment	Deploy Navy
	Manage Navy IT Landscape
	Organizational Flexibility
	Deliver Mobile Personnel and Organizational Management
	Provide Collective Naval Training

FIGURE 11

Illustration of defense Navy Process Map with Process Groups structured based on the process areas.

As no sequence constraints are implied by Process Groups, they cannot be captured in anything more than a Process Map (a process list) and no structured model will represent relationships between or among a set of Process Groups (as it is defined by the structure of the process areas).

The Nature of Process Groups

Process Groups may be tagged and categorized based on the nature of the valued output of each Process Group. They can also be tagged in terms of their financial impact and their value contribution, as well as whether they are strategic, tactical, or operating in nature or in terms of their lifecycle in the delivery of value: provision, planning, delivery, and decommissioning of their ability to create value.

国防海军流程地图	
流程域	**流程组别**
国防海军管理局	操作控制和分析
	执行海军预算
	赠款管理——受赠人
	管理员工资源
	管理职位计划
	制定预算
	区域管理
	管理车辆舰队
	提供医疗卫生服务
	进行分析
	管理财务
	管理人力资本
	提供公司服务
	提供操作支持
海军规划	制定战略计划
	制定海军需求
	制定海军目标
	制定海军计划
军兵力生成	创建海军主数据
	创建海军基本组织
	管理海军人员
	管理海军装备
	发展组织
	进行个人培训
海军部队就业	部署海军
	管理海军IT环境
	组织灵活性
	提供移动人员和组织管理
	提供集体海军训练

图11 国防海军流程地图图示,其中流程组按流程域排列

由于流程组不受任何序列约束,它们不能以流程地图(流程清单)之外的任何方式被捕获,并且任何结构化模型都不能表示两组流程组之间或多组流程组之间的关系(由流程域的结构定义)。

5.流程组的性质

流程组可以根据每个流程组的价值输出的性质进行标记和分类。它们还可以根据其财务影响和价值贡献、是否具有战略/战术或操作性质,或在价值交付的生命周期中进行标记:提供、规划、交付和结束它们创造价值的能力。

PROCESS

What Is a Process?

Although we say that a set of processes is decomposed from a Process Group, this does not mean that a process is simply more detail about the actions needed to complete the goal of a Process Group.

A process, like a progress group, is conceptual; it contains no choices, but will be a member, along with other processes, of the chain of dependent work within the complete Process Life Cycle presented in Figure 12, which is necessary and sufficient for the output of the Process Group to be realized. Each Process Group will consist of the set of processes required to plan, prepare, and deliver its valued output.

In each case, a process will produce a single, usable, and complete business object: a product, a control object, or information object, any of which can then be consumed as a single thing, which makes it essential to fulfill the requirements to complete something that is needed by the enterprise as a means to act. Within transactional work, this will include the set of work required to complete any business object directly related to the purpose of the progress group, and within

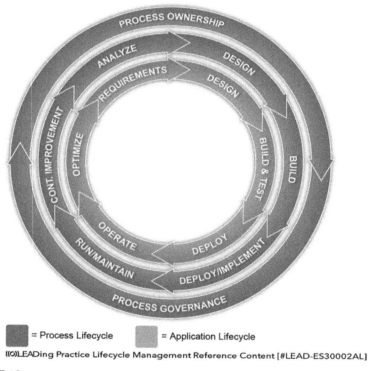

■ = Process Lifecycle   ■ = Application Lifecycle

((©)LEADing Practice Lifecycle Management Reference Content [#LEAD-ES30002AL]

FIGURE 12

Process Life Cycle.

2.4.9　流程

1.什么是流程?

尽管我们说一组流程是从一个流程组中分解出来的,但这并不意味着一个流程只是关于完成一个流程组的目标所需的动作的更详细的描述。

一个流程,就像一个流程组一样,是概念性的,它不包含任何选择,但将与其他流程一起成为如图12所示的完整流程生命周期内依赖工作链的成员,这对于实现流程组的输出是必要的和足够的。每个流程组将包括计划、准备和交付其有价值的输出所需的一组流程。

流程在每种情况下都将生成一个单一的、可用的和完整的业务对象,包括:产品、控制对象或信息对象,其中任何一个对象都可以作为一个单独的对象来使用,这就使得满足需求以完成企业作为一种行为手段所需要的事情变得至关重要。在事务性工作中,这将包括完成任何与流程组的目的直接相关的业务对象所需的工

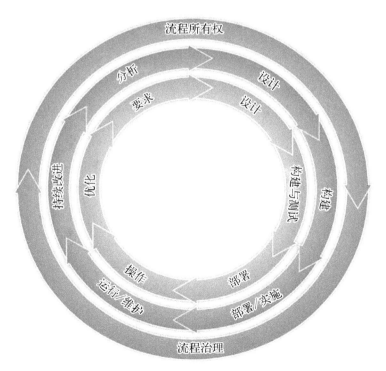

外圈、中圈:流程生命周期　　　　内圈:应用程序生命周期

领导实践生命周期管理参考内容

图12　流程生命周期

transformational work, anything that leads to the creations of the product or economic service that is central to its purpose or mandate. Although the output of a Process Group, the thing of value, and the output of one of the processes in a Process Group may appear to be identical, the purpose of the Process Group is to produce the thing of value, and the purpose of all other work and their associated business objects is to be the means by which that value is produced in a controlled, effective, and efficient method.

The process lifecycle is actually another classification scheme for categorizing processes, distinguishing among those that are part of specifying the conception, planning, and arrangement of processes, the processes that are required to take what was designed and bring it to operation, when it is then monitored, and finally those processes necessary to analyze or assess the performance of one or more processes. (See more on process lifecycle in the BPM Process Life Cycle chapter.)

How to Identify Processes

Each process will produce a single, complete, and meaningful result that contributes to the completion of the valued output necessary for the conclusion of the work of a Process Group. This output may be a transformation of a business object (through transformational work), the result of a complete set of transactions that produce a control object (achieved by transactional work), or the creation of new knowledge or insight (tacit work).

When creating a process, the name given the process will be derived from its goal. The process goal is an atomic statement describing the result of the successful completion of the process based on the process type and output produced.

The output of the process is the resource whose goal it is for the process to change the state of. In better English, we could say that it is the goal of a process to change one or more inputs into a specific output each time it is executed.

Processes are named by referring to the business object that is completed by the process, suffixed with a verb that gives a sense of completion. Verbs such as "determine," "complete," and "answer" provide this sense of completion, whereas "draft," "develop," and "propose" lack the finality needed to convey the needed completeness and will result in candidate processes that will not stand up to further decomposition.

How Are Processes Documented?

Documentation of business processes is captured in a Process Map. An example of such a Defense Process Map is illustrated in Figure 13.

Because processes are a logical view of work and at a level of abstraction where they simply identify what the result of the expended effort is, there is no means to identify the impact of the process not succeeding. That being said, processes have preconditions that must be met or dependencies that must be satisfied. Therefore, whereas a model map may be used to show the inventory of processes, a process model will show the dependency chain required.

作组合,以及在变革性工作中任何导致产品或经济服务的创造的、对其目的或任务至关重要的工作。尽管流程组的输出、有价值的事物和流程组中某个流程的输出看起来可能是相同的,但流程组的目的是产生有价值的事物,而所有其他工作及其相关业务对象的目的是以受控、有效和高效的方法产生该价值。

流程生命周期实际上是对流程进行分类的另一种分类方案,区别于那些规定流程的概念、计划和安排的部分,以及当流程被监视时,需要采取设计并使其运行的流程,最后分析或评估一个或多个流程的性能是必要的。(有关流程生命周期的更多信息,请参见BPM流程生命周期一节。)

2.如何识别流程?

每个流程将产生一个单一、完整和有意义的结果,有助于完成流程组工作所需的有价值的输出。此输出可能是业务对象的转换(通过转换工作)、生成控制对象的完整事务集的结果(通过事务性工作实现)或创建新知识或洞察力(隐性工作)。

创建流程时,给定流程的名称将从其目标派生。流程目标是一个微小语句,根据流程类型和生成的输出描述流程成功完成的结果。

流程的输出是一种资源,其目标是使流程更改状态。更准确地说,我们可以说,流程的目标是在每次执行时将一个或多个输入转化为特定的输出。

流程是通过引用流程完成的业务对象来命名的,该业务对象的后缀是一个提供完成感的动词。诸如决定、完成和回答之类的动词提供了这种完成感,而起草、发展和建议缺乏传达所需的完整性和最终性,并将导致候选流程无法经受进一步的分解。

3.如何记录流程?

业务流程记录在流程地图中,如国防流程地图(图13)。

因为流程是工作的逻辑视图,并且在抽象层次上,它们只是简单地确定所花费工作的结果是什么,所以没有办法确定流程失败带来的影响。话虽如此,流程具有必须满足的先决和依赖条件。因此,虽然模型图可用于显示流程清单,但流程模型将同时显示所需的依赖关系。

Defense Navy Process Map		
Process Area	**Process Group**	**Defense Navy Business Process**
Force Generation	Create Navy Master Data	Create Reference Force Elements
		Create Jobs
		Maintain Material Planning Objects
		Create Material Container
		Maintain Equipment Package
	Create Basic Navy Organization	Create Unit
		Define Stock Elements
		Define Provision Elements
		Maintain Support Relationship
		Maintain Authorized Personnel
		Maintain Real Estate Requirements
		Activate Unit
		Connect to Functional Area Services
		Measure Handling
		Close Unit
	Manage Navy Personnel	Post Personnel
		Move Personnel
	Manage Navy Equipment	Compare Authorized and Actual Material
		Request Equipment
		Finalize Request of Equipment
		Convert Request to Order
		Loan Equipment
		Issue Equipment to a Person
		Return Personal Equipment
	Develop Organization	Create Working Organizations
		Transfer Technical Objects
		Loan Equipment
		Change Supply Relationships
		Change Maintenance Relationship
		Delimit Structures
		Reassign Structures
	Conduct Individual Training	Provide Academic Services
		Grade
		Audit Degree
		Advise Information

FIGURE 13

Illustration of a defense Navy Process Map with business processes structured by Process Groups and process areas.

Any set of these processes may, at any time, be of interest for any reason. These may be assembled into a scenario that is captured in a Process Map. Figure 14– Example Process Map presents an example of process relationship for the processes to provide repaired cars in response to requests provided by customers.

国防海军流程地图		
流程域	**流程组别**	**国防海军业务流程**
军兵力生成	创建海军主数据	创建参考力元素
		创造就业机会
		维护物料计划对象
		创建物料容器
		维护设备包
	创建海军基本组织	创建单元
		定义库存元素
		定义供应元素
		维护支持关系
		维护授权人员
		维护房地产需求
		激活单元
		连接到功能区服务
		措施处理
		封闭单元
	管理海军人员	岗位人员
		调动人员
	管理海军装备	比较授权材料和实际材料
		请求设备
		最终确定设备请求
		将请求转换为订单
		贷款设备
		向某人发放设备
		归还个人设备
	发展组织	创建工作组织
		转移技术对象
		贷款设备
		更改供应关系
		更改维护关系
		界定结构
		重新分配结构
	进行个人培训	提供学术服务
		等级
		审计程度
		建议信息

图13　国防海军流程地图图示，其中业务流程按流程组和流程域排列

这些流程中的任何一组在任何时候都可能因任何原因而受到关注。可以将这些集合到流程图中的场景中。图14展示了响应客户的请求、提供维修车辆的流程关系。

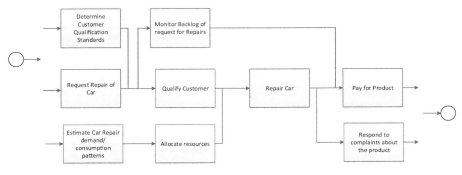

FIGURE 14

Example of a Process Map.

The graphic shows the sequencing relationships between each of the processes of interest for an example set of processes organized to show the sequence of dependency for the completion of each process.

The narrative for this fragment of a larger Process Map would be "… once the request for repair of car" has been received, the customer is qualified based on the customer qualification standards. At the same time, using the estimates of car repair demand/consumption patterns, resources are allocated and available to carry out repairs to cars. Once the request for repairs has been accepted, until the car is repaired the backlog of work will be monitored. When a car has been repaired, it must be paid for and there may

> This example only shows the processes and their flows for the set of processes included in the list in Figure 14. A well-formed Process Map will always include the necessary oversight, planning, provision, delivery, and deregistration processes.

be a requirement to respond to complaints about some aspect of the work.

Processes are named by combining the noun phrase that identifies the name of the completed output, the thing that is the result or purpose of the work with a verb that expresses the idea that the work to create in output results in something that is in its final, complete, and finished form. The product of a process is a complete, well-formed business object. The business object may a thing; a car or a person; a control object (used to record or track the progress to complete the valued output of a Process Group) such as a budget or plan; or an information object employed within the Process Group.

A Process Map will only contain the processes within a Process Group. The Process Map will show the chain of dependency for the completion of the processes necessary to plan, provide, and deliver the valued output as well as the processes necessary to monitor or oversee the work and to dispose of or deregister, resources upon the decision to terminate execution of the Process Group.

THE NATURE OF PROCESSES

When tagging and categorizing processes they may be characterized as Main, Management, or Supporting, or as being Strategic, tactical, or operational in nature.

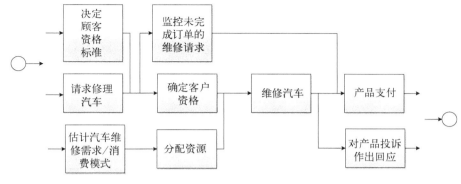

图14　流程图示例

图14显示了每一个流程之间的顺序关系，例如，一组组织起来的流程显示了每个流程完成时的依赖序列。

更大流程图的这一部分的叙述将是"……一旦收到汽车维修请求"，将根据客户资格标准对客户进行资格认证。同时，利用对汽车维修需求/消耗模式的估计，分配资源并用于对汽车进行维修。一旦维修请求被接受，直到汽车被修理，积压的工作将被监控。汽车修好后，必须支付修理费，这样就可以对工作的某些方面的投诉作出回应。

> 这个示例只显示了图14的列表中包含的流程集的流程及其工作流。形成良好的流程地图将始终包括必要的监督、规划、提供、交付和注销流程。

流程的命名方法是将用于标识已完成输出的名称的名词短语、工作的结果或目的与表示在输出中创建的工作将产生最终、完整和已完成形式的内容的动词组合在一起。流程的产品是一个完整的、格式良好的业务对象。业务对象可以是一个事物、一辆汽车或一个人、一个控制对象（用于记录或跟踪完成流程组的有价值输出的进度），如：预算或计划，或流程组中使用的信息对象。

流程地图将只包含流程组中的流程。流程地图将显示完成规划、提供和交付有价值输出所需流程的依赖链，以及监控或监督工作、在决定终止流程组执行时处置或注销资源所需的流程。

2.4.10　流程的性质

当标记和分类流程时，它们可以被描述为主要的、管理的或支持的，或者本质上是战略的、战术的或操作性的。

PROCESS LIFECYCLE VERB TAXONOMY

Because we see a lot of inconsistency in Process Maps and in existing process reference content such as the Supply Chain Operations reference Model (SCOR) or American Productivity & Quality Center (APQC), we created a process verb taxonomy to help you describe and classify processes throughout the entire process lifecycle. As illustrated in Figure 15, this enables the process expert, engineer, or architect to describe and classify

Process Lifecycle Verb Taxonomy

ANALYZE	DESIGN	BUILD	DEPLOY/IMPLEMENT	RUN/MAINTAIN	CONT. IMPROVEMENT
Analyze	Aim	Accept	Accomplish	Administer	Adjust
Appraise	Align	Adapt	Achieve	Assign	Alter
Approximate	Arrange	Assemble	Activate	Audit	Amend
Ascertain	Begin	Assure	Apply	Calculate	Boost
Assess	Blueprint	Build	Assimilate	Chronicle	Change
Capture	Categorize	Chart	Carry out	Communicate	Condense
Clarify	Characterize	Check	Cause	Conserve	Convert
Collate	Classify	Codify	Close	Control	Coordinate
Collect	Cluster	Combine	Complete	Engage	Correct
Consider	Commence	Compile	Conclude	Exchange	Decrease
Count	Compare	Compose	Conduct	Govern	Diminish
Demand	Convene	Configure	Conform	Handle	Eliminate
Detain	Describe	Confirm	Deliver	Keep	Enhance
Detect	Design	Constitute	Deploy	Maintain	Escalate
Diagnose	Determine	Construct	Do	Manage	Improve
Discover	Devise	Craft	Educate	Measure	Incorporate
Estimate	Display	Create	Employ	Monitor	Moderate
Evaluate	Draft	Customize	Evolve	Operate	Modernize
Examine	Draw	Define	Execute	Oversee	Modify
Explore	Drive	Develop	Finish	Preserve	Optimize
Find out	Enter	Enact	Generate	Process	Realign
Forecast	Enumerate	Enlarge	Get done	Oversee	Reassess
Formulate	Establish	Erect	Implement	Promote	Reconsider
Gage	Form	Expand	Include	Protect	Redevelop
Gather	Format	Extend	Initiate	Reconcile	Redirect
Gauge	Found	Fabricate	Instigate	Record	Redraft
Identify	Idea	Increase	Integrate	Recover	Reduce
Inspect	List	Itemize	Interlink	Register	Reevaluate
Investigate	Negotiate	Make	Launch	Reintroduce	Reexamine
Judge	Obtain	Manufacture	Migrate	Report	Reform
Learn	Organize	Match	Perform	Respond	Refresh
Observe	Outline	Pilot	Present	Retain	Regulate
Recognize	Plan	Procure	Progression	Retire	Renew
Reflect on	Plot	Provide	Put into action	Run	Renovate
Research	Prepare	Purchase	Put into operation	Save	Reorganize
Review	Prioritize	Raise	Put into service	Service	Reprioritize
Revise	Propose	Rank	Realize	Set up	Restore
Search	Quantify	Scan	Reallocate	Supervise	Restructure
See	Recommend	Secure	Set off	Support	Revert
Seek out	Select	Shape	Shift	Revolutionize	
Study	Sketch	Systemize	Teach	Turn on	Rework
Survey	Start	Test	Train	Update	Standardize
Think about	Suggest	Translate	Transfer	Uphold	Transfigure
Understand	Verify	Unify	Transition	Withdraw	Transform

© LEADing Practice Business Process Reference Content [#LEAD-ES20005BP]

FIGURE 15

Process verb taxonomy to help describe and classify processes.

2.4.11   流程生命周期的动词分类

因为我们在流程地图和现有流程参考内容（如SCOR或APQC）中看到了很多不一致的地方，所以我们创建了一个流程动词分类法来帮助您描述和分类整个流程生命周期中的流程。如图15所示，这使流程专家、工程师或架构师能够描述和

流程生命周期动词分类

分析	设计	构建	部署/实施	运行/维护	继续改进
分析	旨在	接受	完成	管理	调整
评价	排列	适应	实现	分配	改变
接近	安排	集合	激活	审计	修正
查明	启动	保证	应用	计算	促进
评估	蓝图	建造	同化	记录	变化
捕获	分类	记录	开展	沟通	压缩
澄清	表征	检查	造成	保存	转换
校对	分类	编纂	关闭	控制	使协调
收集	集群	联合	完成	雇用	修正
考虑	着手	编译	断定	交换	减少
计算	比较	撰写	断定	使固定	减少
需求	召开	安置	顺应	治理	消除
留置	描述	确认	交付	处理	增强
查明	设计	构成	部署	保持	逐步升级
诊断	确定	构建	执行	维护	改进
发现	发明	精心制作	教育	管理	包含
估计	显示	创造	雇用	测量	缓和
评价	草案	订制	进化	监视	现代化
检查	描绘	定义	执行	操作	修改
探索	驱动	发展	完成	监督	优化
发现	进入	制定	生成	保存	重新排列
预报	列举	扩大	完成	加工	重新评估
制定	建立	建立	实施	监督	重新考虑
量规	形式	展开	包括	促进	再开发
聚集	格式化	延伸	启动	保护	重定向
量规	发现	制作	使开始	调和	重新起草
识别	构思	增加	整合	记录	减少
检查	列清单	逐项列举	互连	恢复	重新评价
调查	谈判	制作	发射	注册	复查
判断	获得	生产	迁移	重新引入	改革
学习	组织	相配	执行	报告	刷新
观察	概述	试行	显示	回应	调节
辨别	计划	设法获得	进展	保持	重新开始
反思	标出	提供	付诸行动	退休	整修
研究	准备	购买	投入运行	运作	重组
回顾	优先化	提高	投入使用	保存	重新排序
修订	提议	分级	实现	服务	恢复
搜索	量化	扫描	重新分配	设置	重组
看见	推荐	保卫	出发	监督	还原
寻找	选择	形成	转移	支撑	革命化
学习	简述	系统化	讲授	打开	返工
调查	开始	试验	训练	更新	标准化
考虑	建议	翻译	转移	支持	易形
理解	验证	统一	过渡	撤回	转换

领导实践业务流程参考内容

图15   流程动词分类以帮助描述和分类流程

processes in the areas of process analysis, process design, process building, process implementation, process maintenance/monitoring, and continuous process improvement.

PROCESS STEP

What Is a Process Step?

A process step exists as an essential part of what is required to control a process. In each case of a process step, one of the two roles involved initiates the sequence by performing one or more aspects of the work and a second role is involved in completing other parts of the process before the output needed to complete the process is achieved.

A process step is part of the journey to completion of a business object that is produced by a particular process. Each process step, as with a process, will contain the name of the business object that is at the center of the work. Each step will also contain a verb or verb phase taken from the steps derived from the pattern of process steps based on the nature of the process. At this level, although the work is specific, there is nothing to indicate or specify how each action will be accomplished.

How to Identify Process Steps

A process step is a unit of work that is related to exactly one object (e.g., human, sheet of paper, purchase order (system)) and that is executed by one role.

The specifics of the steps involved in a process are based on two factors. The first is the nature of the process, which then permits the core steps to be defined. There are five process types:

1. Respond to request (prepare request, submit request, receive request, act on request, provide response, accept response)
2. Provide or publish an output (prepare output, provide output, receive output)
3. Provide or publish an output and confirm receipt (prepare output, provide output, receive output, verify conformance of output to requirement, confirm receipt of output, receive confirmation of receipt of output)
4. Collaborate to produce a shared output (collaborate on output)
5. Monitor and respond (observe conditions, assess conditions, determine action required, provide direction/assessment, receive direction/assessment)

Superficially, these may look like work that already exists within a process, but this is not the case.

Consider that the sequence of work within the example process includes "request repair for car," "qualify customer," and "repair car." Although this appears to be redundant, it pays to keep in mind the motivation or purpose of each level of work. When we consider these processes, it is important to remember that they are each about the creation of value, whereas at the level of steps the concern has to do with control. Therefore, whereas "request repair for car" is within the continuum of the production of a product (in this case, a repaired car), when we explore how that value is created and controlled, we find a number of viable choices. Two actors can collaborate to determine what the request for the repair of the car should look like to provide to a third party, or one actor can request assistance from another in describing the requirement, or one role can provide the specification of the requirements to a second, who must then act. This decision is central to the idea of achieving control.

分类流程分析、流程设计、流程构建、流程实施、流程维护/监控和持续流程改进（continuous process improvement, CPI）领域的流程。

2.4.12　流程步骤

1. 什么是流程步骤？

流程步骤是控制流程所需的基本部分。在流程步骤的各种情况下，所涉及的两个角色中的一个通过执行工作的一个或多个方面来启动序列，第二个角色在完成流程所需的输出之前完成流程的其他部分。

流程步骤是完成由特定流程生成的业务对象的过程的一部分。与流程一样，每个流程步骤将包含位于工作中心的业务对象的名称。每个步骤还将包含一个动词或阶段动词，该阶段来自基于流程性质的流程步骤模式派生的步骤。在这个层次上，尽管工作是特定的，但是没有什么可以指示或指定每个操作将如何完成。

2. 如何识别流程步骤

流程步骤是与一个对象［如人、纸张、采购订单（系统）］相关且由一个角色执行的工作单元。

流程中涉及的步骤的细节基于两个因素：首先是流程的性质，其次是允许定义核心步骤。有五种流程类型：

（1）响应请求（准备请求、提交请求、接收请求、根据请求采取行动、提供响应、接受响应）；

（2）提供或发布输出（准备输出、提供输出、接收输出）；

（3）提供或发布输出并确认接收（准备输出、提供输出、接收输出、验证输出符合要求、确认接收输出、接收输出确认）；

（4）协作以生成共享输出（协作输出）；

（5）监控和响应（观察条件、评估条件、确定所需行动、提供指导/评估、接受指导/评估）。

从表面上看，这些看起来像已经存在于流程中的工作，但事实并非如此。

考虑到示例流程中的工作顺序包括"请求汽车维修""确认客户资格"和"修理汽车"。尽管这似乎是多余的，但记住每一级工作的动机或目的是值得的。当我们考虑这些流程时，重要的是要记住它们都是关于价值创造的，而在步骤的层次上，关注与控制有关。因此，尽管"请求汽车维修"是在产品（在本例中是一辆修理过的汽车）生产的连续范围内，但当我们探索如何创造和控制该价值时，

The second factor contributing to describing a complete process is driven by seven process-oriented policies.[14] These independent policies will dictate the steps required to flesh out or be added to a process to complement and complete the set core process. The relevant process policies (with the options for each identified) are:

1. When are payments for the output made? (n/a, before, now, later)
2. When is value provided relative to the request? (now, later)
3. Is a profile of the service partner (customer or supplier) maintained? (yes, no)
4. How is price for the output established? (n/a, negotiate, standard pricing)
5. Are the rights to the output transferred? (yes, no)
6. Is the output tracked after the transaction? (yes, no)
7. Is the output prefabricated or made to order? (inventoried, built to order)

Collectively, the answers to these questions and steps identified through the process type complete the list of steps necessary to capture the specification of work at this level. For example if the answer to Question 4, "How is price for the output established?" where that the price is determined by standard pricing, the process step "look up product price on price list" would have to be added. If the answer to this question indicated that prices are determined via negotiation, the process step "negotiate product price" would be needed.

From these patterns, we can see that a process step might legitimately be named in the following manner:

- For a process intended to "respond to (a) request," such as the process to answer a request for an appointment:
  - Provide request for appointment
  - Receive request for appointment
  - Determine appointment
  - Provide response to request for appointment
  - Receive response to request for appointment
  - Examine response to request for appointment
- For a process intended to "provide or publish an output," such as the process to:
  - Provide invoice
  - Receive invoice
  - Accept invoice

Both of these example sets of process steps may be incomplete owing to the influence of the process policies in place for each process, but the additional process steps "create request for appointment" must be added in the first example and "produce invoice" in the second because in either case the particular business object is available except by being "made to order" (Policy 7).

我们发现了许多可行的选择。两个参与者可以协作确定汽车维修请求应该向第三方提供什么样的服务，或者一个参与者可以向另一个参与者请求帮助来描述需求，或者一个角色可以向第二个角色提供需求的规范，然后由第二个角色者来执行。这个决定是实现控制的核心思想。

　　第二个有助于描述完整流程的因素是由七个流程导向策略驱动的[14]。这些独立的政策将规定充实或添加到流程中所需的步骤，以补充和完成设定的核心流程。相关的流程政策（包括每个已确定的选项）包括以下几点。

　　（1）什么时候支付产出的款项？（不适用，以前，现在，以后）

　　（2）何时提供与请求相关的值？（现在，以后）

　　（3）是否维护服务合作伙伴（客户或供应商）的配置文件？（是，否）

　　（4）如何确定产出的价格？（不适用，协商，标准定价）

　　（5）输出的权利是否转移？（是，否）

　　（6）业务处理后是否跟踪输出？（是，否）

　　（7）输出是预制的还是按订单生产的？（库存，按订单生产）

　　总的来说，这些问题的答案和通过流程类型确定的步骤完成了获得此级别工作规范所需的步骤列表。例如，如果政策（4）"如何确定产出的价格？"的答案是价格是由标准定价确定的，则必须添加过程步骤"在价目表上查找产品价格"；如果这个问题的答案表明价格是通过协商确定的，那么需要"协商产品价格"的流程步骤。

　　从这些模式中，我们可以看到一个流程步骤可以通过以下方式合法地命名。

- 对于打算"响应（一个）请求"的流程，如响应预约请求的流程：
  - 提供预约请求；
  - 接收预约请求；
  - 确定预约；
  - 响应预约请求；
  - 接收对预约请求的响应；
  - 检查对预约请求的响应。
- 对于打算"提供或发布输出"的流程，如该流程：
  - 提供发票；
  - 接收发票；
  - 接受发票。

由于每个流程的流程策略的影响，这两个流程步骤的示例集可能都不完整，

Similarly, in both cases the process owner has the discretion to decide whether a profile of the service partner (customer or supplier) is maintained (Policy 3). In cases where the profile exists, another process step, "obtain address for appointment/invoice," is required, and in cases where it is not there would be no additional step. A similar set of decisions is required for each of the process policies in an additive process in which the only point at which the set of process steps making up a particular process will be known is once all the questions have been addressed.

How Are Process Steps Documented?

Process steps may be documented in a structured model that identifies the events (things outside the work that initiate, and terminate the actions described in the model) and the steps required to move the business object to completion.

This modeling approach requires no sub-process, no transaction activity, and no call activity, and only the most basic form of work.

It should be noted that the analysis of activities can only be carried out after either the physical roles are determined from the organization design, from policy, or through practice associated with project- or team-based roles.

In Figure 16 – Example Process Step Model, we see the set of process steps based on the "provide or publish an output" pattern, with no requirement for a payment or other policy-driven steps, except for "develop customer qualification standards," or "built to order," which is added for clarity.

When the decision is made for the product to be delivered not by obtaining it from inventory but by a "build to order" mechanism, a decision must be made as to how the product is to be built. A standard set of choices, where each requires a different set of process steps, is available (not all are required in all cases). Options for building the product are:

- In a product[15,16] manufacturing context: project,[17] jobbing,[18] batch,[19] assembly,[20] or continuous.[21]
- In an economic service[22] context: project, jobbing, batch, or assembly

It is at the level of the process step we see business objects being connected to work. Resources are consumed and output products are produced. In information-centric work the control objects that are consumed and created can be identified and linked to the work that uses or produces them.

While the information flow associated with the control object can be represented separately, in an information flow model, the Process Step Model in Figure 16 shows the sequence of work and where the information structures fit into and connect with the behavior part of the larger equation, which is the complete design and specification of the business to show how control of the work will be maintained and tracked. Figure 17 provides a clearer picture of the work, identifying not just the sequence of work, but the business objects involved in the work and their respective flows.

但第一个示例中必须添加附加的流程步骤"创建预约请求",第二个示例中必须添加"生成发票",因为在这两种任何一种情况下,特定业务对象都是可用的,除非是"按订单生产"[政策(7)]。

同样,在这两种情况下,流程所有人有权决定是否维护服务合作伙伴(客户或供应商)的文件[政策(3)]。如果配置文件存在,则需要另一个流程步骤"获取预约/发票地址";如果没有,则不会有其他步骤。对于附加流程中的每个流程策略,都需要一组类似的决策,在附加流程中,一旦所有问题都解决了,组成特定过程的步骤就会清晰可知。

3. 如何记录流程步骤?

流程步骤可以记录在一个结构化模型中,该模型标识事件(启动和终止模型中描述的操作的工作之外的事情)和将业务对象完成所需的步骤。

这种建模方法不需要子流程、业务活动或支持活动,只需要最基本的工作形式。

应注意的是,只有在从组织设计、政策或通过与项目或基于团队的角色相关的实践确定物理角色之后,才能对活动进行分析。

在图16示例的流程步骤模型中,我们看到了基于"提供或发布输出"模式的流程步骤集,除了"开发客户要求标准"或"按订单构建",不需要支付或其他策略驱动的步骤,这是为了清晰起见而添加的。

当决定交付产品时,必须决定如何构建产品,这不是从库存中取得的,而是通过"按订单生产"的机制,其中标准的选择集是可用的(并非所有情况下都需要)。构建产品的选项包括如下两个。

- 在产品[15,16]制造语境中:项目[17]、工作[18]、批量[19]、组装[20]或连续[21];
- 在经济服务[22]语境中:项目、工作、批量或组装。

在流程步骤的级别上,我们看到业务对象与工作相连:消耗资源生产出产品。在以信息为中心的工作中,可以识别并链接控制对象到使用或生成它们的工作上。

虽然与控制对象相关的信息流可以单独表示,但在信息流模型中,图16中的流程步骤模型显示了工作的顺序以及信息结构与更大方程的行为部分的匹配和连接位置,这是总体的完整设计和规范,以显示如何维护和跟踪工作控制。图17提供了更清晰的工作画面,不仅标识了工作的顺序,还标识了工作中涉及的业务对象及其各自的流程。

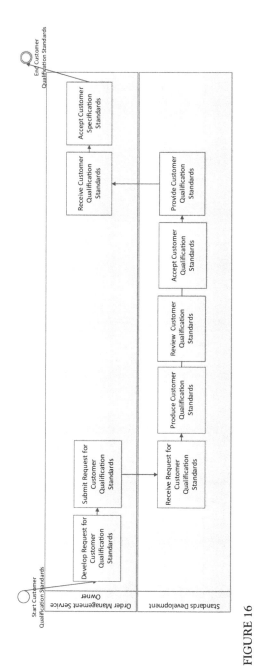

FIGURE 16

Example of a process step model.

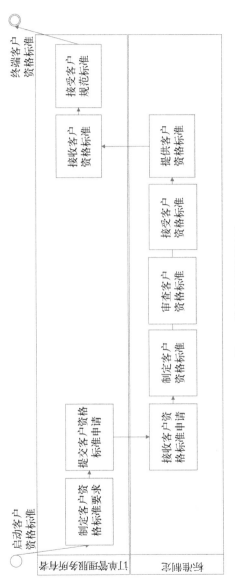

图 16　流程步骤模型示例

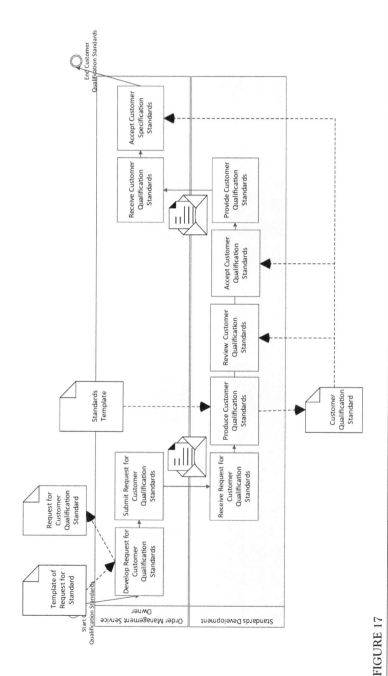

FIGURE 17

Example of a Process Step model with information flows.

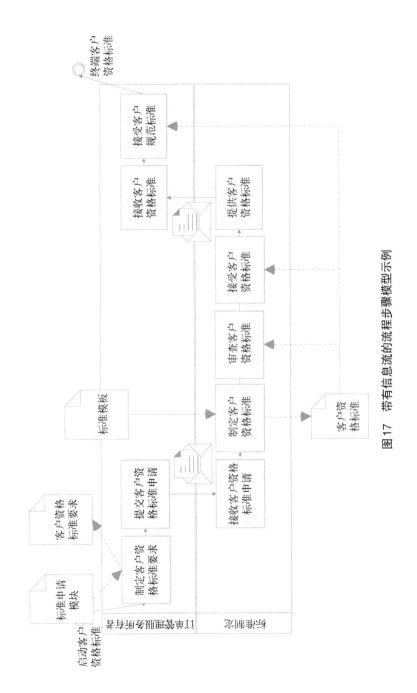

图 17　带有信息流的流程步骤模型示例

The above example Process Step model shows that in the top lane the actor in the initiating role (Order Management Service Owner) would "develop request for customer qualification standard" and then "submit the request for customer qualification standard" to the actor in the receiving role (Standards Writer), represented by the lower lane, who would then first "receive request for customer qualification standard" and then "produce/review/approve customer qualification standard" before "providing the customer qualification standard" back to the requestor, who will first "receive customer qualification standard" and then "accept the customer specification standard," thus complete all steps for the process "produce customer specification standard." Again, this model could be modified to include additional process steps for any one or more of the remaining process policy decisions or to more specifically expose the actual details of the means used to produce or develop the standards, likely project, or jobbing type work.

In this model, the Process Steps are augmented within information flows that show that four business objects are involved in the process: the template for requesting a standard, the request for customer qualification standard, a standards template, and the customer qualification standard itself.

Notice that in these models we are dealing only with the sequence of actions required for the successful, normal behavior to perform the work and therefore complete the process. These steps do not expose the results of decisions, only the activity needed to obtain a decision, and at this level the sequence of work does not change. On the other hand, every one of the steps making up a process may be affected by a set of policies that will affect the sequence of work. Exception conditions, which may result in the work product moving in some manner other than the ideal path seen at the level of the step, are exposed at this level. These exception conditions may occur because of one or more of the following failure types:

1. Execution failure
2. Deadline expiration
3. Resource unavailability
4. External triggers
5. Constraint violation

Each and any of these failure types may occur during any step. Although the complete specification of the rules for handling each failure may be required, care should be taken in performing, documenting, or executing actions in conditions where these failure modes exist, because the result can be a significant explosion in the options of how this work may flow. Of course, this complexity makes process models using existing, standard techniques complex to model and virtually impossible to verify. By separating the failure modes that are applied to a set, from the choices that are about the application of rules within the step, accidental complexity is reduced.

The nature of each of these failure modes is as follows:

1. **Execution failure:** Execution failures during the execution of a step will typically mean the work item is unable to progress and no further steps are possible within the normal flow. This may result in escalation of the execution of work to a different role or that handling of the work by other roles is required.

　　上述流程步骤模型示例表明,在顶层通道中,发起角色的参与者(订单管理服务所有者)将"制定客户要求资格标准",然后向接收角色的参与者(标准编写者)"提交客户资格标准的请求",由下层"通道"表示,在"制定客户要求资格标准"之前,谁先"收到客户资格标准的要求",然后"制定/审查/批准客户资格标准",谁先"收到客户资格标准",然后"接受客户规范标准",从而完成流程"生产客户要求资格标准"的所有步骤。同样,可以修改此模型,使其包含任何一个或多个剩余流程决策的附加流程步骤,或者更具体地公开用于生产或开发标准、可能的项目或者工作类型的方法的实际细节。

　　在这个模型中,流程步骤在信息流中进行了扩充,这些信息流显示流程中涉及四个业务对象:需求标准的模板、客户资格标准的要求、标准模板和客户资格标准本身。

　　请注意,在这些模型中,我们只处理成功的、正常的行为所需的行为序列,以执行工作,从而完成流程。这些步骤不会公开决策的结果,只公开获得决策所需的活动,并且在这个级别上,工作顺序不会改变。另一方面,组成流程的每个步骤都可能受到一组影响工作顺序的策略的影响。异常条件(可能导致工作产品以某种方式移动,而不是步骤级别上看到的理想路径)在此级别暴露。由于以下一种或多种故障类型,可能会发生这些异常情况:

　　(1)执行失败;

　　(2)期限到期;

　　(3)资源不可用;

　　(4)外部触发因素;

　　(5)违反约束条件。

　　这些故障类型中的每一种都可能在任何步骤中发生。尽管可能需要处理每个故障的完整规范,但在这些故障模式存在的情况下,在执行、记录或执行操作时应小心,因为结果可能会对工作流程的选择产生重大影响。当然,这种复杂性使得使用现有标准技术的流程模型变得复杂,甚至几乎无法验证。通过将应用于某个集合的故障模式与步骤中有关规则应用的选择分开,可以降低意外的复杂性。

　　这些故障模式的性质如下。

　　(1)执行失败:执行步骤期间的执行失败通常意味着工作无法进行,并且在正常流程中无法执行下一步的步骤。这可能导致工作执行转移到其他角色,或者需要其他角色来处理工作。

2. **Deadline expiration:** Commonly there is a requirement for a step to be completed in or by a particular time constraint. If this cannot happen, exception handling involving other roles different from the above case may be required.

3. **Resource unavailability:** Often the execution of work requires access to one or more resources during its execution, and fails due to either no resource is found to do the work, or the resource becoming unavailable during the course of the work. If these are not available to the work item at initiation, it is usually not possible for the work item to proceed. This may require some combination of other activities actions to:

   a. Obtain the necessary resources
   b. Reallocate resources
   c. Abandon the execution of work
   d. Re-specify the execution of work

   Examples of this type of failure include having in adequate information or the incorrect information necessary to perform a transaction.

4. **External trigger:** Events external to an activity may affect the ability to act on the execution of work and may therefore require some alternative form of handling. Events may occur in activities that are not directly linked to the work in question being executed. They may occur anywhere within the process model or even in other process models. Addressing the impact of these external triggers will typically mean that current execution of work needs to be halted or possibly undone and some alternative action taken. Examples of an external trigger include the requestor who initiates a process cancelling the request, a systematic breakdown of the process, such as a failure of enabling automation.

5. **Constraint violation (including inadequate authority to approve):** Constraints in the context of an activity are typically found within the information integrity requirements needed and operational consistency of the business. On-going monitoring is generally required to ensure that they are enforced. Implementation of routines to identify and handle constraint violations detected within the context of a step is similar to the issue of dealing with external triggers. Typically, the constraint will detect and need to deal with the violation, although there is no reason why the constraint could not be specified and handled at a block or process level.

6. To address its failure condition, each exception case should be modeled separately as a separate overlay to the activities in the activity baseline. These conditions can lead to the creation of multiple paths, any number of which can result in failure of the activity, and therefore the affected parent process, to complete. Then it is at the level of the steps where work occurs and the possibility of failure exists.

Each instance of a failure mode can be seen in Figure 18 as triggering an event that moves the workflow from the current, planned, sequence to a new, exception flow. Each exception condition is terminated by an intermediate event, which can

（2）期限到期：通常需要在特定的时间限制内或通过特定的时间限制完成某个步骤。如果不能做到这一点，则可能需要处理涉及与上述情况不同的其他角色的异常。

（3）资源不可用：工作的执行常需要在执行流程中使用一个或多个资源，但由于找不到资源来执行工作或在工作流程中资源变得不可用而失败。如果启动时工作项无法使用这些内容，则通常无法继续进行工作项。这可能需要一些其他活动的配合，以：

a. 获得必要的资源；

b. 重新分配资源；

c. 放弃工作执行；

d. 重新规定工程的执行。

此类故障的例子还包括信息不足或执行业务所需的信息不正确。

（4）外部触发因素：活动外部的事件可能影响对工作执行的操作能力，因此可能需要某种替代处理形式。事件可能发生在与正在执行的工作没有直接联系的活动中。它们可能发生在流程模型中的任何地方，甚至在其他流程模型中。解决这些外部触发因素的影响通常意味着当前的工作执行需要暂停或可能撤销，并采取一些替代措施。外部触发因素的示例包括启动取消请求的流程的请求者、流程的系统性故障，如启动自动化功能失败。

（5）违反约束条件（包括批准权限不足）：活动前后关系中的约束通常在所需的信息完整性要求和业务的操作一致性中找到。这通常需要进行持续监控，以确保相关措施得到执行。用于识别和处理在流程中检测到的冲突实例的方法类似于处理外部触发器的问题。通常，约束将检测并处理冲突，尽管可以在模块或流程层中指定和处理约束。

（6）为了解决导致失败的条件，每个异常情况都应该单独建立模型。这些条件可能导致多个失败因素产生，每个因素都可能导致活动失败，从而导致受影响的父流程无法完成。随后，到达工作发生和存在故障可能性的步骤层级。

图18中可以看到故障模式的每个实例都触发了一个事件，该事件将工作流从当前计划的序列移动到一个新的异常流。每个异常条件都由一个中间事件终止，

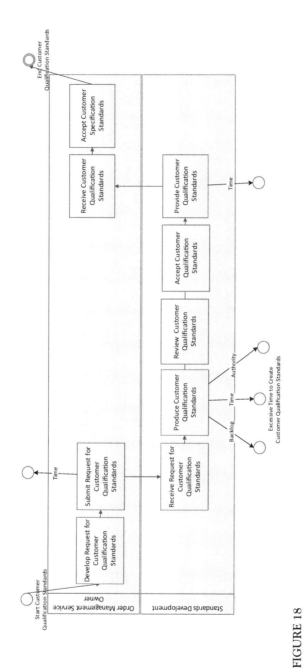

FIGURE 18

Example of a Process Step with Exceptions.

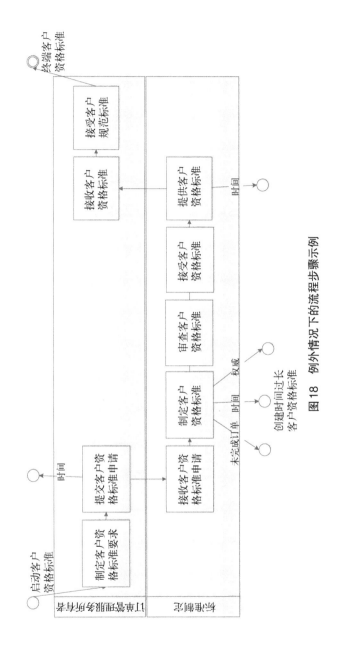

图 18 例外情况下的流程步骤示例

then in turn be recognized as the triggering event for the appropriate workflow; i.e., if the Standards Developer takes too long to "Develop the Customer Qualification Standard," this will trigger the workflow that transfers the work from the current workflow to a new workflow that is not on this model and therefore can be addressed as a single bundle of behavior, capturing each escalation path and its rules and behavior as a discrete package. Because each exception condition and its associated work is now both well-bounded and focused on a single aspect of the work, the ability to develop and validate the specification is now standardized and repeatable.

The decision to establish and enforce any exception condition is wholly within the discretion of any actor engaged in the process. In the example, the time from the "request for customer qualification standards" and their actual receipt is monitored and if the intervening process to "submit the request for customer qualification standards" takes too long, the process may be interrupted and completed through a separate exception process. Similarly, the step to "provide customer qualification standards" may also be escalated or subject to exception handling based on time criteria. Finally, the process to "develop the customer qualification standards" themselves may be escalated if it remains in the backlog too long, the standards development resource has an issue with the authority available to do the work, or they take too long with this work. In each case, the exception condition flows to an event that is then mirrored by an equivalent event that indicates the start of the applicable process designed to address the implied work. Of course, the potential exists for the control policies for every process step to have the potential to experience failure for any combination of the failure conditions.

THE NATURE OF PROCESS STEPS

When tagging or categorizing process steps, they may be identified as simple/static, generic/hybrid, or complex/dynamic.

ACTIVITY

What Is an Activity?

The activities of a process are classified within each of the process steps. Activities describe work. Each activity specifies one of the inputs consumed within the activity and exposes it as work separate from the work needed to transform it, or is the transaction that produces the output identified as being the result of completing the step.

Process activities provide the details of the complete set of actions required to produce an output from a process step. The flow between activities exposes the impact of failure of the work to complete successfully, i.e., the content of a data field is not available where it is required or does not conform to the specification, or a part is not of acceptable quality, and so on.

然后该中间事件又可以被识别为适当工作流的触发事件,即如果标准开发人员"开发客户资格标准"花费的时间太长,则这将触发将工作从当前工作流传输到不在此模型上的新工作流的工作流,因此可以作为单个行为组合处理,将每个升级路径及其规则和行为捕获为一个离散包装。由于每个异常条件及其相关的工作现在都有很好的边界,并且集中在工作的一个方面,开发和验证规范的能力现在是标准化的和可重复的。

建立和执行任何例外条件的决定完全由参与该流程的参与者自行决定。在本例中,从"客户资格标准申请"到实际接收的时间受到监控,如果介入流程"提交客户资格标准申请"花费的时间太长,则可以通过单独的例外流程中断并完成该流程。同样,"提供客户资格标准"的步骤也可能会升级,或者根据时间标准进行异常处理。最后,"开发客户资格标准"本身的流程可能会升级,如果它积压太久,标准开发资源与可用的权限有问题,或他们花了太长的时间来做这项工作。在每种情况下,异常条件都会流到一个事件,然后由一个等效事件镜像该事件,该事件指示设计用于处理隐含工作的适用流程的开始。当然,每一个流程步骤的控制策略都有可能在任何故障条件组合下遇到故障。

2.4.13　流程步骤的性质

标记或分类流程步骤时,可以将其标识为简单/静态、通用/混合或复杂/动态。

2.4.14　活动

1. 什么是活动?

流程的活动在每个流程步骤中进行分类。活动描述工作,每个活动指定活动中使用的一个输入,并将其显示为独立于转换活动所需工作的工作,或者是生成标识为完成步骤结果的输出业务。

流程活动提供了从流程步骤生成输出所需的整套操作的详细信息。活动之间的流动暴露了工作未能成功完成的影响,即数据字段的内容在需要或不符合规范时不可用,或者部分质量不可接受等。

An activity is the part of the actual physical work system that specifies how to complete the change in the form or state of the inputs, oversee, or even achieve the completion of an interaction with other actors and seeing the decisions that must be made, who makes them, and under what circumstances.

At this level, we are capable of exposing and capturing the individual atomic questions that a worker must ask and answer to complete the step.

How to Identify Activities

At this level, we are capable of exposing and capturing the interaction with the individual atomic objects that a worker in a role can view, access, and/or manipulate within the work.

Activities will be of one of two types:

1. There will be one activity will always recognize the work to convert some set of inputs into an output (transformational work), complete a transaction within rule-based work, or, in situations of high ambiguity, reach a judgment or conclusion (tacit work)
2. The remaining activities are each associated with one of the inputs consumed in creating the output

How Are Activities Documented?

Process activities are documented in a structured model that identifies the events (things outside the process that initiate, terminate, or happen during the course of the action described in the model), and the set of activities required within the scope of the business step.

In the example shown in Figure 19 the work becomes much more concrete. The context of the activity is well-bounded by the process step within which it is classified. The events that show that start and end of the process activity model are the

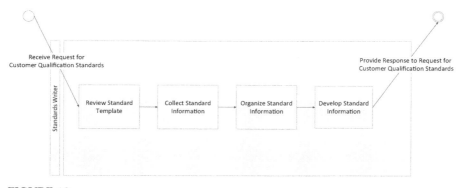

FIGURE 19

Example of a process activity model.

　　活动是实际物理工作系统的一部分,它指定如何完成输入的形式或状态的更改、监督甚至完成与其他参与者的交互,并查看必须做出的决定、谁做出的决定以及在什么情况下做出的决定。

　　在这个层次上,我们能够暴露和捕获单个的微小问题,这些问题是工作人员完成这一步骤必须询问和回答的。

2. 如何识别活动

　　在这个级别上,我们能够公开和捕获与单个微小对象的交互,角色中的工作人员可以在工作中查看、访问或操作这些微小对象。

　　活动将分为两类:

　　(1)将有一个活动始终识别工作,将一些输入转换为输出(转换工作),在基于规则的工作中完成事务,或在高度模糊的情况下,做出判断或结论(隐性工作);

　　(2)其余的活动都与创建输出所消耗的输入之一相关联。

3. 如何记录活动?

　　流程活动记录在一个结构化模型中,该模型标识事件(流程之外的事件,这些事件在模型中描述的操作过程中启动、终止或发生),以及业务步骤范围内所需的一组活动。

　　在如图19所示的示例中,工作变得更加具体。活动的前后关系被分类的流程步骤很好地限定。显示流程活动模型开始和结束的事件是流程和后续步骤的名称,流程活动本身是单独的、离散的操作,完成这些操作以解决特定问题,或者在完

图19　流程活动模型示例

names of the proceeding and following steps and the process activities themselves are individual, discrete actions that are complete to address a specific problem, or challenge on the way to completing the step. The focus of each item of work is on accomplishing or completing a different thing. Whereas all activity is carried out by a single role in most cases, including this example, this is not always the situation. In collaborative work, multiple roles may again appear at this level, but the purpose of each activity can be described by a single verb.

THE NATURE OF ACTIVITY

Process activities may be tagged or categorized as simple/static, generic/hybrid, or complex/dynamic and may be of the type: create, maintain, transmit, receive, track, find, store, pay, observe, assess, or destroy.

THE WORK SYSTEM
What Is the Work System?

The work system is composed of the combination of human work performing an activity aided by one or more resources that provide capability in performing this work; i.e., each resource acts as a package of capabilities to enable the work or act as a multiplier on what the human could reasonably accomplish without the aid of the resource.

Work system design seeks to optimize the contribution of a range of capabilities in the performance of a set of work to appropriately deliver the desired value and cost combination. Conception and execution of the work system will of necessity require that the designers address the problem through the interplay of the components to create a holistic understanding of the parts in relation to the whole such that all the parts of a work system relate (directly or indirectly) to the creation of a single specific result.

The nature of the activities will depend on the features of the capabilities available to enable the work. While it seems complicated, this is a simple concept, let us illustrate this simplistic example, where the work system for digging a hole with a shovel (which draws on the capability of a small tool, a shovel), is different from digging a hole with a back end loader (which will draw on the different capabilities in this package of equipment). Similarly, the activity for entering process, resources, work tasks, and data into SAP will, for example, be different from executing the equivalent work in PeopleSoft.

How to Determine the Work System

Only one business role (which may or may not be fulfilled by multiple actors performing that role and which is supported by a resource providing particular capabilities) will be involved in the activities needed to complete an activity. Each activity may be enabled through any number of capabilities.

成步骤的过程中遇到挑战。每项工作的重点是完成相同或完成不同的事情。尽管在大多数情况下所有活动都由一个角色执行，包括这个例子，但情况并非总是如此。在协作工作中，多个角色可能会再次出现在此级别，但每个活动的目的可以用单个动词来描述。

2.4.15　活动的性质

流程活动可以标记或分类为简单/静态、通用/混合或复杂/动态，并且可以是以下类型：创建、维护、传输、接收、跟踪、查找、存储、支付、观察、评估或销毁。

2.4.16　工作系统

1. 什么是工作系统？

工作系统由执行一项活动的人力工作组合而成，该活动由一项或多项资源辅助，这些资源提供执行该项工作的能力。即，每项资源都充当一组能力或充当一个乘数，以使该项工作得以开展，使人力在没有资源帮助的情况下能够合理完成工作。

工作系统设计旨在优化一系列能力在一组工作中的贡献，以提供所需的价值和成本组合。工作系统的概念和执行必然要求设计人员通过各组成部分的相互作用来解决问题，对与整体相关的部分进行整体理解，从而使工作系统的所有部分（直接或间接）与创建单个特定结果相关。

活动的性质将取决于可用于开展工作的能力的特点。虽然看起来很复杂，但这是一个简单的概念。让我们来举例说明这个简单的例子：用铲子挖洞的工作系统（利用小工具"铲子"的能力）不同于用装载机挖洞（利用设备中的不同能力）。同样，例如，将流程、资源、工作任务和数据输入 SAP 的活动将不同于在 PeopleSoft 中执行等效工作。

2. 如何确定工作系统

只有一个业务角色（可能由执行该角色的多个参与者完成，由提供特定功能的资源进行支持）将参与完成活动所需的活动，每个活动可以通过任何数量的功能来启用。

One challenge of developing the work system is that it requires a mode of thinking at odds with much current practice, which focuses on disciplines that are analysis-centric. Analysis, by definition, is about the decomposition of a problem into its parts to formulate an understanding of the constituent parts of the problem so as to study the parts and their relationships, and to reach a conclusion. Rather than taking the reductionist approach of traditional design of work, the work system design involves understanding how things influence each other within a whole. Although thinking about each part is critical to understanding, thinking about the whole is critical to creating a work environment that cost-effectively optimizes the tradeoffs between human and machine capabilities to create a result that supports the larger enterprise strategy and cost.

How Is the Work System Documented?

The work of the human is in one "swim lane" and the work of each capability is each captured in a separate one.

One example of the method for documenting work might be to capture transactional work enabled by a software system. Such a specification requires a clear understanding of the work involved, as set by the step that provides the context to the work, the nature of the work (transformational, transactional, or tacit), as well as that of the enabling capabilities to create a work system that finds the optimal balance between the manual and automated parts.

PROCEDURE

What Is a Procedure?

Whereas Process Groups, processes, and process steps are conceptual and process activities describe tangible work, procedure actually specifies how the work is done and will typically represent common agreed-upon practice controlled at a supervising (human or machine) level. Using the specification of the design of software or other capabilities is the means to capture the behavior and information aspects of systems or machines at the lowest externally observable level. Procedures are captured within the design of the product and within the manuals of operations and are an essential part of the product behavior.

A clear understanding of the context of the work is required to document a work system. Every domain of human endeavor and the standards for describing work in each domain has its own requirements as to what is needed. Within those domains enabled by software-based systems, different tools to describe activity are often also necessary, with distinctions based on the solution architecture being employed so as to both cleanly separate process or business logic from software implementation and provide a complete specification of application behavior and therefore a specification of the application interface, or the features of other capabilities and the work to be performed by each role when using the system, providing input to training aides, application manuals, and other products.

　　开发工作系统的一个挑战是，它需要一种与当前许多实践不符的思维模式，这种模式侧重于以分析为中心的学科。从定义上讲，分析是将一个问题分解成各个部分，以形成对问题组成部分的理解，从而研究各个部分及其关系，并得出结论。工作系统设计不是采用传统工作设计的简化方法，而是理解事物在整体中是如何相互影响的。尽管思考每一个部分对于理解都是重要的，但是思考整体对于创造一个工作环境是至关重要的，这个工作环境能够经济有效地优化人和机器能力之间的权衡，从而产生支撑更大企业战略和更高成本的结果。

　　3. 如何记录工作系统？

　　人类的工作是在一条"通道"上进行的，每一种能力的工作都是在一条单独的"通道"中进行的。

　　用于记录工作的方法的一个示例可以是捕获由软件系统启用的事务工作。这样的规范要求清楚地理解所涉及的工作，由为工作提供背景的步骤、工作的性质（转换、交易或隐性）以及启用功能来创建一个工作系统，找到最优的平衡之间的手动和自动部分。

2.4.17　程序

　　1. 什么是程序？

　　虽然流程组、流程和流程步骤是概念性的，流程活动描述的是有形的工作，但程序实际上指定了工作是如何完成的，并且通常代表在监督（人或机器）级别控制的共同商定的实践。使用软件设计规范或其他能力是在最低的外部可观测水平上，捕获系统或机器行为和信息方面的手段。程序包含在产品设计和操作手册中，是产品行为的重要组成部分。

　　记录工作系统需要对工作环境有清楚的理解。人类努力的每个领域和描述每个领域中工作的标准都有自己的需求。在基于软件的系统所支持的那些领域中，描述活动的不同工具通常也是必要的，使用基于解决方案体系结构的区别，以便将流程或业务逻辑与软件实现完全分离，并提供应用程序行为的完整规范，因此，应用程序接口的规范或其他功能的特性，以及每个角色在使用系统时要执行的工作，为培训助手、应用程序手册和其他产品提供输入。

How to Identify Procedures

Each procedure will reference a single object that the role can view, access, or manipulate in a transactional application with a graphical user interface. The objects might include menu items, buttons, fields, scroll bars, a specific tool such as a steering wheel or a particular switch on the steering column, and so on.

How Are Procedures Documented?

The language of specification of procedures will vary based on the specifics of the capabilities and combinations of capabilities being used. For example, even with software design, the specification for each of a report, interface, conversion, enhancement, or form may be different.

CONNECTING THE WORK SPACES

To understand how these views of work connect, it is useful to view them in an integrated manner. This is shown in Figure 20.

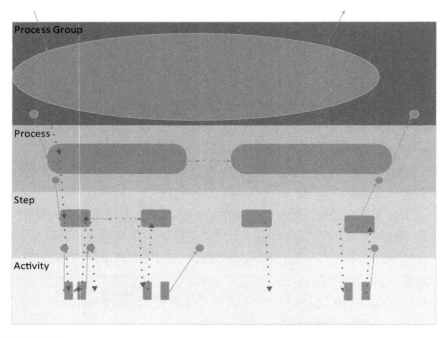

FIGURE 20

Integrated view for process hierarchy.

2. 如何识别程序

每个程序都将引用一个对象,相关角色可以在具有图形用户界面(UI)的业务性应用中查看、访问或操作该对象。对象包括:菜单项、按钮、字段、滚动条、特定工具(如方向盘或转向柱上的特定开关)等。

3. 如何记录程序?

程序规范的语言将根据所使用的能力和能力组合的具体情况而有所不同。例如,即使有了软件设计,报表、接口、转换、改善或表单的每一个规范也可能不同。

2.4.18　连接工作空间

为了理解这些工作视图是如何连接的,以集成的方式查看它们是很有用的。如图20所示。

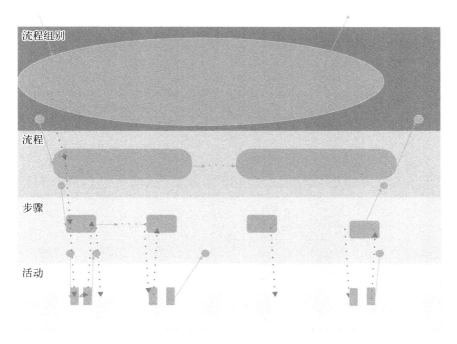

图20　流程层次结构的集成视图

The view of work in which value is delivered occurs within the Process Group and at this level appears to be seamless. To expose the work needed to provide the product that delivers this value I must look into the processes, and then if I want to understand how the delivery of the processes is controlled I will look into the relevant process steps, and from there, if I want to actually understand the work involved I need to examine the process activities. What this means is that, as one enters a process, the work goes through a series of steps that start with the desire to create value and end with the work required to provide the control. In the course of this, the product needed to obtain the value are produced. In the case of software enabled transactional work, this occurs when the operator, working within an activity, presses the "Enter" key to complete the final activity needed to complete the final piece of work. This triggers the event that ends the activity flow, which provides the final piece of control needed to complete the product, necessary to obtain the value, all completing instantaneously.

PROCESS SCENARIOS

Process scenarios express the user story of how work is performed to achieve a particular result, using reusable design objects at the appropriate level to express the concern through the exploration of the process hierarchy. Although the hierarchy captures work as level, well-bounded components with a purpose in terms of the creation of value, production of useable products, maintenance of control, or the actual execution of work, it does not actually expose how the business uses or consumes these processes in the pursuit of a specific business purpose; process scenarios do this extremely well.

To develop a meaningful scenario, three questions must be asked and answered.

1. What is the purpose of the scenario; i.e., is it intended to show production of products or how control will be maintained, or is it to show the actual execution of work.
2. What is the defined objective of the work?
3. Where is the birth of the process relative to the scope of the analysis?

A Process Map may be used to capture the processes within a single Process Group, or be within a scenario drawn on the processes of multiple Process Groups to show a more complex set of linkages. For example, a Process Groups scenario map may call on processes within the process management Process Group, budgeting Process Group, procurement process, and payment groups to show the processes that cross a wide swath of an enterprise to fully execute an integrated procurement cycle, not only including the processes directly implicated in such a body of work, but showing all necessary oversight, planning, provision, delivery, and deregistration processes.

　　工作视图的价值出现在流程组中,并且在此级别上似乎是无缝的。为了公开提供具有此价值的产品所需的工作,我们必须研究流程。如果想了解如何控制流程的交付,应研究相关的流程步骤,如果想从实际中了解所涉及的工作,需要检查流程活动。这意味着当一个人进入一个流程时,工作需要经历一系列的步骤:从创造价值的愿望开始,到提供控制所需的工作结束。在此流程中,生产出获得价值所需的产品。在启用软件的业务性工作的情况下,当操作员在活动中工作,按"Enter"键完成最终工作所需的最终活动时,就会发生这种情况。这将触发结束活动流的事件,活动流提供完成产品所需的最后一个控件以及获取值所必需的控件,所有这些控件都会立即完成。

2.4.19　流程方案

　　流程方案表示如何执行工作以获得特定结果的工作计划,使用适当级别的、可重复使用的设计对象可通过探索流程层次结构来表达关注点。虽然层次结构将工作视为水平的、有界限的组件,其目的在于创造价值、生产可用产品、维护控制或实际执行工作,但它实际上并未揭示企业如何使用或消费,这些流程是为了追求特定的商业目的,流程方案在此方面非常合适。

　　为了开发一个有意义的方案,必须提出并回答三个问题。

　　(1)该方案的目的是什么,即它是为了展示产品的生产或如何保持控制,还是为了展示工作的实际执行。

　　(2)工作的既定目标是什么?

　　(3)与分析范围相关的流程应该在哪里产生?

　　流程地图可用于总览单个流程组中的流程,或位于多个流程组的流程上绘制的场景方案中,以显示一组更复杂的链接。例如,流程组方案图可以调用流程管理流程组、预算流程组、采购流程和付款组中的流程,以显示跨越企业范围的流程,以及完全执行集成的采购周期,而不仅仅包括与这一工作机构直接相关的流程,而且显示了所有必要的监督、规划、提供、交付和注销流程。

PROCESS TYPE

Tagging processes according to their type categorizes the processes based on their role within the enterprise in terms of whether they are essential management, process supporting processes, or represent a primary process within the enterprise. This is identified and correctly used in competency modeling, business model design, as part of the value chain, accountability, or operating model view.

Management processes will appear in the accountability view and may be subject to decisions about how activities are designed and implemented. Management processes are engaged in planning, budgeting, control, oversight, and monitoring of main or supporting processes.

Main processes are processes within a process that deliver the output.

Supporting processes are processes that are necessary to ensure the main process is given everything it needs to meet the purpose for which it was designed, deployed and is operated.

As you can see Figure 21, management processes control the organization. Main processes produce the products/services for customers. Support processes provide resources for main processes. Together, they constitute the hierarchy or process architecture of an organization.

While most organizations are familiar with categorizing process types in their organizations, we feel compelled to point out the mistakes regarding tagging process types. The most common mistake in tagging process types is not understanding the difference between categorizing process types and competency types. The tagging of process types into main processes, also called core processes, does not recognize the difference between the nature of processes and the role of competency; where the latter concept is generally used in literature and by most

FIGURE 21

Example of process types.

2.4.20　流程类型

根据流程类型对流程进行标记,根据流程在企业中的角色对流程进行分类,包括基本管理、流程支持体系或代表企业内的主要流程。这在能力建模、业务模型设计中被识别并正确使用,作为价值链、问责制或运营模型视图的一部分。

管理流程将出现在问责制视图中,并可能取决于有关如何设计和实施活动的决策。管理流程涉及主流程或支持流程的规划、预算、控制、监督和监控。

主流程是提供输出的流程中的流程。

支持流程是确保主流程得到满足其设计、部署和运行目的所需的一切所必需的流程。

如图21所示,管理流程控制着组织,主流程为客户生产产品/服务,支持流程为主流程提供资源,它们一起构成组织的层次结构或流程架构。

虽然大多数组织都对其组织中的流程类型分类很熟悉,但我们不得不指出标记流程类型的错误。标记流程类型最常见的错误是不理解分类流程类型和能力类型之间的区别。将流程类型标记为主流程(也称为核心流程)并不能识别流程的性质和能力之间的差异,后者的概念通常在文献中使用,大多数管理人员都会使用它来识别组织对外部环境的响应能力[23]。在文献中,战略竞争力和差异化被定义

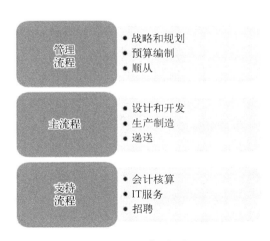

图21　流程类型示例

executives to identify the organization's ability to be responsive to its external environment.[23] Strategic competitiveness and differentiation have been defined in the literature as an organization's ability to identify major changes in the external environment to quickly commit one's competencies to new courses of action, and to act promptly when it is time to halt or reverse them.[24] Organizations in every industry face competitive forces and develop strategies to define their direction. With the ability to classify competencies into core-differentiated competencies, core competitive competencies, or non-core competencies, it enables sorting of the role played by each competency in the execution of the strategy, the creation of value, and the link to the relevant business model discipline. Through renewal of the core competitive and core-differentiated competencies, one enables renewal in terms of business model innovation and transformation.[25] However, this requires understanding one's competencies according to the value they provide and where and how they provide it to the enterprise.

When decomposing process type classification to recognize management processes, main (core) processes, or supporting processes, it is possible for a process of any of these types to be within any one of core differentiating, core competitive or the non-core competencies. This means a non-core competency will have management processes, main (core) processes or supporting processes, so will core differentiating competency as well as a core competitive competency. Therefore, while they are two very different tagging mechanisms, one is applied within process modeling and another within competency/business modeling. Except that both are tagging mechanisms, they have nothing to do with each other and we actually understand how the mix-up came about. As shown in Figure 22, a main process will go across non-value creating as well as value creating aspects. So will a management process and a supporting process; they are all part of various service flows to the outcome, e.g., customer value creation.

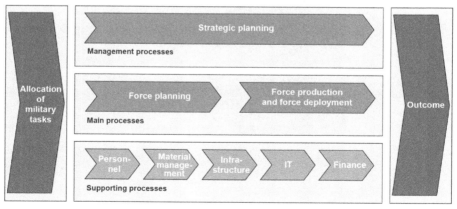

FIGURE 22

Process types in a defense example.

为一个组织能够识别外部环境中的重大变化、迅速将自己的能力投入到新的行动方案中,并在需要停止或逆转这些行动方案时迅速采取行动的能力[24]。每个行业的组织都面临着竞争,并制定战略来确定他们的方向。凭借将能力分为核心差异化能力、核心竞争能力或非核心能力的能力,它能够对每个能力在战略执行、价值创造和与相关商业模式规程的链接中所扮演的角色进行分类。通过更新核心竞争能力和核心差异化能力,我们可以在商业模式创新和转型方面实现更新[25]。然而,这需要根据一个人提供的价值和在哪里以及如何向企业提供能力来理解他自己的能力。

当分解流程类型以识别管理流程、主(核心)流程或支持流程时,这些类型的流程可能处于核心差异化能力、核心竞争能力或非核心能力中的任何一种。这意味着非核心能力将具有管理流程、主(核心)流程或支持流程,核心差异化能力和核心竞争能力也将如此。因此,尽管它们是两种非常不同的标记机制,但一种应用于流程建模,另一种应用于能力/业务建模。除了这两个都是标记机制,它们彼此之间没有任何关系。我们实际上了解了混合是如何产生的,如图22所示,主流程将跨越非价值创造和价值创造两个方面,管理流程和支持流程也是如此,它们都是通向结果的各种服务流的一部分,如客户价值创造。

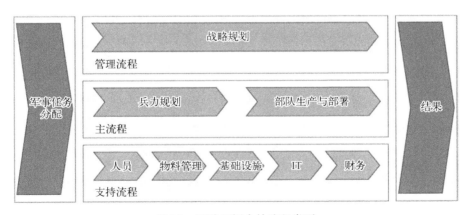

图22　国防示例中的流程类型

Therefore, while process also can be viewed as a "value chain" in which each process, step, and activity contributes to the result, some management, main, and supporting activities directly contribute value, while others may not. However, we all know that all processes have and consume enterprise resources independent of the value creation and realization. The challenge for managers is to eliminate steps that do not add value and to improve the efficiency of those that do. Within the main processes, there will be aspects that are non-value creating, also aspects that do.

A more detailed version of the Process Map defense example is found in Figure 23.

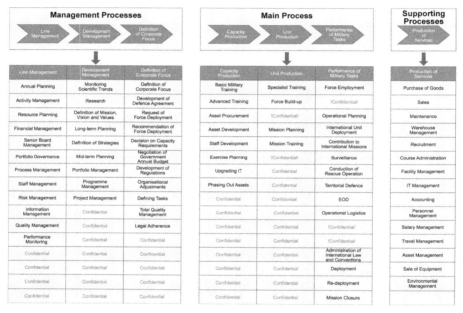

FIGURE 23

Process types in a more detailed Process Map defense example.

PROCESS TIER

In addition to process tagging in terms of process classification and process categorization, processes can be tagged according to their strategy, tactics, and operational tiers. The reason this applies to all processes is that all processes exist within the strategic, tactical, or operational aspects of the organization.

- **Strategic aspects:** This tier affects the entire direction of the firm. An example may be the mission, vision, strategic business objectives, and specific business performance indicators and business plans. The strategic tier has long-term, complex decisions made by executives and senior management, and the measurement reporting view used is for the most scorecards.

因此，虽然流程也可以被视为一个价值链，其中每个流程、步骤和活动都对结果有贡献，但一些管理活动、主要活动和支持活动直接贡献价值，而其他活动则可能不会。然而，我们都知道，所有流程都拥有和消耗企业资源，而这些资源独立于价值的创造和实现。管理者面临的挑战是消除那些不增值的步骤，并提高那些增值步骤的效率。在主流程中，会有一些方面是不创造价值的，同样需要采取上述方法。

国防流程地图示例的更详细版本如图23所示。

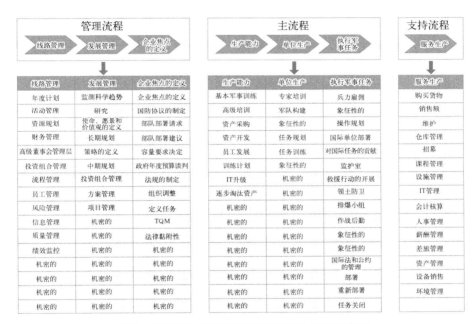

图23 更详细的国防流程地图示例中的流程类型

2.4.21 流程层级

除流程分集和流程分类方面的流程标记之外，流程还可以根据其战略、战术和操作层进行标记。这适用于所有流程，其原因是所有流程都存在于组织的战略、战术或运营方面。

- 战略方面：这一层影响公司的整个方向。如使命、愿景、战略业务目标以及特定的业务绩效指标和业务计划。战略层由业务主管和高级管理层做出长期、复杂的决策，所使用的度量报告视图适用于大多数记分卡。

- **Tactical aspects:** The aspects at this tier are more medium-term, subjects of less complex decisions and primarily performed by middle managers. They follow from strategic decisions and aim to meet the critical success factors. The way to do this is for governance, evaluation, reports, control, and monitoring, and the measurement reporting view used is, for the most dashboards.
- **Operational aspects:** At this tier, decisions are made day-to-day by operational managers. They are simple and routine, and the measurement reporting view used is, for the most cockpits.

Figure 24 illustrates an example of the "Enterprise Tiers" and relevant process context.[26]

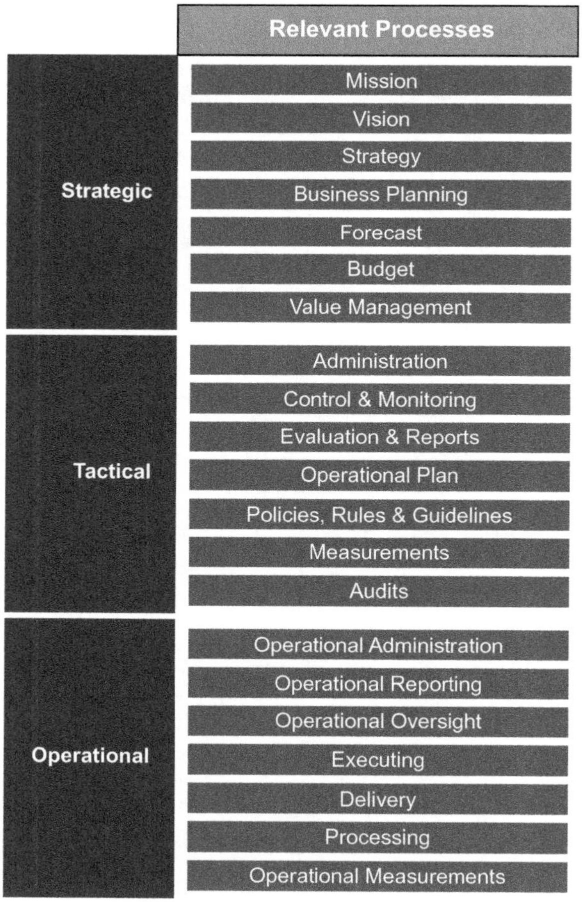

FIGURE 24

Example of "Enterprise Tiers" and relevant process context.

- 战术方面：这一层更为中间，决策不太复杂，主要由中层管理人员执行。他们遵循战略决策，旨在满足关键的成功因素。实现的方法是治理、评估、报告、控制和监控，而使用的度量报告视图大多是仪表盘。
- 运营方面：在这一层，运营经理每天都会借助驾驶舱使用评估报告视图做出简单和常规的决策。

图24说明了企业层和相关流程前后关系的示例[26]。

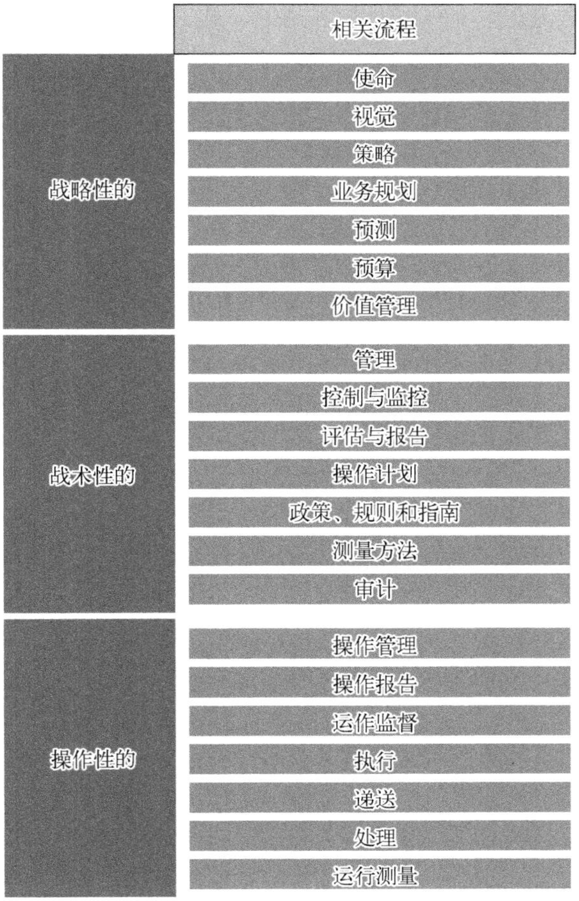

	相关流程
战略性的	使命
	视觉
	策略
	业务规划
	预测
	预算
	价值管理
战术性的	管理
	控制与监控
	评估与报告
	操作计划
	政策、规则和指南
	测量方法
	审计
操作性的	操作管理
	操作报告
	运作监督
	执行
	递送
	处理
	运行测量

图24　企业层级和相关流程关系示例

As demonstrated in Figure 24, the enterprise tiers represent tagging possibilities that link the processes, the goal and objective view, decision making, and the system measurement and reporting view. Beside the ability to classify processes according to strategic, tactical, and operational processes, and relate the processes to the right accountability level, ideal for process ownership, process governance, process analytics, and reporting, tier tagging enables information or the service flow between the enterprise tiers, making the tier process tagging a powerful tool Figure 25.

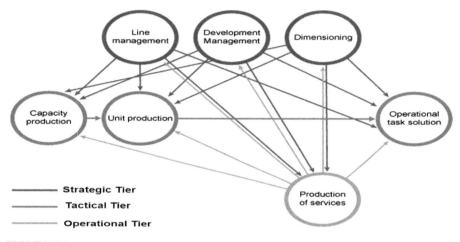

FIGURE 25

Example of enterprise tiers and the relations across tiers.

Tiers	Business Process	Process Step	Business Process Activity	Solution Area	WorkFlow
1. Strategic	Develop and manage strategies	Develop HR strategy	Identify strategic HR needs	Portal	End-to-end Process Integration
2. Tactical	Manage treasury operations	Manage treasury policies and procedures	Develop and confirm internal controls for treasury	ERP	Human capital management
2. Tactical	Manage IT knowledge	Develop IT knowledge management strategy	Plan IT knowledge management actions and priorities	ERP	Corporate services
2. Tactical	Dispose of assets	Dispose of product/service assets	Perform sale or trade	Portal	User Product/serviceivity Enablement
2. Tactical	Develop product and services	Design, build, and evaluate product and services	Build prototypes	ERP	Corporate services
2. Tactical	Develop and maintain information technology solutions	Develop the IT development strategy	Establish sourcing strategy for IT development	ERP	Corporate services
3. Operational	Develop and maintain information technology solutions	Develop the IT development strategy	Define development processes, methodologies, and tools standards	ERP	Corporate services
3. Operational	Manage taxes	Develop tax strategy and plan	Develop foreign, national, state, and local tax strategy	ERP	Corporate services

FIGURE 26

Detailed example of sorting processes according to enterprise tiers.

如图24所示，企业的层级表示标记可能性，它们链接流程、目标和目标视图、决策制定以及系统测量和报告视图。除了根据战略、战术和运营流程对流程进行分类，并将流程与正确的问责制级别相关联，这是流程所有权、流程治理、流程分析和报告的理想选择，分层标记还能够实现企业层之间的信息或服务流，使层级流程标记成为一个强大的工具（图25）。

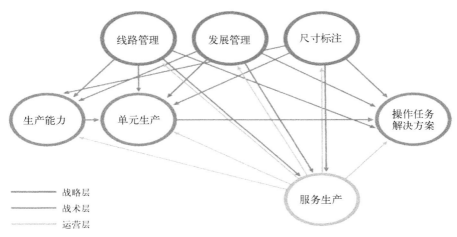

图25　企业层和跨层关系示例

层级	业务流程	流程步骤	业务流程活动	解决领域	工作流
1.战略性的	制定和管理策略	制定人力资源战略	识别战略人力资源需求	入口站点	端到端流程集成
2.战术性的	管理国库业务	管理财政政策和程序	制定和确认财政部的内部控制	企业资源计划(ERP)	人力资本管理
2.战术性的	管理IT知识	制定IT知识管理战略	计划IT知识管理行动和优先级	ERP	公司服务
2.战术性的	处置资产	处置产品/服务资产	进行销售或贸易	入口站点	用户产品/服务可实施性
2.战术性的	开发产品和服务	设计、构建和评估产品和服务	构建原型	ERP	公司服务
2.战术性的	开发和维护信息技术解决方案	制定IT发展战略	建立从IT发展获得的策略	ERP	公司服务
3.操作性的	开发和维护信息技术解决方案	制定IT发展战略	定义开发过程、方法和工具标准	ERP	公司服务
3.操作性的	管理税收	制定税收战略和计划	制定外国、国家、州和地方税收战略	ERP	公司服务

图26　根据企业层分类流程的详细示例

In Figure 26 is a more detailed example of how the tagging of the processes according to strategic, tactical, and operational views is an ideal way to go across end-to-end processes and see connections in the different tiers.

PROCESS NATURE

Not all processes are created equal. Some processes are simple and highly repeatable, and involve the same transactional and transformation work that can be done over and over; others are more complex or dynamic, involved in cases where the rules either are not clear, or are chaotic or subject to dynamic change. Understanding the nature and importance of processes is therefore central to effective process management and the basis for developing a successful business process management basis. Characteristics of the tagging process are as follows:

1. Simple and static processes are that are well understood, highly repeatable and are carried out multiple times in exactly the same way.
2. Generic and hybrid processes have mixed components and objects-like resources, tasks, measures, or rules, but similar properties.
3. Characteristics of complex and dynamic processes are that they are difficult and challenging, the way they are carried out changes over time, and the rules, practices, and procedures of this class of process are subject to ongoing evolution.

The ability to tag processes and categorize them according to their simple, hybrid, and complex nature enables the link to process drivers, reporting, process automation, and even risk, the operating model, as well as cost and value.

- **Process drivers and process nature:** Understanding the factors that lead to the drivers, e.g., performance and value drivers of a particular process, provides a means not only of understanding what to consider when evaluating the operations of a process, but also of developing the approach to improving it. When we consider the factors that influence process design, we see that many organizations are looking at performance drivers when modelling processes. However, we do know that the focus cannot not only about performance; by the nature of many of the process drivers, we must be concerned with value creation and realization.
- **Cost and process nature:** When we consider the factors that influence process design, we see that many organizations are looking at processes with a cost-centric view. Although it should not be this way, for some organizations BPM is all about cost. This is where the identification of cost factors in the processes is an ideal tool. In the simple and hybrid processes, cost cutting

图26是一个更详细的例子，说明了根据战略、战术和运营视图标记流程是一种跨越端到端流程并查看不同层级连接的理想方式。

2.4.22　流程性质

并非所有流程的创建都是平等的。有些流程简单且具有高度可重复性，并且涉及可反复完成的相同事务和转换工作；另一些流程更为复杂或动态，涉及规则不明确、混乱或动态变化的情况。因此，了解流程的性质和重要性是有效流程管理的核心，也是建立成功BPM基础的基础。标记流程的特征如下。

（1）简单和静态的流程被很好理解，高度可重复，并且以完全相同的方式执行多次。

（2）通用流程和混合流程具有混合的组件和对象，如资源、任务、度量或规则，但具有相似的属性。

（3）复杂和动态流程的特点是，它们是困难和具有挑战性的，它们的执行方式随着时间的推移而变化，这类流程的规则、实践和程序都会不断演变。

能够根据流程的简单、混合和复杂特性标记流程并对其进行分类的能力使得链接到流程驱动因素、报告、流程自动化甚至风险、操作模型以及成本和价值成为可能。

- 流程驱动因素和流程性质：了解驱动因素，如特定流程的绩效和价值驱动因素，不仅提供了了解评估流程操作时应考虑的因素，还提供了开发改进流程的方法。当我们考虑影响流程设计的因素时，我们发现许多组织在建模流程时都在考虑绩效驱动因素。然而，我们知道，关注的焦点不仅仅是绩效，根据许多流程驱动因素的性质，我们必须关注价值创造和实现。

- 成本和流程性质：当我们考虑影响流程设计的因素时，我们发现许多组织都在以成本为中心的观点看待流程。虽然不应该这样做，但对于某些组织来说，BPM完全是关于成本的。在这一点上，识别流程中的成本因素是一个理想的工具。在简单和混合的流程中，通过标准化来削减成本是一个巨

through standardization is a huge potential, whereas human-based cost cutting, which is generally achieved by training or other means that improve human performance, must occur for processes that are more complex. More straightforward standard cost cutting is normally applied for the less complex processes.

- **Value aspects and process nature:** Once the enterprise truly understands which processes are of value, it is appropriate to invest in optimization of performance to include all factors of the nature of the process and thereby the simple, hybrid, and complex processes that are a part of value creation and realization.

- **Process reporting and process nature:** Reports are periodic accounts of the activities of the enterprise, whereas the other reporting tools are real-time or near real-time communications tools for providing information in such a way as to connect the activities to their strategic and tactical tiers. Where scorecards are a summary record of events of the execution, dashboards provide at-a-glance views of key performance indicators relevant to a particular objective, and cockpits provide real-time/actual measurement of dynamic and especially complex processes. In this case, because the concerns are just about control, all that is required here is the application of sound standards and guidelines.

- **Process automation and process nature:** Human may be to replaced by machine or automation achieve a lower cost of operations. First, aspects of human labor were replaced with simple process automation. Today, both simple and complex processes are automated across the enterprise. For processes that tend to be either more complex or more valuable, it makes sense to seek improvements through exploring and developing methods and practices to achieve greater value before automating them. Generally, transaction volumes aside, investment in automation or other capabilities should receive low priority when the process is cost focused and simple; the business case will not be there unless the transaction volumes are significantly larger than elsewhere in the process landscape. Understanding the dimensions of a particular process nature is necessary to its automation design.

- **Risk aspects and process nature:** When considering the effects of uncertainty on process design and operations, risk increases as complexity increases. Whereas the probability and impact of variation in the performance of a process are consistent within processes of the same nature, when considering other than simple and static processes the reason for the risk varies. Variations within simple processes are generally not material to their performance and therefore not of concern to the extent that explicit risk management-centric oversight is required. On the other hand, when considering bottlenecks in processes, a phenomenon in which a single or limited number of components, aspects, or resources affect or

大的潜力,而对于更复杂的流程,通常通过培训或其他提高人员绩效的手段来实现的基于人员的成本削减则必须发生。更直接的标准成本削减通常适用于不太复杂的流程。

- 价值方面和流程性质:一旦企业真正了解哪些流程是有价值的,就应该对流程性质的所有因素进行投资和绩效优化,包括简单、混合和复杂的流程,这些流程是价值创造和实现的一部分。

- 流程报告和流程性质:报告是企业活动的"定期账户",而其他报告工具是实时或接近实时的通信工具,用于提供信息,从而将活动与其战略和战术层连接起来。当记分卡是执行事件的汇总记录时,仪表盘提供与特定目标相关的KPI的一目了然的视图,而驾驶舱提供动态和特别复杂流程的实时/实际测量。在这种情况下,因为关注的仅仅是控制,所以这里需要的只是应用健全的标准和指导方针。

- 流程自动化和流程性质:用机器或自动化代替人工可降低操作成本。首先,人类劳动的各个方面被简单的流程自动化所取代。如今,简单和复杂的流程在整个企业中都是自动化的。对于那些趋向于更复杂或更有价值的流程,在实现自动化之前,通过探索和开发方法和实践来寻求改进是有意义的。通常,不考虑事务发生数量,当流程以成本为中心且简单时,对其自动化或其他功能的投资应获得低优先级;除非其事务发生数量明显大于流程环境中的其他地方,否则业务案例将不存在。了解特定流程性质的维度对于其自动化设计是必要的。

- 风险方面和流程性质:当考虑不确定性对流程设计和操作的影响时,风险随着复杂性的增加而增加。当考虑除简单和静态流程外的其他流程时,同一性质的流程中,流程性能变化的概率和影响是一致的,但是风险的原因是不同的。简单流程中的变化通常对其绩效没有实质性影响,因此在需要明确的以风险管理为中心的监督的情况下,对此并不需要担心。另一方面,当考虑流程中的瓶颈(一个单一或有限数量的组件、方面或资源影响或

otherwise limit the capacity, affecting performance and value, the same bottleneck would have a different impact on a high-value process than it would on a lesser-value process. Therefore, considering the process nature, one distinguishes between pain points, weakness cluster, and value clusters.

- **Operating Model and process nature:** An operating model is an abstract representation of how an organization operates, or could operate, across process, organization, and technology domains to achieve its purpose and execute its strategy. The operational model is orthogonal to the previous models. Whereas all other models examine strategic importance and process complexity within a particular context, the operating model considers the implications of relative process standardization and integration irrespective of its value or relative complexity.

MISCATEGORIZATION AND MISCLASSIFICATION

It is important to mention that, while the logical clustering of categorization and classification is important and brings many benefits, there are also some common mistakes that are made. For the most part these errors occur due to a logical fallacy in which diverse and dissimilar objects, concepts, entities, etc. are grouped (categorization) or sorted by order, based upon illogical common denominators, or common denominators and relations based on over generalization of perceived underlying patterns are made. It is therefore relevant as illustrated in this document, to have the right decomposed levels and relationship in place as well as the right tier, type and the nature identified.

CONCLUSIONS

While this chapter should be seen and used as a description of what process tagging is and how it can be applied, but it does not show all aspects of where the concepts can be useful. It attempted to build a basis of a structured way of thinking, working, modeling, and implementation of process classification and categorization. It endeavored to provide a standardized terminology, build common understanding, and make available the standardized and integrated classification and categorization tags to processes, enabling process practitioners to use the process tagging reference content to:

- Identify the relevant process tags
- Identify the process nature and enable understanding
- Specify process complexity
- Decompose the relevant process levels into the smallest parts that can, should, and need to be modeled, and then compose the entities to the right content through mapping, simulation, and scenarios

以其他方式限制容量、影响绩效和价值的现象)时,同一瓶颈对高价值流程的影响与对低价值流程的影响不同。因此,基于流程的性质,我们可以区分痛点、弱点集群和价值集群。

- 操作模型和流程性质:操作模型是一个组织如何在流程、组织和技术领域内操作或可以操作以实现其目的和执行其战略的抽象表示。操作模型与前面的模型是正交的。尽管所有其他模型都在特定环境中检查战略重要性和流程复杂性,但操作模型考虑了相关流程标准化和集成的影响,而不管其价值或相对复杂性如何。

2.4.23 错误分类和错误分集

值得一提的是,虽然分类和分集的逻辑聚类很重要,并带来许多好处,但也存在一些常见的错误。在大多数情况下,这些错误是由逻辑谬误造成的,其中各种不同的对象、概念、实体等被分组(分类)或按顺序排序,基于不合逻辑的共同特性,或基于感知到的潜在模式的过度泛化的共同特性和关系。因此,如本书所示,适当的分解层级和关系以及适当的层级、类型和性质是相关的。

2.4.24 结论

虽然本节应该被看作对什么是流程标记以及如何应用它的描述,但是它并没有显示概念在哪些方面有用。它试图建立一种结构化的思维、工作、建模及流程分集和分类实现方法的基础。它努力提供标准化术语,建立共识,并向流程提供标准化和一体化的分集和分类标记,使流程从业者能够使用流程标记参考内容:

- 识别相关流程标记;
- 识别并能够理解流程的性质;
- 详述流程的复杂性;
- 将相关流程级别分解为可以、应该和需要建模的最小部分,然后通过映射、模拟和场景将实体组合到正确的内容中;

- Model the relevant process meta objects through the architectural layers (process architecture relevant)
- Visualize and clarify the process tags with the process templates by using maps, matrices, and models (alternative representation of information)
- Reduce and/or enhance complexity of process modeling, engineering, and architecture principles applying a common logical clustering
- Provide a structured process content that can be used for blueprinting and implementation

The beauty and power of this approach is that it enables understanding of the nature of a process, its levels, the relationships, enabling that the requirements for analysis are explicit, the starting and ending points are abundantly clear, and the level of analysis is set in each case by the context of the view of perspective of the containing description of work.

End Notes

1. Cohen H. and Lefebvre C., eds., *Handbook of Categorization in Cognitive Science* (Elsevier, 2005).
2. http://oll.libertyfund.org/index.php?option=com_staticxt&staticfile=show.php%3Ftitle=166&Itemid=99999999.
3. Cohen H. and Lefebvre C., eds., Ibid.
4. Ross, Jeanne; Weill, Peter; Robertson, David C. Enterprise Architecture As Strategy: Creating a Foundation for Business Execution. Harvard Business Review Press (2006).
5. Ibid.
6. Ibid.
7. Dr M. Hammer, *Business Process Re-engineering*.
8. H. Smith and P. Fingar, *Business Process Management*.
9. BPMN.
10. http://www.modelio.org/documentation/metamodel/Metamodel_HTML/90.html.
11. http://wiki.scn.sap.com/wiki/display/ModHandbook/Process+Hierarchy.
12. Ibid.
13. Within a single role. A swim lane is a graphical tool for showing a collection of steps, or activities for which a specific role is responsible.
14. Business information analysis and integration technique (BIAIT): the new horizon, Walter M. Carlson IBM, ACM SIGMIS, Volume 10 Issue 4, Spring 1979.
15. Products are tangible and discernible items.
16. Process choice is demand driven. Three primary questions bear on the selection of the production process step:
   a. How much variety in products or services must be provided?
   b. What degree of equipment flexibility will be needed?
   c. What is the expected volume of output?
17. Initiate, plan, execute, close out.
18. Specify, design, obtain materials, set up, create components, assemble, finish, test, accept, take down.
19. Set up, create components, assemble, finish, test, accept, take down.

- 通过架构层对相关流程元对象建模（与流程架构相关）；
- 通过使用地图、矩阵和模型（信息的替代表示），使用流程模板可视化和澄清流程标签；
- 减少和/或增强流程建模、工程和架构原则的复杂性，应用通用逻辑集群；
- 提供可用于蓝图编制和实施的结构化流程内容。

这种方法的优点和威力在于，它能够理解一个流程的性质、层次、关系，使分析的要求明确，起点和终点充分清晰，并且在每种情况下，分析的层次都是由包含工作描述的透视图的上下文来设置的。

参考文献

［ 1 ］ Cohen H. and Lefebvre C., eds., Handbook of Categorization in Cognitive Science (Elsevier, 2005).
［ 2 ］ http://oll.libertyfund.org/index.php?option=com_staticxt&staticfile=show.php%3Ftitle=166&Itemid=99999999.
［ 3 ］ Cohen H. and Lefebvre C., eds., Ibid.
［ 4 ］ Ross, Jeanne; Weill, Peter; Robertson, David C. Enterprise Architecture As Strategy: Creating a Foundation for Business Execution. Harvard Business Review Press (2006).
［ 5 ］ Ibid.
［ 6 ］ Ibid.
［ 7 ］ Dr M. Hammer, Business Process Re-engineering.
［ 8 ］ H. Smith and P. Fingar, Business Process Management.
［ 9 ］ BPMN.
［10］ http://www.modelio.org/documentation/metamodel/Metamodel_HTML/90.html.
［11］ http://wiki.scn.sap.com/wiki/display/ModHandbook/Process+Hierarchy.
［12］ Ibid.
［13］ Within a single role. A swim lane is a graphical tool for showing a collection of steps, or activities for which a specific role is responsible.
［14］ Business information analysis and integration technique (BIAIT): the new horizon, Walter M. Carlson IBM, ACM SIGMIS, Volume 10 Issue 4, Spring 1979.
［15］ Products are tangible and discernible items.
［16］ Process choice is demand driven. Three primary questions bear on the selection of the production process step:
a. How much variety in products or services must be provided?
b. What degree of equipment flexibility will be needed?
c. What is the expected volume of output?
［17］ Initiate, plan, execute, close out.
［18］ Specify, design, obtain materials, set up, create components, assemble, finish, test, accept, take down.
［19］ Set up, create components, assemble, finish, test, accept, take down.

20. Produce, test, accept.
21. Ibid.
22. An economic service is the production of an essentially intangible benefit, either in its own right or as a significant element of a tangible product, which through some form of exchange satisfies an identified need.
23. Hamel and Prahalad, 1994; Sull, 2009.
24. Shimizu and Hitt, (2004).
25. Burgelman, 1983.
26. LEADing Practice-Categorization & Classificaion Body of Knowledge, 2014.

［20］ Produce, test, accept.

［21］ Ibid.

［22］ An economic service is the production of an essentially intangible benefit, either in its own right or as a significant element of a tangible product, which through some form of exchange satisfies an identified need.

［23］ Hamel and Prahalad, 1994; Sull, 2009.

［24］ Shimizu and Hitt, (2004).

［25］ Burgelman, 1983.

［26］ LEADing Practice-Categorization & Classificaion Body of Knowledge, 2014.

Why Work with Process Templates

Mark von Rosing, Maria Hove, Henrik von Scheel, Ulrik Foldager

INTRODUCTION

In the Business Process Management (BPM) Ontology and Semantics chapter, we provided you with a detailed and extensive description of the concept of ontology and semantics: what they are, what their purpose is, and perhaps most importantly of all, how to use them effectively in the BPM way of thinking, the BPM way of working, and the BPM way of modeling.

Although we see an immense amount of literature on the adoption and implementation of BPM in a rapidly growing market, what surprises us is the low level of maturity of standards in terms of consistent and integrated templates to describe process. Because we see this as one of the main reasons for high-cost and low-value creation around process analysis, process mapping, process documentation, and process governance, we have chosen to focus in this chapter on process templates.

The chapter will specify what process templates are, the relationship between BPM ontology and semantics and how it links directly to the concept of templates, why they are needed, where they can be applied, and the benefits of applying them.

We believe that the principles of process templates are relevant to any organization, independent of industry, business model, or operating model.

THE RELATIONSHIP BETWEEN BUSINESS PROCESS MANAGEMENT ONTOLOGY AND PROCESS TEMPLATES

When an organization decides to make use of ontology and semantics to lay the foundation of what we call "process things," it is done for a vast variety of purposes (we will be naming a few of them throughout this chapter), but the most important one is that once you have established a specific and clear definition for a meta object, for example, this definition will be available to all relevant employees across organizational boundaries of the enterprise after it has been documented and published for use. This means that a common understanding and consensus has been reached within the organization for what name a particular meta object has for whenever you are referring to that particular meta object. Of course, this makes it a lot more practical for organizations to handle objects in the bigger picture: not just for documenting, but also for using them when modeling, engineering, and architecting process concepts and solutions, regardless of the business unit and/or business requirement. In the sense of semantics, then, it allows you to accurately describe how a particular object relates to another

The Complete Business Process Handbook. http://dx.doi.org/10.1016/B978-0-12-799959-3.00009-4

2.5　为何使用流程模板

Mark von Rosing, Maria Hove, Henrik von Scheel, Ulrik Foldager

2.5.1　介绍

在2.3节BPM本体论和语义中,我们向您详细而广泛地描述了本体论和语义的概念:它们是什么、它们的目的是什么,但最重要的是,如何在BPM思维方式、BPM运行方式和BPM建模方式中有效地使用它们。

虽然我们看到了大量有关在快速增长的市场中采用和实施BPM的文献,但令人惊讶的是,标准在用于描述流程的一致性和集成模板方面的成熟度很低。因为我们认为这是围绕流程分析、流程映射、流程文档和流程管理进行高成本和低价值创建的主要原因之一,所以我们选择在本节中集中讨论流程模板。

本节将详细说明什么是流程模板,BPM本体和语义之间的关系,以及它如何直接链接到模板的概念,为什么需要它们、在哪里可以应用它们及应用它们的好处。

我们认为流程模板的原则与任何组织都相关,独立于行业、业务模型或操作模型。

2.5.2　BPM本体与流程模板的关系

当一个组织决定利用本体和语义来为我们所说的"流程事物"("process things")奠定基础时,它是为各种各样的目的(我们将在本节中命名它们中的一些)而做的,但最重要的是,一旦您为元对象建立了明确而具体的定义,例如,此定义在记录并发布供使用后,将可供跨企业组织边界的所有相关员工使用。这意味着在组织内部就特定元对象的名称达成了共识和一致,无论您何时引用该特定元对象。当然,这使得组织更实际地处理全局中的对象:不仅用于文档化,也在建模、工程和架构流程概念及解决方案时使用它们,而不管业务部门或业务需求如何。从语义的意义上讲,它允许您准确地描述一个特定对象与另一个特定

particular object (regardless of object type or hierarchical location). This has to be defined as well, of course, but just like the ontology definitions, an organization must also reach a common understanding and consensus in semantic relationships regarding how exactly each object relates to another. This is meticulous work and takes time and effort, but it is nevertheless extremely important to avoid common pitfalls.

Thus, we know what to call a particular object. In our case, we choose to use the driver meta object (through the creation of our ontology), and we know how the driver meta object relates to a business process meta object (because we have also defined a set of semantics that accurately describe how they influence and relate to each other). If we would then create a process template in which the relationship to a value driver is relevant, we would be able to use the business process meta object and place it in the process template, for both information and documentation purposes, as well as the ability to relate it to other aspects further on. We would most likely be identifying and listing (for example, in columns in an Excel spreadsheet) values such as the name of each business process meta object, where it is located, what resources it uses, etc. Maps are always used within the concept of the BPM way of thinking, which is the starting point, and where the conceptual aspects are covered. With this planner's view we generate and describe business concepts, document important and essential information regarding the business, and create a general overview of more or less anything of importance.

Continuing from this path, we would next create a process matrix for the purpose of relating the value driver meta object (in a row next to the columns in the Excel spreadsheet) to the relevant business process meta object. Matrices are almost always created within the concept of the BPM way of working, because here we begin to actually take action and relate objects to each other. Keep in mind that whenever you are creating a matrix, you are actively using the information provided to you through the previous creation of a process map. The map provides you with the information you need to create an efficient process matrix. By creating a process matrix, we then allow ourselves to identify directly and accurately which kind of value driver (whether internal or external) has an impact on the business process (regardless of impact type, although it has to carry some importance because we expect to note down information that affects the business somehow, and bearing that in mind that it is therefore worth documenting) on the business processes of the organization (regardless of business unit). Not only do we describe which value driver affects which business process, we can also identify exactly how the value driver affects the business process, where the impact occurs, what the consequences are, and who is responsible (role object) and who is accountable (owner object) for acting upon this knowledge.

Last but not least, we could—if deemed necessary and/or beneficial—create a process model to build a visual representation of how these value drivers would affect the business processes of the organization. Process models, as the name implies, are mostly used within the concept of the BPM way of modeling. Here we visually illustrate behaviors, relationships, connectivity, location, function, and purpose. Keep in mind, however, that a model always makes use of both the process map and the process matrix. The map and the matrix are your source of information; the model is how you would visualize this information.

对象的关系(无论对象类型或层次位置如何)。当然,这也必须被定义,但就像本体论的定义一样,一个组织也必须在语义关系中就每个对象与另一个对象的确切关系达成共识。这是一项细致的工作,需要时间和精力,但避免常见的陷阱仍然非常重要。

因此,我们知道一个特定的对象是什么。在我们的例子中,我们选择使用驱动元对象(通过创建本体),并且我们知道驱动元对象如何与业务流程元对象相关(因为我们还定义了一组语义,精确描述了它们如何影响和相互关联)。如果我们随后创建一个流程模板,其中与价值驱动因素的关系是相关的,那么我们将能够使用业务流程元对象,将其放置在流程模板中用于信息和文档的使用,并将其与其他方面进一步关联。我们很可能正在识别和列出(如在Excel电子表格的纵列中)价值,如每个业务流程元对象的名称、它所在的位置、它使用的资源等。映射总是在BPM思维方式的概念中使用,这是一个起点,并且涵盖了概念方面。根据该流程计划专员的观点,我们生成和描述业务概念,记录有关业务的重要和基本信息,并概述任何重要的内容。

沿着这条路径,我们接下来将创建一个流程矩阵,以便将价值驱动元对象(在Excel电子表格中纵列旁边的一行中)与相关的业务流程元对象相关联。矩阵几乎总是在BPM工作方式的概念中创建的,因为在这里我们开始采取实际行动并将对象相互关联。请记住,无论何时创建矩阵,您都应尽可能地使用以前创建流程图时提供给您的信息。流程地图为您提供了创建高效流程矩阵所需的信息。通过创建一个流程矩阵,我们就可以直接而准确地确定哪种价值驱动因素(无论是内部还是外部)对组织(不考虑业务部门)的业务流程产生影响(无论影响类型如何,虽然它必须带有一定重要性,但因为我们希望记下以某种方式影响业务的信息,所以它是值得记录的)。我们不仅可以描述哪些价值驱动因素影响哪些业务流程,还可以准确地确定价值驱动因素如何影响业务流程、影响发生的位置、后果是什么、谁承担责任(角色对象)以及谁负有责任(所有者对象),并根据这些知识采取行动。

最后,如果认为必要或有益的话,我们可以创建一个流程模型来构建这些价值驱动因素将如何影响组织的业务流程的可视化表示。顾名思义,流程模型主要用于BPM建模方式的概念中。在这里,我们直观地说明了行为、关系、连接性、位置、功能和目的。但是,请记住,模型总是同时使用流程地图和流程矩阵。流程地图和矩阵是您的信息源,而模型是您将如何可视化这些信息。

As you can see, this is why the BPM ontology and semantics have real business value, because you have put down definitions of *what* (ontology) the objects are and *how* (semantics) they relate to other objects. As you can imagine, this is an essential piece of information for any process expert (process modeler), process engineer, and/ or process architect in daily work. This is also the foundation and the reason why our process templates are 100% standardized and integrated with each other, enabling the ability to share process objects across various process templates.

WHAT ARE PROCESS TEMPLATES?

A process template is a documentation product such as a process map, process matrix, or process model. Process templates are created to describe some aspect of a process, a process landscape, process flow, process solution, or state. In enterprise architecture, these would be called artifacts. Templates enable the capture and relation of objects within the same template or across multiple templates, each of which promotes its own view. Process templates enable the capture and relation of process-centric objects within the same template or across multiple templates, each of which promotes its own view of a process.

The purpose of having process templates that address the various process concepts is to set out or describe how to organize and structure the viewpoints and process objects associated with the various disciplines and bring them together to create a common understanding. Standard process templates are important because they establish the elements of the artifacts, i.e., the relevant process objects to be addressed when the template is used.

Within the set of templates presented in this chapter, each template is part of an overarching ontological and semantically based specification that ensures that all of the objects are appropriately related. Reuse of the content of one process template or view for another is therefore ensured. Without this standardization and integration, the process templates create more work and cost, and are actually of little value. What many practitioners and organizations do not realize is the importance of having such integration and standardization across the landscape of this work, and therefore the value of the result.

Years of research in the Global University Alliance[1] have identified the semantic relations of the various process objects and how they can be applied within different contexts. These relationships are built into the process templates, e.g., process maps, process matrices, and/or process models.

PROCESS MAPS

A process map is intended to be an accurate list and representation of a set of decomposed and/or composed process objects. The purpose of this map is to inventory and create a list of all processes in the enterprise.

The content of a process map is based on which objects/elements can be related so the columns of the map conform to the semantic rules within the context in which they are being used.

正如您所看到的,这就是为什么BPM本体和语义具有真正的业务价值,因为您已经放下了对象是什么(本体)以及它们如何与其他对象相关的定义(语义)。正如您可以想象的那样,对于任何流程专家(流程建模者)、流程工程师或流程架构师来说,这都是日常工作中必不可少的一部分信息。这也是我们的流程模板100%标准化和相互集成的基础和原因,它使得流程对象能够在不同的流程模板之间共享。

2.5.3　什么是流程模板?

流程模板是一个文档,如流程地图、流程矩阵或流程模型。创建流程模板是为了描述流程的某些方面、流程布局、流程流、流程解决方案或状态。在企业体系结构中,这些称为制品(artifacts)。模板可以捕获同一模板内或跨多个模板的对象并建立对象关系,每个模板都可以展示自己的视图。

拥有处理各种流程概念的流程模板的目的是制定或描述如何组织和构造与各种规程相关联的视角和流程对象,并将它们结合在一起以形成共同的理解。标准流程模板很重要,因为它们建立了制品的元素,即使用模板时要处理的相关流程对象。

在本节介绍的一组模板中,每个模板都是本体总体和基于语义的规范的一部分,该规范确保所有对象都适当相关。因此,可以确保将一个流程模板或视图的内容重新用于另一个流程模板或视图。如果没有这种标准化和一体化,流程模板将产生更多的工作和成本,而且实际上几乎没有价值。许多从业者和组织没有意识到,在这项工作的各个方面进行这种整合和标准化的重要性,因此结果的价值也是如此。

全球大学联盟[1]多年的研究已经确定了各种流程对象的语义关系,以及它们如何在不同的环境中应用。这些关系构建在流程模板中,如流程地图、流程矩阵和/或流程模型。

2.5.4　流程地图

流程地图旨在准确列出和表示一组分解或组合的流程对象。这个地图的目的是创建和存储企业中所有流程的列表。

流程地图的内容基于哪些对象/元素可以关联,因此图中的列符合使用它们的语境中的语义规则。

This list helps us to understand the breadth of functionality provided by each of the processes. It will also provide a centralized and official overview and record of the key processes in the enterprise, each situated within the specific process area and process group in which it participates as well as linking in the channel, stakeholder, owner, and role/resource (including the manager) involved. Table 1 is an example of such a process map.

Members of the BPM team carry out the tasks necessary to complete the process map in the manner described in Table 2.

Mapping a process enables the following, among others:

- Identify relevant processes, including the name of the process
- Specify a unique process identifying number or ID
- Specify the level of process detail (see process levels in process tagging chapter)
- Link the involved business units and stakeholders to the relevant process
- Detail the process owners, in terms of which process is owned by whom
- Other process roles involved this can include:
  - Process roles

Table 1 *Example of a Process Map*

Process#	Business Process area	What Specification:				Who Specification:			
		Process Groups	Business Process	Process Steps	Process Activities	Stakeholder Involved	Process Owner	Managers Involved	Roles/Resources Involved
#									

Table 2 *Example of How a Process Map is Based on Semantic Rules and Tasks*

The "what" specification: Identify, select, and categorize the business processes.	
Rules	Process relates directly to business construct, i.e., when collecting the inventory of processes within a sliver or slice of the enterprise
Tasks	• Identify and categorize the process areas related to the business areas or the end-to-end flow areas. • Specify and categorize the process groups based on the related business groups or the end-to-end flow groups. • Select, label, and categorize the business processes according to the groups. • Spot the process steps related to the business process. • Identify the process activities related to the process steps and business processes.
The "who" specification: Identify the relevant stakeholder, owners, and managers involved.	
Rules	Process relates directly to role and to resource (stakeholder) and owners, i.e., when collecting the inventory of processes related to a community within the enterprise
Tasks	• Identify and categorize the stakeholders linked to the business processes. • Specify the business process owners and categorize them according to their business process ownership. • Spot and categorize the business process-related managers. • Recognize, classify, and label the roles/resources of the business processes.

　　此列表帮助我们了解每个流程提供的功能的广度。它还将提供企业中关键流程的集中和正式的概述和记录，每个流程位于其参与的特定流程区域和流程组内，并在涉及的渠道、利益相关方、所有人和角色/资源（包括经理）中进行链接。表1就是这样一个流程地图的例子。

　　BPM团队的成员应该按照表2中描述的方式执行完成流程地图所需的任务。

　　绘制流程地图可以实现以下功能。

- 识别相关流程，包括流程名称；
- 指定唯一的流程标识号或ID；
- 指定流程详细信息的级别（请参阅流程2.4.5小节中的流程级别）；
- 将相关业务部门和利益相关者与相关流程联系起来；
- 详细说明流程所有人，就流程所有人而言。
- 涉及的其他流程角色包括：
  - 流程角色；

表1　流程地图示例

流程 #	什么的规范：					谁的规范：			
	业务流程区域	流程组	业务流程	流程步骤	流程活动	利益相关方	流程所有人	相关管理人员	涉及的角色/资源
#									

表2　流程地图如何基于语义规则和任务的示例

关于什么的规范：识别、选择和分类业务流程。	
规则	流程直接与业务结构相关，即在企业的一个条块或切片内收集流程的库存时。
任务	• 识别和分类与业务领域或端到端流程领域相关的流程领域。
	• 根据相关业务组或端到端流程组指定和分类流程组。
	• 根据组别选择、标记和分类业务流程。
	• 找出与业务流程相关的流程步骤。
	• 识别与流程步骤和业务流程相关的流程活动。
关于谁的规范：确定相关的利益相关方、所有人和相关管理人员。	
规则	流程直接与角色、资源（利益相关方）和所有人相关，即收集与企业内社区相关的流程库存时。
任务	• 识别和分类与业务流程相关的利益相关方。
	• 指定业务流程所有人，并根据其业务流程所有权对其进行分类。
	• 识别和分类与业务流程相关的管理人员。
	• 识别、分类和标记业务流程的角色/资源。

- Process approver, in terms of who approves the process and or the work
- Process checker, in terms of who checks the work

As we have already mentioned, the above reasons determine the design and content of what is within such a process map and therefore how such a process map looks.

PROCESS MATRIX

Process matrices show the relationship between two specific sets of decomposed (broken down) objects in a process-centric context. The core idea of the process matrices is that they each consist of a set of process objects that semantically have primary and therefore direct natural relations to each other. The result is that these are always in the form of two lists (a row and a column) in which the process objects with which they share a relationship are each rated according to them within the body of the matrix. Within the process matrix, this allows one to relate the unfamiliar to the familiar, thus connecting process objects in the different layers (composition).

Table 3 is a process matrix illustrating the columns of the process map combined with performance indicators. Using this template would result in the content of every column having a minimum of one process indicator.

The process-performance indicator matrix's capture should be based on enterprise modeling and architecture rules outlined in Table 2. In addition to those rules and tasks, the rules and tasks outlined in Table 4 are applied when completing Table 3.

Reasons for creating process matrices can include:

1. Link processes to business goals and strategy
2. Create value-oriented process relations
3. Relate business competencies and processes
4. Understand the end-to-end process flow
5. Identify process problems and pain points—fixing a defective or inefficient process
6. Specify which business objects, information objects, and or data objects are involved

Table 3 *Example of a Process Matrix Showing How Processes Relates to Performance Indicators*

	Indicators Process Number	What Specification:			Who/Whose Specification:			
		Business Process	Process Steps	Process Activities	Stakeholder Involved	Process Owner	Managers Involved	Roles/ Resource Involved
Performance Indicator 1	#							
Performance Indicator 2	#							
Performance Indicator N	#							

- 流程批准人（就批准流程或工作的人员而言）；
- 流程检查员（依据谁检查工作）。

正如我们已经提到的，上述原因决定了这样一个流程地图中的内容和设计，以及这样一个流程地图的全局。

2.5.5　流程矩阵

流程矩阵显示了以流程为中心的环境中两个特定的分解（瓦解）对象集之间的关系。流程矩阵的核心思想是：它们每个都由一组流程对象组成，这些流程对象在语义上具有彼此之间的直接自然关系。结果是它们总是以两个列表（行和列）的形式存在，其中与它们有共享关系的流程对象在矩阵的主体内根据它们进行评级。在流程矩阵中，这允许将不熟悉的与熟悉的关联起来，从而将不同层别中的流程对象（组合）连接起来。

表3是一个流程矩阵，说明了结合绩效指标的流程地图的各列。使用此模板将使每个列的内容至少具有一个流程指标。

PPI矩阵的获得应基于表2中概述的企业建模和体系结构规则。除了这些规则和任务，表4中概述的规则和任务在完成表3时也适用。

创建流程矩阵的原因包括：

（1）将流程与业务目标和战略联系起来；

（2）建立以价值为导向的流程关系；

（3）关联业务能力和流程；

（4）了解端到端流程；

（5）识别流程问题和症结——修复有缺陷或效率低下的流程；

（6）指定涉及哪些业务对象、信息对象和/或数据对象；

表3　显示流程与绩效指标之间关系的流程矩阵示例

	指示流程编号	什么规格：			谁/谁的规格：			
		业务流程	流程步骤	流程活动	利益相关方	流程所有人	相关管理人员	涉及的角色/资源
绩效指标1	#							
绩效指标2	#							
绩效指标N	#							

Table 4 *Relationship of Process Objects to Performance Indicators and Tasks Associated With It*

Performance Indicator: A metric used by an enterprise to indicate its overall success or the success of a particular area in which it is engaged.	
Rules	(D) Process relates to performance (performance indicator).
Tasks	• Associate and tie the performance indicator(s) to the business processes. • Associate and tie the performance indicator(s) to the process steps of the business process. • Associate and tie the performance indicator(s) to the process activities of the business process. • Associate and tie the performance indicator(s) to the stakeholders involved in the business process. • Associate and tie the performance indicator(s) to the process owners of the business process. • Associate and tie the performance indicator(s) to the managers involved in the business process. • Associate and tie the performance indicator(s) to the roles/resources involved in the business process.

---

7. Connect performance indictors to processes
8. Improve the operating model
9. Reduce process cost
10. Associate relevant rules to the processes
11. Identify and relate compliance aspects
12. Process automation
13. Process measurements and reporting as part of the organizational analytics and decision making
14. Service model improvement

As we have mentioned, these reasons determine the design and content of what is within such a process matrix and therefore how such a process matrix should look. For example, reporting would require process matrices with the following relations: other relations' performance indicators (measures), business goals, and who would receive what report.

PROCESS MODEL

Once information has been collected and organized in the process maps and/or process matrices, a process model may be crafted to enable the complex set of resulting information to be used in different disciplines, and within this to be communicated more easily to stakeholders, management, and leadership. The fully integrated and standardized process templates enable the practitioner to work and model with the process objects throughout all aspects of the enterprise (business, application, and technology) with more confidence in the completeness and alignment of their information. Their semantic relations and connection are governed not only by the

表4　流程对象与绩效指标及相关任务的关系

绩效指标：企业用来表示其整体成功或某一特定领域成功的指标。

规则	（D）流程与性能（绩效指标）有关。
任务	• 将绩效指标与业务流程关联起来。 • 将绩效指标与业务流程的流程步骤联系起来。 • 将绩效指标与业务流程的流程活动联系起来。 • 将绩效指标与业务流程中涉及的利益相关方联系起来。 • 将绩效指标与业务流程的流程所有人联系起来。 • 将绩效指标与业务流程中相关的管理人员联系起来。 • 将绩效指标与业务流程中涉及的角色/资源联系起来。

（7）将绩效指标与流程连接起来；

（8）改进操作模式；

（9）降低流程成本；

（10）将相关规则与流程关联起来；

（11）识别和关联合规性方面；

（12）流程自动化；

（13）作为组织分析和决策的一部分的流程度量和报告；

（14）服务模式改进。

正如我们所提到的，这些原因决定了这样一个流程矩阵中的设计和内容，以及这样一个流程矩阵应该是什么样子的。例如，报告需要具有以下关系的流程矩阵：其他关系的绩效指标（度量）、业务目标以及谁将收到什么报告。

2.5.6　流程模型

一旦在流程地图或流程矩阵中收集和组织了信息，就可以创建流程模型，以使复杂的结果信息集能够在不同的学科中使用，并在此基础上更容易地与利益相关方、管理层和领导层沟通。完全一体化和标准化的流程模板使从业者能够在整个企业的各个方面（业务、应用程序和技术）使用流程对象进行工作和建模，并对其信息的完整性和一致性更有信心。它们的语义关系和连接不仅受对象的控制，还受流程建模规则和任务的控制，这些规则和任务可以确保流程模板如何以及在何

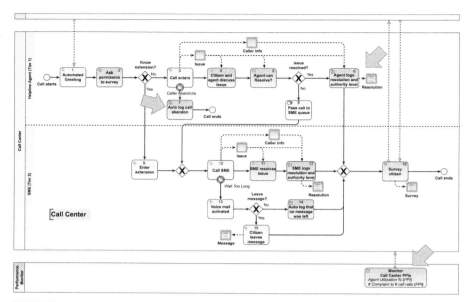

FIGURE 1

Example of process model, with measurements and reports specified within notations (Example modeled in iGrafx®).

objects, but also by the process modeling rules and tasks, which ensure how and where the process templates interlink and share common process objects.

An example of such a process model is illustrated in Figure 1, which demonstrates processes, roles involved, interactions, data aspects, and measurements and reporting aspects.

In this example, through a process such as described in Table 1, the stakeholders, managers, and owners identified in the process map are related to the performance indicators through the process matrix (Table 3). If such a vertical alignment of measurements across levels is ignored, it is possible that activities or processes will be measured in ways that do not contribute to the overall success of the organization. The danger to an organization might even be that the performance indicators could lead to conflicts in strategy or value creation. Together, the process map in Table 1 and process matrix in Table 3 and the process model in Figure 1 provide a good example of how it is possible to relate the relevant process information.

The process maps, process matrices, and process models that specify the semantic relations in this illustration are just examples. As such, they do not show all possible relations that exist. Other views of other information are possible. We could, for example, choose to organize this same information by processes, measurements, or the data involved. An example of an alternative view is found in Figure 2, in which the sales call center from Figure 1 shows a sales analysis and a cost of goods sold analysis by process.

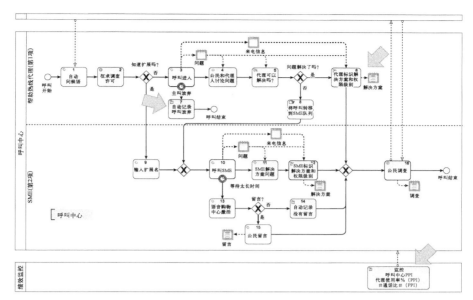

图1　流程模型示例,在符号中指定测量和报告(在 iGrafx® 中建模的示例)

处相互链接和共享公共的流程对象。

图1展示了这种流程模型的一个示例,它演示了流程、涉及的角色、交互、数据方面以及度量和报告方面。

在本例中,通过如表1所述的流程,流程地图中标识的利益相关方、经理和所有人通过流程矩阵与绩效指标相关(表3)。如果忽略跨级别测量的这种垂直校准,那么活动或流程的测量方式可能不会对组织的整体成功起到促进作用。对组织的危险甚至可能是错误的绩效指标导致战略或价值创造方向的冲突。表1中的流程地图和表3中的流程矩阵以及图1中的流程模型一起为如何关联相关流程信息提供了一个很好的示例。

流程地图、流程矩阵和流程模型(在例证中指定语义关系)只是示例。因此,它们并不显示所有可能存在的关系。其他信息的其他视图是可能的。例如,我们可以选择通过流程、度量或涉及的数据来组织相同的信息。图2中提供了一个备选视图的示例,其中图1中的销售呼叫中心按流程显示了销售分析和销货成本分析。

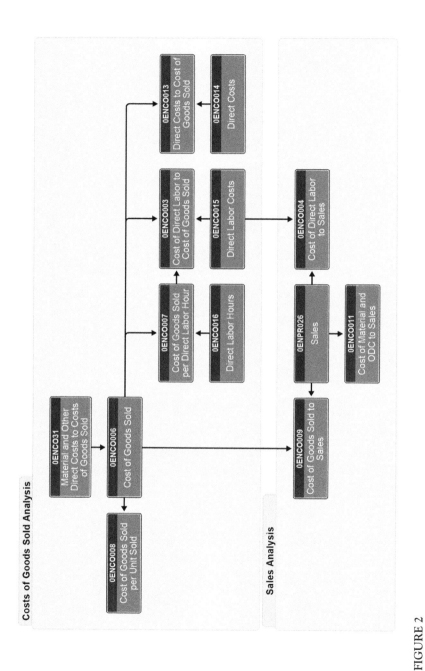

FIGURE 2

Example of automated process measure, e.g., SAP system measurement with the relevant processes, data object queries, transaction codes, and process flow relations. [2]

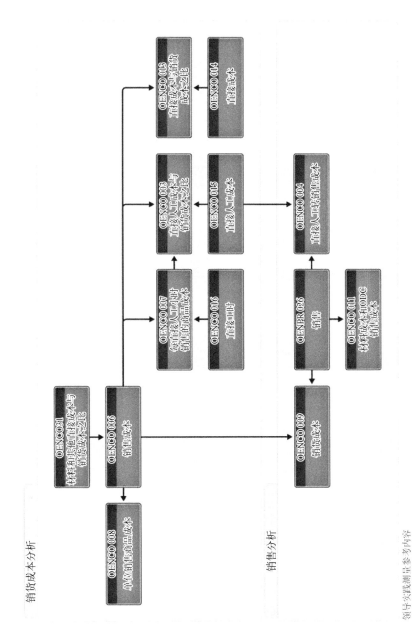

图2 自动化流程度量示例，如SAP系统度量与相关流程、数据对象查询、业务代码和流程关系[2]

THE MOST COMMON PROCESS TEMPLATES

As we have just explained, the process templates consist of process maps, matɪ and models that capture the relevant process elements (meta objects). Each of thɪ process templates is based on a specific view with particular stakeholder concerns to enable process identification, creation, and realization in achieving the outlined needs and wants. For this, the process templates identify the relevant stakeholders, their requirements and concerns, the process descriptions and their semantic rationale, and the corresponding tasks to the specific views and viewpoints. Each of these process templates is thereby built to support a particular need and want.

Figure 3 illustrates an overview of the most common process templates. All of the process templates listed are fully integrated and standardized, enabling full reusability of shared aspects between process templates, where 1 in Figure 3 shows the objects in the process maps, 2 shows the objects in the process matrices, and 3 shows those of the process models. The specific process templates therefore not only show which objects are within what template, specifying whether it is a map, matrix, or model, it furthermore shows where the object of one template can be reused in another: where the objects have and should be integrated and standardized, because they are the same. That most organizations do not have such integrated and standardized process templates is the single source of the high costs of modeling, engineering, and architecture and the low maturity of output.

BENEFITS OF PROCESS TEMPLATES

One of the strangest things we have heard is that "real process experts do not use templates," or "templates are a substitute for a real subject matter experts." However, we have learned through hard experience that there are times when using one is not only the most appropriate choice, but frequently the sole choice that addresses the problem. Some benefits to using process templates are that:

- They ensure consistency with project artifacts
- All subject matter experts work in a standardized way
- They ensuring cross-integration of templates
- They save time with templates
- They are reusable
- They enable better governance
- They are less expensive
- They are faster to populate across various teams
- They develop routine
- They maintain consistency among various team members and artifacts
- They immediately lift the artifacts to maturity level 3 and 4
- They develop a standard in your BPM Center of Excellence (CoE)
- If you are a non-designer, use templates to give a more professional edge to your own marketing materials.
- They get things done faster

2.5.7　最常见的流程模板

正如我们刚才所解释的,流程模板由流程地图、矩阵和捕获相关流程元素(元对象)的模型组成。这些流程模板中的每一个都基于一个特定的视图,具有特定的利益相关方关注点,以实现流程识别、创建和实现所概述的需要和需求。为此,流程模板识别相关的利益相关方、他们的需求和关注点、流程描述及其语义基础,以及特定视图和观点的相应任务。因此,这些流程模板中的每一个都是为支持特定的需求而构建的。

图3说明了最常见的流程模板的概述。所列的所有流程模板都是完全一体化和标准化的,能够充分重用流程模板之间的共享方面,其中图3中的"1"显示流程图中的对象,"2"显示流程矩阵中的对象,"3"显示流程模型中的对象。因此,特定的流程模板不仅显示哪些对象在哪个模板中,指定它是流程图、矩阵还是模型,而且还显示了一个模板的对象可以在另一个模板中重用的位置:对象已经并且应该一体化和标准化的位置,因为它们是相同的。大多数组织没有这样的一体化和标准化的流程模板,这是建模、工程和体系结构成本高、输出成熟度低的单一原因。

2.5.8　流程模板的好处

我们听到的最奇怪的事情之一是"真正的流程专家不使用模板"或者"模板是真正的学科专家的替代品"。然而,我们通过艰苦的经验了解到,有时使用模板不仅是最合适的选择,而且往往是解决专业问题的唯一选择。使用流程模板的一些好处如下。

- 它们确保与项目制品的一致性。
- 所有学科专家以标准化方式工作。
- 它们确保模板的交叉集成。
- 使用模板节省时间。
- 可重复使用。
- 它们能够实现更好的治理。
- 它们能够价格更低。
- 它们能够更快地在不同的团队中应用。
- 它们制定规范。
- 它们保持不同团队成员和制品之间的一致性。
- 它们立即将制品提升到成熟度3级和4级。
- 它们在您的BPM CoE制定了一个标准。
- 如果您不是设计师,请使用模板为自己的营销材料提供更专业的优势。
- 它们能更快地完成工作。

Process Objects \ Templates	Forces & Drivers (FD)	Vision, Mission & Goals (VMG)	Requirement (Rq)	Stakeholder (ST)	Strategy (S)	Balanced Scorecard (BSC)	Performance (Pe)	Measurement & Reporting (MR)	Competency/Business Model (BC)	Revenue (Rev)	Cost (Co)	Operating (Op)	Information (I)	Role (Ro)	Owner (O)	Organizational Chart (OC)	Object (Ob)	Workflow (WF)	Rule (Ru)	Risk (RS)	Security (SC)	Case (CS)	Process (P)	BPM Notations (BPMN)	Service (Se)	Application (A)	Application Service (AS)	Application Rules (AR)	System Measurements/Reporting (AM)	Application Interface (AI)	Application Screen (Asc)	Compliance (C)	Data (D)	Platform (Pl)
Process Area (categorization)	1,2	2	1,2	1,2,3	1,2,3		2,3	2,3	2			1		2,3	2,3	2,3		2,3					1,2,3	2										
Process Group (categorization)	1,2	2	1,2	1,2,3	1,2,3		2,3	2,3	2			1		2,3	2,3	2,3		2,3					1,2,3	2										
Business Process	2		1,2						2	2	2	1						2,3	2,3	2	2	2,3	1,2,3	2,3	2	2								
Process Step			2,3							2	2							2,3	2,3	2	2		1,2,3	3	2	2								
Process Activity			2,3							2	2							2,3	2,3				1,2,3	3	2	2								
Events			2,3																2,3		2	2	2,3	3	2									
Gateways			2,3																2,3				2,3	3										
Object (Business & Information & Data)			2,3										1,2,3				1,2						2,3	2,3	2,3		2,3							
Process Flow (incl. Input/output)			2,3						1,2,3	3	3			1,2			2	1,2,3					3	3	2,3		2,3			3				
Process Roles			2,3														2,3						2,3	2,3	1,2		2				2,3			
Process Rules										2,3	2,3								1,2,3				2,3	2,3				2,3				1,2		
Process Measurement (PPI)						2,3	1,2,3	1,2,3		2,3	1,2	2,3											2	2,3	2,3				1,2					
Process Owner	2	2	1,2,3		2,3		2,3	2,3				2,3			1,2								1,2,3	2,3	2,3								1	1

1 = Map　2 = Matrix　3 = Model

FIGURE 3

The most common process templates.[1]

	模板																																	
流程目标	部队和潜水员(FD)	愿景、使命与目标(VMG)	需求(Rq)	利益相关者(ST)	策略(S)	平衡计分卡(BSC)	绩效(Pe)	测量和报告(MR)	能力/商业模式(BC)	收入(Rev)	成本(Co)	操作(Op)	信息(I)	角色(Ro)	所有者(O)	组织图表(OC)	目标(Ob)	工作流(WF)	规则(Ru)	风险(RS)	安全性(SC)	案例(CS)	流程(P)	BPM符号(BPMN)	服务(Se)	应用(A)	应用服务(AS)	应用规则(AR)	系统测量/报告(AM)	应用接口(AI)	应用程序屏幕(Asc)	遵从(C)	数据(D)	平台(PL)
流程领域(分类)	1,2	2	1,2	1,2,3	1,2,3		2,3	2,3	2			1		2,3	2,3	2,3		2,3					1,2,3	2										
流程组别(分类)	1,2	2	1,2	1,2,3	1,2,3		2,3	2,3	2			1		2,3	2,3	2,3		2,3					1,2,3	2										
业务流程	2		1,2	1,2,3					2	2	2	1						2,3	2,3	2	2	2,3	1,2,3	2,3	2	2								
流程步骤			2,3							2	2							2,3	2,3				1,2,3	3	2	2								
流程活动			2,3							2	2							2,3	2,3				1,2,3	3	2	2								
事件			2,3																2,3	2	2	2	2,3	3	2		2,3							
网关			2,3																2,3				2,3	3	2		2,3							
目标(业务、信息和数据)			2,3										1,2,3				1,2	1,2,3	2,3				2,3	2,3	2,3									
流程流(包括输入/输出)			2,3							3	3						2	1,2,3					3	3	2,3		2			3	2,3			
流程角色			2,3						1,2,3		2			1,2			2,3						2,3	2,3	1,2								1	1
流程规则			2,3																1,2,3				2,3	2,3				2,3				1,2		
流程测量(PPI)			2,3			2,3	1,2,3	1,2,3		2,3	2,3												2	2,3					1,2					
流程所有者	2	2	1,2,3	1,2,3	2,3		2,3	2,3	1,2,3	1,2	1,2	2,3			1,2								1,2,3	2,3	2,3									

1=图 2=矩阵 3=模型

图3 最常见的流程模板[3]

- They maintain artifact consistency
- They simplify updates and changes
- Process templates can be used by the various people who work with processes, i.e., process experts, process engineers, and process architects. As a matter of fact, aspects of the templates can be reused across the various process roles.
- There are many ways to personalize them without sacrificing the benefits and consistency of the process templates. If your team has unusual needs, you can customize a process template and then create the BPM project.

Remember that the process templates we have illustrated are designed based on a complete view of the enterprise semantics and are therefore fully integrated and standardized with each other. This means that we know which aspects of one process template can be reused in another template. A further advantage is that the process templates are designed to meet the needs of most process experts/teams in many different settings and to be fully integrated into the BPM lifecycle, BPM roles, BPM governance, and BPM change management. Working with the various process templates we present in this book is a smart idea and will lift your maturity and save significant amounts of time and money.

CONCLUSION

In this chapter, we have focused on process templates and why they are important within organizations working with their processes. The subject is therefore relevant for BPM CoE, BPM teams, process experts, and other subject matter experts working with processes.

We covered what process templates are, how they can be used, and where they can or should be applied to draw on the ontology and semantic-based process relations standardized to ensure reusability and replication of success in outlining the correct connection points based on a common relationship pattern of the process objects.

We furthermore detailed the differences between process templates in terms of process maps, process matrices, and process models, and ended with the benefits and value of process templates.

We showed that by using process templates to manage the different kinds of highly connected information and relations the process creation is ensured and that:

- The process map (which lists the various related objects to capture the decomposed unrelated objects) is a critical design tool
- The process matrix (which is composed in terms of relating specific objects together) provides the continuity for and interconnection between a process map (a representation of decomposed and/or composed objects)
- A process model (a representation of interconnected and related objects) is critical to integrating and standardizing the process templates and tools of the practitioner.

Furthermore, it is an essential part of supporting, integrating, and standardizing the practitioner's way of thinking, working, and modeling.

- 它们保持制品的一致性。
- 它们简化了更新和更改。
- 流程模板可供处理流程的各种人员使用,即流程专家、流程工程师和流程架构师。事实上,模板的各个方面可以跨各种流程角色重用。
- 在不牺牲流程模板的好处和一致性的情况下,有许多方法可以对其进行个性化设置。如果您的团队有不寻常的需求,您可以定制流程模板,然后创建BPM项目。

请记住,我们所演示的流程模板是基于企业语义的完整视图设计的,因此它们彼此完全一体化和标准化,这意味着我们知道一个流程模板的哪些方面可以在另一个模板中重用。另一个优势是,流程模板的设计能够满足大多数流程专家/团队在许多不同设置中的需求,并且能够完全集成到BPM生命周期、BPM角色、BPM治理和BPM变更管理中。使用我们在本书中介绍的各种流程模板是一个明智的想法,它将提高您的成熟度并节省大量的时间和金钱。

2.5.9 结论

在本节中,我们重点介绍了流程模板以及为什么它们在使用其流程的组织中很重要。因此,该主题与BPM CoE、BPM团队、流程专家和其他处理流程的学科专家相关。

我们介绍了什么是流程模板,如何使用它们,以及在哪里可以或应该应用它们来利用标准化的本体论和基于语义的流程关系,以确保在基于流程对象的公共关系模式的基础上,成功地概括出正确连接点的可用性和复制性。

我们进一步详细说明了流程模板在流程图、流程矩阵和流程模型方面的差异,并以流程模板的好处和价值为结束。

我们表明,通过使用流程模板来管理不同类型的高度关联的信息和关系,可以确保流程创建,并且:

- 流程地图(列出各种相关对象以捕获分解的无关对象)是一个关键的设计工具;
- 流程矩阵(由相关特定对象组成)提供了流程地图(分解对象或组合对象的表示)之间的连续性和互联性;
- 流程模型(互联和相关对象的表示)对于整合和标准化从业者的流程模板和工具至关重要。

此外,它是支持、整合和标准化从业者的思维、工作和建模方式的一个重要部分。

As already shown, the illustrated process map, process matrices, and process models and the specification of semantic relations are just examples. Because they were examples, not all possible relations were specified. However, all of the possible relationships for the various process templates will be illustrated with detailed examples in the various chapters across Volumes 1 and 2 of *The Complete Business Process Handbook.*

End Notes

1. http://www.globaluniversityalliance.net/.
2. Enterprise Cost Model, LEADing Practice Measurement Reference Content [#LEAD-ES20014PG].
3. Common Process Templates Overview, LEADing Practice Business Process Reference Content [#LEAD-ES20005BP].

如前所示的流程地图、流程矩阵和流程模型以及语义关系的规范只是示例。因为它们是示例,所以并没有指定所有可能的关系。但是,各种流程模板的所有可能的关系将在《完全流程手册》第1卷和第2卷的各个章节中以详细的示例加以说明。

参考文献

[1] http://www.globaluniversityalliance.net/.
[2] Enterprise Cost Model, LEADing Practice Measurement Reference Content [#LEAD-ES20014PG].
[3] Common Process Templates Overview, LEADing Practice Business Process Reference Content [#LEAD-ES20005BP].